THE ENVIR♻NMENT

the essentials of...

THE ENVIR♲NMENT

Joseph Kerski and Simon Ross

Hodder Arnold

A MEMBER OF THE HODDER HEADLINE GROUP

First published in Great Britain in 2005 by
Hodder Education, a member of the Hodder Headline Group,
338 Euston Road, London NW1 3BH

http://www.hoddereducation.com

Distributed in the United States of America by
Oxford University Press Inc.
198 Madison Avenue, New York, NY 10016

The advice and information in this book are believed to be true and
accurate at the date of going to press, but neither the authors nor the publisher
can accept any legal responsibility or liability for any errors or omissions.

British Library Cataloguing in Publication Data
A catalogue record for this book is available from the British Library

Library of Congress Cataloging-in-Publication Data
A catalog record for this book is available from the Library of Congress

ISBN-10 0 340 81632 5

ISBN-13 978 0 340 81632 5

 2 3 4 5 6 7 8 9 10

Typeset in 8/10pt New Baskerville by Dorchester Typesetting Group Ltd
Printed and bound in Spain

What do you think about this book? Or any other Hodder Education title?
Please send your comments to the feedback section on www.hoddereducation.com

Contents

Preface

When the the first astronauts ventured into space during the 1960s, they looked down upon the Earth and remarked how fragile and isolated the planet looked. The blue planet, as it became known, with its swirling white clouds looked vulnerable to these early space pioneers. For the first time we, the people of planet Earth, became aware of how much we needed to care for our collective home and prevent it from being damaged. Our future depended upon it.

Concern for the environment grew during the last few decades of the twentieth century for three reasons. First, information and technological revolutions allowed us to monitor the Earth as never before. Second, as people learned more about how the Earth's environment worked and how it was changing, environmental awareness, sensitivity and activism increased to where the environment was discussed and debated more than nearly every other issue. We began to realise that humans are responsible for much of the environmental change occurring, and many fought for awareness, programmes and laws that sought to reverse the harm that was becoming more and more evident. Third, the population explosion and results of two centuries of industrialisation made it evident that areas of environmental concern around the world were not isolated pockets, but enormous areas, some of them global in scale. Rainforests were found to be disappearing at a staggering rate, wiping out hundreds of species of plants and animals. Alarmingly, some species were becoming extinct before scientists even had the chance to discover them in the first place. Acid rain, resulting from a chemical cocktail of industrial pollutants, was found to be harming lake and forest ecosystems in pristine environments ranging from Scandinavia to North America. Rivers and underground aquifers across the world were becoming increasingly polluted as a result of industrial and agricultural waste. Diseases resulting from dirty water, such as cholera, were endemic in some regions. We were over-exploiting and wasting our natural resources with little thought to the legacy that we would leave behind for our children and grandchildren. The future looked bleak indeed.

Perhaps the most significant concern of recent years has been associated with global warming. There seems little scientific doubt – despite the scepticism of some politicians with a vested interest in maintaining economic growth through industrialisation – that global warming is taking place and that it will have an impact upon our environment. Exactly what that impact will be, nobody is absolutely sure, but it seems possible that climates across the world will change and that ecosystems will alter too. Sea levels may rise, threatening low-lying island chains such as the Maldives and vast deltaic plains like southern Bangladesh. Storms may become more frequent in the mid-latitudes, and landslides and avalanches may become more commonplace in mountain regions. For many, the impact of global warming fuelled by people's thoughtless pollution of the atmosphere represents the major environmental issue of the present day.

As the twentieth century drew to a close, the environment started to take centre-stage politically with more and more pressure groups and ordinary people expressing their concern for the future. International conferences were held to discuss the major issues of poverty and environmental degradation, perhaps the most influential being the 1992 Rio 'Earth Summit'. Western democratic governments began to put the environment high on their political agendas in recognition of the concerns that people (the voters) had about the state of the planet. In response to these concerns and pressure from the international community, governments have begun to take action. Industries have been forced to reduce the emission of harmful gases by installing filters. There are stricter guidelines on water quality and polluters risk heavy fines. People and local authorities are being encouraged to recycle and reduce waste and industries are being urged to conserve energy. Renewable energy sources such as wind, water and biofuels are being developed to replace the less sustainable exploitation of fossil fuels. In 1997, the Kyoto Protocol paved the way for international agreement on reducing emissions of gases thought to be responsible for global warming.

Essentials of the Environment aims to provide readers with a quick reference guide to the major environmental issues facing society today. Written and researched by experts in a variety of fields, it aims to be comprehensive, accurate and also thought-provoking. Most importantly of all, however, *Essentials of the Environment* aims to generate a greater awareness of the importance of conserving and preserving our environment for the benefit of generations to come. As the saying goes, we should take nothing but pictures, leave nothing but footprints and kill nothing but time.

Several themes run through *Essentials of the Environment*. The environment is wonderfully complex and interrelated. One cannot understand biodiversity, for example, without understanding species differentiation, climate or life zones. Therefore, even though the book is arranged alphabetically, the entries are all related, at a variety of scales. Through research and tools such as geographic information systems, remote sensing and field probes, we have learned a great deal about environmental processes and forces, though it is equally clear that we have much yet to learn. Despite the fact that environmental awareness and knowledge have never been greater in the history of civilisation, the capacity to harm the environment continues to threaten the very thing that we are trying to protect.

Finally, although it is easy to become discouraged about the environmental harm that we have inflicted and continue to inflict, the book is filled with success stories – people and actions that have saved species, protected lands and encouraged sustainability and conservation. It is our hope that this book will contribute to awareness, understanding and positive action that will protect that which we all know, love and depend on for our very existence.

> In the end, we will conserve
> Only what we love
> Love only what we understand
> Understand what we are taught.
> *Baba Dioum*

Simon Ross
Joseph Kerski
January 2005

List of contributors

Joseph Kerski Joseph Kerski is Geographer at the US Geological Survey's Rocky Mountain Mapping Center in Denver, Colorado, USA. He creates curricula based on geography and Geographic Information Systems (GIS); supports the implementation of science, GIS and geography at all levels of society through the provision of technical, educational and materials support and articles, web resources and books; seeks and nurtures educational partnerships; teaches 40 workshops annually; and conducts research in the effectiveness of geographic technologies in teaching and learning. Joseph holds three geography degrees and encourages everyone to explore, study and value the Earth.

Simon Ross Simon Ross is Head of Geography and Director of Co-Curriculum at Queen's College, Taunton, UK and Chairman of the Wellington Branch of the Geographical Association. He is the author of several widely-used school textbooks including the *Geography 21* series (Collins Educational) and *Basic Mapskills* and *Essential AS Geography* (Nelson Thornes). In 2003, his book *Essential Mapskills* (Nelson Thornes) was awarded the prestigious Gold Award by the Geographical Association. A leading authority on natural hazards, Simon lectures regularly to sixth-form students and has acted as a geography consultant to the BBC. He is passionate about the environment and the need to educate young people in order to conserve and preserve it for generations to come.

Ann Bowen Roseberry Sports and Community College, Co. Durham, UK

Andrew Hignell formerly at Wells Cathedral School, Somerset, UK

Alison Rae Roedean School, Sussex, UK

David Armstrong University of Colorado – Boulder, USA

Robert Larkin University of Colorado – Colorado Springs, USA

Suzanne Larsen University of Colorado – Boulder, USA

Bob Coulter Missouri Botanical Garden, USA

Jim Brey University of Wisconsin – Fox Valley, USA

Acknowledgements

I thank our publisher Hodder Arnold for publishing a book that has such global consequence as this. I am grateful to all who made contributions, and most especially to my colleague Simon Ross for his diligence every step of the way.

To the educators across the world that I have worked with over the past decade, my sincere gratitude and appreciation for your dedication to teach the concepts in this book to students.

This book makes it clear that humans have the single greatest impact on the environment. This impact does not have to be negative. Indeed, success stories of people making a positive difference in the environment abound. I challenge all who read this book to look for these successes, be inspired by them, and take action in your own community to protect what we all so dearly love. Make your own success story!

I wish to dedicate this book to Janell, for her excellence, thoughtfulness, and faithfulness, and to Emily and Lilia – may you and others of your generation never tire of learning about, enjoying, valuing, and protecting our environment.

Joseph Kerski

Like Joseph, I would like to thank all the authors who contributed to this book.

I would also like to thank my family for yet again putting up with me shutting myself away in my study to work on this book when I should really have been entertaining them! Thanks to my pupils – past and present – at Queen's College for inspiring me with a desire to promote environmental awareness amongst the young, on whose shoulders the future of the planet rests. A special thanks to Sheila Moose and Rosie for keeping me on the straight and narrow.

Simon Ross

A

ACID RAIN

All rain is slightly acidic because the falling water combines with carbon dioxide as it passes through the lower atmosphere, to form carbonic acid. The term acid rain was increasingly used by environmentalists from the 1970s onwards to embrace the presence of atmospheric pollutants, such as sulphur dioxide and nitrogen oxides.

The apparent increasing presence of these substances in rainfall has become an emotive issue, but many people forget that it is not a new discovery. As far back as the nineteenth century, the presence of acids in the rainfall over the industrial centre of Manchester, north-west Lancashire, was identified by Angus Smith, an industrial chemist, who was employed as the UK's first Inspector of Factories.

The causes of acid deposition

Acid deposition is a far more precise term than acid rain to describe the way polluting substances mix with precipitation. Researchers have shown that acid deposition has two components: wet deposition and dry deposition.

Wet deposition refers to all types of precipitation – snow, hail and fog, as well as rain – that have a pH lower than the global average of 5.6. The acidity comes from two pollutants – sulphur dioxide and nitrogen oxides – which react with water vapour in clouds, sunlight and oxygen, producing sulphuric acid and nitric acid. Rain and other forms of precipitation subsequently 'wash' these acids from the air and deposit them back on the surface of the EARTH.

Dry deposition occurs when these particles are deposited in solid form, usually blown by wind, on to buildings or vegetation or any other surface, before particles are washed off by rain. This makes the resulting run-off even more acidic, because the dry deposition adds to the wet deposition.

As a contemporary issue, acid rain, or more precisely acidic deposition, came to the forefront after the first United Nations Environment Conference in Stockholm in 1972. Evidence for the detrimental effect of acidic precipitation was presented by a group of scientists, who cited especially the decline in fish populations in lakes and rivers in Sweden and a decrease in forest growth.

The effects of acid deposition have been widely experienced across Europe and North America for many years, and in the past few years they have also been identified in the 'tiger economies' of South-East Asia as these nations rapidly industrialise. In addition to the well publicised acidification of rivers and lakes, and the killing of fish and other aquatic life, scientists have shown that acid deposition can accelerate rates of chemical weathering, especially on buildings made of limestone and sandstone.

It has also been shown that people can become seriously ill, and can even die from the effects of acid deposition. The substances cause serious respiratory problems, including asthma, as well as dry coughs, headaches and throat infections. Other indirect effects occur after acid rain has been absorbed by either plants or animals, so that when people eat these contaminated plants and animals, the toxins within them cause brain damage, kidney problems and possibly Alzheimer's disease.

This growing awareness of the negative effects on humans has prompted a number of policies aimed at reducing the rates of wet and dry deposition. These include washing coal before it is burnt and using devices called scrubbers to remove chemically the sulphur dioxide from the gases being discharged through chimneys (see **FLUE GAS DESULPHURISATION**). Another way is to encourage power stations to use fuels that produce smaller amounts of sulphur dioxide, rather than burning coal, while other people favour switching over to alternative and renewable energy sources.

The long-term solution will therefore require global action and international agreements. Acid deposition is not contained by international frontiers. In simple terms, there is a need for all governments to play their part in reducing, if not eliminating, a danger that threatens all forms of life in both developing and developed countries.

ADIABATIC HEATING AND COOLING

The environment is constantly subjected to many influences, but one of its chief influences is invisible – the air. The air's chemical balance, the pollutants it holds, its storms, clouds and precipitation, and its many other characteristics all affect the environment. This is primarily because air moves from place to place, carrying with it its distinctive characteristics, which are then modified en route.

Air is constantly rising and falling due to changes in its temperature and density. In the same way, a hot air balloon pilot fires the burner to heat the air inside the balloon to make it rise. Air parcels that are lifted to higher altitudes or caused to descend go through a process called adiabatic cooling and heating. In this process, the temperature of the air changes, but the change is due to external pressure changes caused by the vertical movement of the air. The air does not lose any of its heat energy in the process, so if it is returned to the original altitude, it will return to the original temperature.

Adiabatic cooling involves the lifting of a parcel of air. Some of the common lifting mechanisms include convection. When a kettle of water is heated on the stove, the water closest to the heat source becomes less dense. When it rises, cooler water takes its place, quickly resulting in a circulation of water in the kettle. Similarly, convection in the atmosphere results from temperature and density differences, and serves to mix the air. Air also mixes when air masses converge at the surface, or when air diverges aloft. Divergence may occur with low pressure because air is

rising during a low. Air masses are frequently forced to rise when their horizontal movement is blocked by a mountain range. As an air parcel is lifted into higher regions with less air pressure, it expands. This expansion cools the air, because the air molecules in it are further apart and collide less frequently.

Lifted air always cools down, because air pressure always falls as one goes higher in the atmosphere. The cooling rate of dry (unsaturated) air is generally 5.5°F per 1000 feet or about 10°C per kilometre. This rate of temperature change is called the dry adiabatic lapse rate.

If the air becomes saturated (when the dew point temperature is reached) condensation occurs forming clouds. As latent heat is released when condensation takes place, the rising air now cools less rapidly, commonly between 2.2°F per 1000 feet (about 4°C) in very warm moist air to the dry rate of 5.5°F (about 10°C) if the air is very dry. This new rate of cooling is termed the saturated adiabatic lapse rate.

If an air 'parcel' descends, it will enter a region that has higher air pressure, which will compress the 'parcel' resulting in an increase in temperature at the dry adiabatic lapse rate. As the descending air is warming all the time, condensation and cloud formation will not occur.

Places on the Earth characterised by descending air are cloud-free and dry. Such places exist at about 30 degrees north and south latitude, where air descends after rising from the heating at the equator. In Africa this wide belt of descending air created the hot, dry, Sahara Desert, above. Another place characterised by descending air is on the leeward side of mountains, opposite to the direction from where predominant winds blow.
From: www.rallyetrain.de

Adiabatic heating and cooling have a great effect on the climate of a region, and therefore on the environment. Places on the Earth characterised by ascending air are typically cloudy and wet. For example, air at the equator is heated to a hotter temperature than air on either side of it. As air ascends, it condenses into clouds, and precipitation falls, giving the equatorial areas of Africa (the Congo, for example) and South America (the Amazon Basin) their wet climate and resulting tropical rainforest vegetation and associated wildlife.

Climate models associated with global warming suggest that the current zones of ascending and descending air may well alter, leading to changes in the world's climates. Regions that may have been previously well watered may see reduced rainfall in the future, which may lead to dire environmental consequences. Previously dry regions may see more rainfall with flooding and soil erosion becoming an issue.

WEBSITES

Adiabatic Processes from Palomar College:
http://daphne.palomar.edu/jthorngren/adiabatic_processes.htm
Stability and Cloud Development from
Lyndon State College:
http://apollo.lsc.vsc.edu/classes/met130/notes/chapter7/index.html

Detailed slides showing adiabatic cooling and heating:
www.phys.ufl.edu/courses/met1010/chapter7-2.pdf
Online Weather Course from the American Meteorological Society:
http://64.55.87.13/amsedu/online/info/

AFFORESTATION

Afforestation is the process of transforming an area into forest, often when trees have not grown there before. This is different from reforestation, the replanting of trees in an area that was once forested but where the trees have been destroyed, for example, as a result of a forest fire. Afforestation may be used as part of a programme to reduce flood hazards in a drainage basin because increased interception by the trees slows down the transfer of water.

Some of the earliest examples of afforestation were associated with the Forestry Commission in the UK. Ennerdale Water in the Lake District was given reservoir status and its capacity increased in the 1930s in order to provide a water supply for the growing cities in the North West, especially Manchester. Extensive tracts of coniferous forest were planted in Ennerdale to improve the hydrology and reduce siltation of the reservoir. The presence of the trees slows the water flow into the reservoir and promotes natural filtration

Case study

Afforestation schemes

Bihar, India

Chami Murmu is a young tribal woman in an impoverished district in Bihar, India. Her tribal group, Sahyogi Mahila, has helped afforest 523 hectares of barren land by planting nearly 2 million saplings of which 90 per cent have survived.

The group began the work at Bhalubasa, a village with 70 families. A barren rocky hillock was planted with 10 hectares of acacia trees. The forest is protected by a village vigilance committee that keeps out cattle likely to destroy the trees, prevents neighbouring villagers felling the trees and has the power to fine anyone caught cutting the trees. The women no longer walk 10 kilometres to collect firewood, but use the dry twigs and leaves collected from the woodland. No one needs to buy firewood any longer. Money collected from the fines has also been reinvested in the communities to provide loans for marriages, medical treatment, improvements to agriculture and to help set up new businesses.

This work was repeated in the region to cover eventually 47 villages. Each village handles everything from the plantation to the collection of fines. *continued over*

Case Study continued

The European Union

Forestry in the EU has been recognised as having a major role to play as part of the Common Agricultural Policy and in the wider context of rural development. Since 1992 grants have been available in the EU for afforestation of agricultural land. This was in response to overproduction in agriculture and an effort to take land out of production to reduce food surpluses. In contrast to earlier schemes that primarily planted conifers, these more recent schemes promote the planting of native deciduous trees and a more diverse range of conifers. In Ireland policies aim to increase afforestation from 8 to 17 per cent of the land area by 2035. This will require 20,000 hectares of forest to be planted each year. Proponents of the scheme argue that forestry is lucrative and can contribute in a significant way to farm incomes. The grants available make afforestation the only guaranteed income in farming for the next 20 years. The annual grant can be up to £500 per hectare planted. It is tax free – and farmers need only complete one form!

Karnataka, India

Coldplay, one of the world's best-known musical groups, is committed to planting trees to improve the world's environment. In conjunction with Future Forests the band is helping to plant 10,000 mango trees in Karnataka in India. The trees provide fruit for trade and local food but will also help to soak up carbon dioxide to improve air quality and reduce deforestation's negative impact on the environment.

of the water. However, such afforestation has been severely criticised by local residents and by environmentalists. They argue that the regimented lines of planted trees are monotonous and that the species planted are not native UK trees. They also argue that the trees contribute to the acidification of the soil and the water, leading to changes in the forest and freshwater ecosystems. Since the original planting, the Forestry Commission has been working to create a more natural-looking forest by varying the species planted and using more native deciduous trees.

Since these early plans, afforestation has spread to many parts of the world. It is seen as the solution to many environmental and economic problems, including reducing soil erosion and flooding, providing timber supplies for local craft industries and improving fuelwood supplies. The Coldplay example quoted above is hoped to counter increased levels of carbon dioxide in the atmosphere to reduce the rate of GLOBAL WARMING and to improve trade and local food supplies.

WEBSITES

Information on the EU grants:
www.irishforests.com/farm/page15.html
Article on afforestation in Ireland, the Netherlands and Greece,

'Local perspectives on European Afforestation', at The Arkleton Centre Crossroads Conference website:
www.abdn.ac.uk/arkleton/conf2000/papers.s html
Information and the opportunity to purchase your tree as part of the Coldplay afforestation scheme:
www.futureforests.com

SEE ALSO
deforestation

AGRICULTURE

A strict definition of agriculture would be the practice of cultivating the soil in order to produce crops. However, in its broadest sense it often also includes pastoral farming, which involves the rearing of animals.

Early people were hunters and gatherers. Relics of this primitive agriculture can still be found practised by rainforest tribes. Gradually husbandry (looking after animals) and crop farming skills improved and small-scale sedentary farming developed. Archaeological dating techniques suggest that cereal farming and domestication of animals was widespread in Mediterranean countries by 7000 BC. Over time, specialised techniques developed; for example, the Egyptians were among the earliest people to develop widespread agriculture using irrigation and manure. By Roman times (200 BC to AD 400) agriculture was commonplace in Western Europe. From the seventeenth century onwards, selective breeding of crops and animals began to improve yields and the use of crop rotation allowed land to be used continuously. The greatest impact on agriculture came with the Industrial Revolution and the development of new farm machinery, leading ultimately to the development of highly intensive and factory farming methods, especially in Europe and the USA.

Today, in many parts of the developing world, agriculture is still very labour intensive, especially where rice cultivation dominates. Overall, about 75 per cent of the world's population is engaged in farming, although this masks great differences. In the UK less than 2 per cent of the population is employed in agriculture, whereas in Bangladesh the number tops 80 per cent. This is largely because in the UK the farms yield high outputs and machinery is widely used.

The type of farming in an area is heavily controlled by the physical conditions in an area, the climate, relief and soils. All crops and animals require a reliable water supply and most plants cease growing at temperatures below 6°C. Crops grow best on deep, fertile free draining soils, while less fertile soils are more suited to pastoral farming. Lowland areas with gentle relief are more easily farmed. Land that is high and steep tends to be used for pastoral farming, although the use of terracing may allow some crop growth, for example where olives are grown in the Mediterranean and on tea plantations in parts of India.

Human factors may also play a part in the type of agriculture practised in an area. Labour is in abundance in some parts of the world, such as in South-East Asia where rice growing employs large numbers. This contrasts with upland areas. For example, in the UK population densities are low and traditional hill sheep farming is adapted to low inputs of labour – often referred to by the phrase 'one man and his dog'. The availability of capital to invest, the spread of diseases such as BSE (bovine spongiform encephalopathy) and foot and mouth, and the availability of markets all play a part. In developing countries governments have

Classification of farm types

Arable, pastoral or mixed
Arable farms grow crops while pastoral farms concentrate on the rearing of animals. Where farms both grow crops and rear animals it is called mixed farming – common in large areas of Western Europe.

Subsistence or commercial
Many farmers, especially in the developing world, produce food only to feed themselves and their family. This is called subsistence farming and contrasts with most developed countries where farming is commercial – the crops and animals are sold in order to make a profit.

Intensive or extensive
Farms with high inputs of labour or capital and with high outputs or yields per hectare are intensive farms such as rice growing in South-East Asia, market gardening in The Netherlands and cereal production in East Anglia, UK. Extensive farms tend to use large areas of land and have low inputs of labour and capital. As a result yields per hectare are lower, such as on hill sheep farms in the UK, shifting cultivation in the Amazon basin and wheat growing in the Prairies of North America.

Sedentary or nomadic
Most farming is sedentary, characterised by permanent settlements and the same land farmed each year. However, there are still small pockets of the world where farming remains nomadic and farmers move around looking for fresh pastureland to graze animals or for new plots of land to cultivate.

received international loans to assist agricultural developments such as the implementation of the Green Revolution and policies such as the Common Agricultural Policy in the EU.

Agriculture can have profound effects upon the environment. Over-cultivation and over-grazing can lead to vegetation being destroyed, making soils vulnerable to erosion by wind and rain. Soil erosion is a major problem throughout the world and has significant local effects on food production. The use of

Case study

Two cotton farmers

Billy Tiller and Mama Idrissou both farm cotton, although a few thousand kilometres and the Atlantic Ocean separate them. Both resemble each other in some ways – they are in their thirties, both followed their fathers into farming, both wear baseball caps and both love to farm.

These two farmers produce the same amount of cotton, but experience great differences in climatic conditions, have different size farms, use differing production techniques, and operate under different political conditions.

Billy Tiller, USA

I farm in west Texas near the border with New Mexico. I own 2146 hectares, of which 1611 hectares are planted with cotton. I don't have irrigation and have to rely on the rainfall if it comes. The cotton is always a gamble because of the harsh climate – sandstorms, tornadoes, floods, droughts, heatwaves and frost are all possible. It is like a slot machine – you put the money in, pull the handle and sometimes nothing comes out. That's what happened in 2003 – erratic rains, strong winds, hail, cold and static storms ruined the cotton. I was left with only 21 hectares of cotton, and that was pretty poor and straggly. I don't worry too much because I get a subsidy from the US Government, but even so I only just broke even in 2003. Now I have started to plant sorghum and keep some cattle.

Mama Idissou, Benin

I farm 30 hectares of land south of Parakou in Benin in western Africa and in 2003 I was able to harvest about 20 hectares of cotton. The cotton grows well in the muggy wet climate – rainfall is not a problem here. The plants are healthy but I have had better – in one year the yields were more than eight times as much as those in Texas. But we are still poor – I find it difficult to sell the cotton because I can't compete with the American growers who are subsidised. It is a pity, because the cotton is helping to develop our country. The money has built schools, roads and clinics and people now live longer, more of them can read and write, and they are better nourished.

Source: Adapted from 'The subsidies gap', *The Guardian*, 8 September 2003.

chemicals in farming causes pollution of water sources and can have damaging effects on ecosystems. Excess nitrates, for example, can lead to rivers becoming choked by algae – a process known as eutrophication. The clearing of land for ranching or crop growing is a major cause of deforestation, especially in countries such as Brazil and Indonesia. Huge fires associated with these clearances can lead to atmospheric pollution.

WEBSITES

General information on agriculture:

www.agriculture.com

The Food and Agricultural Organisation of the UN:

www.fao.org

Information on cotton farming:

www.cottonfarming.com/home/main.ihtml

www.agr.state.tx.us/education/teach/ mkt_cotton.htm

SEE ALSO

Common Agricultural Policy, Green Revolution, extensive farming, intensive farming, organic farming, rice, soils

AIR MASSES AND FRONTS

Air masses are large parcels of air that sit over a particular region for a long time: so long, in fact, that they ultimately take on a particular set of characteristics from the place over which they are sitting. Air masses may be as large as continents. They form most easily in places where there is little pressure difference or where no low pressure exists that would mix and carry the parcel away before it could be modified by the surface over which it lies.

Air that stagnates over cold, dry polar regions becomes cold and dry and is called a 'continental polar' air mass. If it is really cold, it is termed 'arctic air'. Air that has its origins over cold northern or southern oceans is cold and moist and is called 'maritime polar'. Similarly, air that sits over a hot, dry desert is called 'continental tropical'. Tropical air that forms over warm seas is termed 'maritime tropical' and is warm and high in water vapour content. Truly hot, moist tropical air is sometimes called 'equatorial air'.

After a suitable period of time over the source region, an air mass takes on the temperature and moisture characteristics of the region. If you are located in the source region, the temperature and moisture conditions of the air mass are 'normal' conditions for you. In most cases, however, the air mass is prompted to move in response to differences in pressure. Low pressure areas (depressions) in the middle latitudes in the northern hemisphere commonly pull air from tropical areas to the eastern side and polar air masses to the western side. The mixing is in the opposite direction in the southern hemisphere. Because air that is dissimilar in temperature and humidity does not easily mix, fairly sharp boundaries are formed between dissimilar air masses. These boundaries are called 'fronts'.

Air masses are an important influence on the environments they pass over. For example, much of the Great Plains of the northern USA and southern Canada are affected by air masses originating in the subarctic region far to the north. This influences the types of crops that can be grown and the natural vegetation, which in turn affects everything from **BIODIVERSITY** to the migrations of birds. Air fronts are also important to environments around the globe. Air is constantly on the move, mixing warm and cold air to moderate temperatures and bringing precipitation and storms to different regions. Prevailing winds that drive air masses influence the type of vegetation that grows on windward versus leeward sides of mountains (**RAIN SHADOW**), the effect of the jet stream and even ocean currents.

Fronts: the air mass battlegrounds

Fronts are the boundaries, at the surface, between air masses. Their advance and retreat reminded early meteorologists of the move-

ment of army troops in wartime and they were given the name 'fronts' as a result. If the air that is advancing is warm, the front is termed a warm front. If cold air is advancing, the boundary is called a cold front. If there is no movement along the boundary then the front is termed 'stationary'. Sometimes a cold front overtakes the warm front at the surface, forcing the warm air aloft. This is called an occluded front.

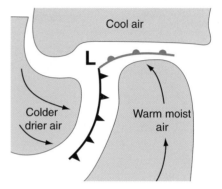

Air masses being moved by low pressure system in the northern hemisphere

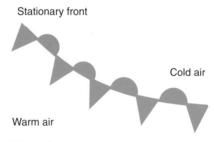

Stationary front map view

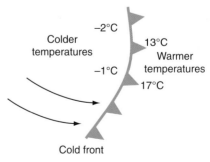

Cold front map view

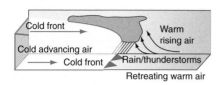

Cold front cross-section

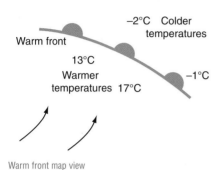

Warm front map view

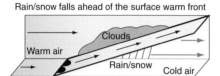

Warm front cross-section

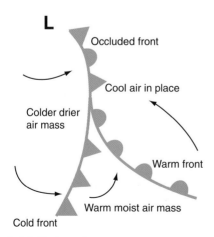

Occluded front map view

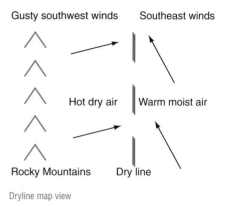

Dryline map view

WEBSITES

University of Illinois WW2010 Project: http://ww2010.atmos.uiuc.edu/(Gh)/guides/home.rxml

MetOffice's Air Masses and Fronts Education Resources: www.met-office.gov.uk/education/training/air.html

Weather Talk from the Naval Meteorology Command: http://pao.cnmoc.navy.mil/pao/Educate/WeatherTalk2/indexnew.htm

Air Masses and Fronts from the University of Wisconsin: http://physics.uwstout.edu/wx/Notes/ch8notes.htm

AMENSALISM

An elephant tramping through the bush has a detrimental effect on numerous populations of animals and plants, with no gain to itself. This is amensalism. Amensalism is a **SYMBIOSIS** in which one species is harmed and the species doing the harming receives no benefit. Amensalism characterises many human actions. For example, humans may use a broad-spectrum insecticide that kills many types of insects – some harmful, some harmless, and some beneficial (and who knows what else). Amensalism might be considered a sort of 'ecological clumsiness'.

One particular form of amensalism is allelopathy, when one species produces and releases chemical substances that inhibit the growth of another species. Allelopathic substances range from acids to bases to simple organic compounds. All these substances are chemicals produced by plants that seem to have no direct use in metabolism. For example, the antibiotic juglone is secreted by black walnut (*Juglans nigra*) trees. This substance inhibits the growth of trees, shrubs, grasses and herbs found growing near black walnut trees.

In the chaparral vegetation of California, USA, certain species of shrubs, notably *Salvia leucophylla* (mint) and *Artemisia californica* (sagebrush) are known to produce allelopathic

Artemisia californica. Photo: Joseph Kerski

Case study

Mould and penicillin

Bacteriologist Alexander Fleming returned to his research laboratory at St Mary's Hospital in London after the First World War. His battlefront experience had shown him how serious a killer bacteria could be, more effective in killing even than enemy artillery. He wanted to find a chemical that could stop bacterial infection and thereby save lives. He discovered lysozyme, an enzyme occurring in many body fluids, such as tears. It had a natural antibacterial effect, but not against the strongest infectious agents, so he kept looking.

Fleming had so much going on in his lab that it was often in disorder. This disorder proved very fortunate. In 1928 he was straightening up a pile of Petri dishes where he had been growing bacteria, but which had been piled in the sink. He opened each one and examined it before tossing it into the cleaning solution. One made him stop and think. Some mould was growing on one of the dishes. This was not too unusual, but all around the mould, the staph bacteria had been killed. This *was* highly unusual. Taking a sample of the mould, Fleming found that it was from the *Penicillium* family, later specified as *Penicillium notatum*.

Fleming published his findings in 1929, but they raised little interest. He found some chemists to help refine and grow the mould. By 1938 Australian Howard Florey was appointed professor of pathology at Oxford University, hiring German refugee Ernst Chain, who became enthusiastic about the search for antibacterial chemicals. He found Fleming's old paper on penicillin. The Oxford team, as Florey's researchers have become known, began experimenting with the penicillin mould, including injecting it first into mice, and then into humans.

By then, the world was at war again, and penicillin was scarce. Florey's connections at the Rockefeller Foundation in the USA funded further research, and an agricultural research centre in Illinois had developed techniques of fermentation needed for penicillin growth. The nutrient base for the penicillin grown there was corn (maize), which was not commonly grown in the UK. The penicillin from corn yielded almost 500 times more penicillin than from other bases. More productive strains of the mould were sought – one of the best came from a rotting cantaloupe from a Peoria, Illinois market. From January to May 1943, only 400 million units of penicillin had been made, but by the time the war ended, US companies were making 650 billion units a month. Fleming, Florey, and Chain were all awarded the 1945 Nobel Prize for Medicine. During 1998 a buyer from Cheltenham UK purchased a small vial of mould used by Fleming, on behalf of an American client – paying £8050 at Christie's in London.

The interaction between the mould, *Penicillium*, and bacteria seems to be an example of amensalism. The mould (a fungus) secretes a substance (penicillin) that adversely affects bacteria. It breaks down their cell walls, but the mould seems to get nothing in return. Humans, however, have benefited immensely from this environmental process.

substances. These chemicals often accumulate in the soil during the dry season, reducing the germination and growth of grasses and herbs in an area up to one to two metres from the secreting plants. This is an amensal relationship.

WEBSITES

Species interactions: competition and amensalism from the University of Alaska: www.geobotany.uaf.edu/teaching/biol474/lesson08notes.pdf

Competition and amensalism from Northern Arizona University: http://jan.ucc.nau.edu/~doetqp-p/courses/env470/Lectures/lec18/lec18.htm

The discovery of penicillin: www.pbs.org/wgbh/aso/databank/entries/dm28pe.html

http://history1900s.about.com/library/weekly/aa062801a.htm

ANTARCTICA

Antarctica, a continent of 14 million square kilometres, is located south of 60° south latitude. The continent is mostly covered by two enormous ice sheets – in fact, just 2 per cent of the land is exposed. Ancient Greek geographers predicted that Antarctica must exist to balance out the Arctic in the north. But it was not until American sealer Captain John Davis landed on Antarctica in 1821 that they were proved right.

Antarctica holds the world record for the coldest temperature – an amazing –89.5°C at Vostok Station on 21 July 1983. Antarctica is also very dry. It is technically a polar desert, and receives less than 2.5 centimetres of water, in the form of snow, per year. As if the cold, dry environment was not enough, Antarctica is also very windy. Local winds, called katabatic winds, develop from dense, cold air flowing down slopes as a result of gravity, and can exceed 320 kilometres per hour. Finally, Antarctica is also comparatively high in elevation – its land averages 1500 metres higher than the rest of the Earth's continents.

Antarctica has no native human populations – the few hundred people living there are conducting research or exploration.

Geographic South Pole. The copper pipe marks the exact spot of 90° south latitude as determined each January using the satellite-based Global Positioning System, or GPS. In the mid-background are flags of the original 12 signatory nations to the Antarctic Treaty. In the background is the geodesic dome housing facilities of the Amundsen-Scott South Pole Station.
Photo: Lynn Teo Simarski, NSF/US Arctic Program

However, Antarctica is not devoid of life. Algae, mosses, lichens, bacteria, mites, and springtails (insects just over 1 millimetre long) inhabit the continent. Other species, such as penguins and seals, live part of their lives on Antarctica, but hunt for food in the sea. Whales, fish and krill (shrimp-like crustaceans that grow up to 5 centimetres long) live in the frigid waters surrounding the continent.

Aside from its rugged beauty, Antarctica needs to be protected, since it plays such an important role in regulating the Earth's environment – its water levels, climate and weather. Cold waters originating in the Antarctic drive the movement of ocean currents worldwide. These currents in turn help to drive major weather systems across the planet.

Antarctica's ice sheets contain about 70 per cent of the Earth's fresh water, and contain enough water to raise sea levels by 70 metres. Global warming threatens to melt portions of these vast ice sheets, causing widespread flooding, particularly of low-lying areas such as the Netherlands. Even though few people live on Antarctica at any given moment, environmental change there has the potential to have an impact on large portions of the population far away.

Antarctica's ice sheets give us clues as to what the Earth was like hundreds of thousands of years ago. For example, scientists studying ice cores have discovered that 18,000 years ago the Earth was windier and dustier than it is today. They know this because dust from different continents is embedded in the ice from that period.

Currently, Antarctica is protected by the Antarctic Treaty of 1959. The Treaty has 45 signatories, 27 of them consultative parties on the basis of being original signatories or by conducting substantial research in Antarctica. Antarctica is administered through meetings of the consultative member nations. Several countries have made territorial claims in Antarctica – interested in the mineral wealth of the continent – but none have been recognised. In 1997 the Environmental Protection Protocol to the Treaty was ratified, designating Antarctica as a 'natural reserve devoted to peace and science'. Despite this protection, Antarctica is not safe from the effects of global environmental issues such as global warming and the 'ozone hole'.

WEBSITES

USGS terraweb resources on Antarctica:
http://terraweb.wr.usgs.gov/projects/Antarctica
CIA World Factbook:
www.odci.gov/cia/publications/factbook/geos/ay.html
Rice University 'Glacier' pages:
www.glacier.rice.edu
Australian Antarctic Division:
www.aad.gov.au
NOAA's role in the South Pole:
www.nmao.noaa.gov/oldies/nspt.html
Antarctic Survey:
www.antarctica.ac.uk

SEE ALSO

ozone

ANTICYCLONES

Anticyclones are areas of higher atmospheric pressure than the areas surrounding them. They occur where the air is sinking towards the Earth's surface and are therefore usually associated with dry and sunny weather conditions. Once the descending air reaches the ground it spreads outwards (diverges), which explains why air moves out from the centre of an anticyclone. Due to the rotation of the Earth, the air spirals outwards in a clockwise direction in the northern hemisphere and anticlockwise in the southern hemisphere.

Does water flowing down a drain really move in the opposite direction in the

---

Antarctica – a scientific research lab

Antarctica provides a unique location for many scientific research projects. Probably the most famous observation to arise from Antarctic research is the discovery of the 'ozone hole'. Every year, starting in September and lasting for a few months, the ozone layer in the stratosphere above the Antarctic thins by as much as about 70 per cent. This thinning is referred to as the 'ozone hole'. During the mid-1990s, the ozone hole was about as large as the continent of North America. This discovery provoked great concern, since a thinned ozone layer lets in increased amounts of hazardous ultraviolet radiation from the sun. This radiation can cause skin cancer, eye cataracts and other health problems in humans.

Many other scientific projects are currently underway on Earth's southernmost continent, such as the studies of clean air conducted by the US National Oceanic and Atmospheric Administration (NOAA). Since the nearest urban area is over 4800 kilometres away, atmospheric influences that are solely from local sources can be avoided. NOAA's site on Antarctica is one of four placed around the world to track changes in composition of the atmosphere and link these changes to climate patterns.

southern hemisphere from in the northern hemisphere? This is a popular myth, but it simply isn't true. What *is* true, though, is that **AIR MASSES** move in opposite directions from high and low pressure zones in the northern versus the southern hemisphere. The Earth is a spheroid, slightly flattened at the poles. It spins each day on its axis. The combination of the spherical shape of the Earth and its rotation causes air to 'turn' as it moves. Air flowing towards a particular place at a given time finds that place in a different location due to the Earth's rotation. This apparent shift to the right in the northern hemisphere and left in the southern hemisphere is known as the Coriolis Effect.

The Coriolis effect is a force of inertia described by the nineteenth-century French engineer-mathematician Gustave-Gaspard Coriolis in 1835. Coriolis showed that an inertial force – acting to the right of the direction of body motion for anticlockwise rotation of the reference frame, or to the left for clockwise rotation – must be included in the equations of motion. The Earth rotates towards the east, which is why the sun appears to move towards the west. Looking down on the North Pole, the Earth rotates in an anticlockwise motion. The effect of the Coriolis force is an apparent deflection of the path of an object that moves within a rotating coordinate system. The object does not actually deviate from its path, but it appears to do so because of the motion of the coordinate system.

For example, if a group of children on a merry-go-round roll a ball back and forth to each other, the ball looks as if it is curving to the children on the merry-go-round. To an observer from above, however, the ball rolls in a straight line. In another example, if a cannon were fired towards the Equator from the North Pole, the projectile would land to the right, or west, of its true path. A similar displacement to the right occurs if the projectile is fired in any direction.

Anticyclones, otherwise known as high pressure systems, are composed of masses of air that have descended from higher in the atmosphere or are pooled near the surface due to density. If this occurs, and the pressure of the air at this location is higher than in the surrounding area, as measured by a barometer, then this place is by definition the location of a centre of high pressure or an

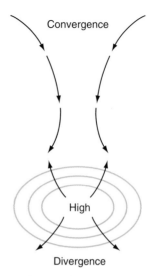

Airflow around an anticyclone in the northern hemisphere. From the University of Wisconsin

anticyclone. The air from the centre of the anticyclone moves outwards to areas of lower pressure.

If the air moving out of the anticyclone is replaced from above or additional chilling continues to make the air denser, the high will persist and may even strengthen. If this does not happen, the air in the centre of the high is depleted and the pressure falls to a point where it is equal to the surroundings and the high disappears.

High-pressure conditions are significant features of the atmosphere, affecting our weather and **CLIMATE**. They also affect the environment over broad areas of the earth. Winds are often quite light around an anticyclone, indicated by wide spacings between isobars (lines of equal pressure) on a weather map. Where anticyclones persist over broad regional areas, we typically find **DESERTS**, for example over northern Africa where the Sahara desert is located. Anticyclones can be identified on weather charts as large areas where pressure is higher than in surrounding areas.

In winter in the UK, the clear, settled conditions and light winds associated with anticyclones can lead to frost and fog. Clear skies allow heat to be lost from the surface of the Earth by radiation. This allows temperatures to fall steadily overnight, leading to air or ground frosts. Light winds and falling temperatures can encourage fog to form; these fogs can linger well into the following morning and be slow to clear. If high pressure becomes established over Northern Europe during winter, this can bring a spell of cold easterly winds to the UK.

Pollution incidents are often more serious during anticyclones. This is because sinking air prevents the dispersal of pollutants, causing a concentration near the ground. In the 1950s, London suffered horrendous smogs during winter anticyclones, resulting in the premature deaths of thousands of people. Cities such as Los Angeles, Mexico City, and Athens suffer particularly bad pollution incidents, mainly resulting from vehicle emissions, during anticyclones.

In summer in the UK, the clear settled conditions associated with anticyclones can bring long sunny days and warm temperatures. The weather is normally dry, although occasionally very hot temperatures can trigger thunderstorms. An anticyclone situated over the UK or near the European continent usually brings warm, fine weather.

Long, hot summers can lead to droughts and water shortages. The UK was particularly badly affected by a drought in 1976 when reservoirs dried up and many people had to have their water supplies rationed. The heavy thunderstorms that frequently mark the end of anticyclonic conditions have the potential to cause serious SOIL EROSION, particularly if vegetation cover has been reduced.

WEBSITES

The UK Environmental Change Network: www.ecn.ac.uk/Education/anticyclones.htm
WW2010 Project from the University of Illinois:
http://ww2010.atmos.uiuc.edu/(Gh)/guides /mtr/fcst/sfc/hgh.rxml
UK Met Office:
www.met-office.gov.uk
Movie and explanations of the Coriolis Force:
http://ww2010.atmos.uiuc.edu/(Gh)/guides /mtr/fw/crls.rxml
and
http://zebu.uoregon.edu/~js/glossary/ coriolis_effect.html

SEE ALSO

pollution

ARTESIAN BASINS

An artesian basin is a large saucer-shaped geological structure in which one or more aquifers are sandwiched between layers of impermeable rock. This formation traps water underground and is one of the most important global sources of fresh water. Artesian basins are relatively easy to exploit because the water is under great pressure. Any well or borehole dug down into the saturated rock acts as a release to that pressure, causing water to rise upwards without any human assistance. The presence of an artesian basin is of huge value, often making the difference between an area being economically viable and not so, as in north-east Australia, south-east England, the North American High Plains and northern India.

London, a developed world city with a population of 7.5 million people, needs a huge water supply. Yet it is located in one of the driest parts of the UK and on low-lying land that is not particularly suitable for reservoir construction. Fortunately London is situated on an artesian basin called the London Basin. The aquifer rock is chalk, which is sandwiched between impermeable clays above and below. The chalk layer comes to the surface on either side of the city in the North Downs and the Chiltern Hills, where it is able to soak up rainfall to recharge its supply. In the past, the water pressure from this artesian structure resulted in a natural fountain in what is today Trafalgar Square. Now, however, the pressure has been reduced by excessive exploitation and the fountain is created artificially. Water demand and supply is unevenly distributed throughout the year. Maximum extraction takes place in summer when demand is highest, followed by winter recharge when rainfall is higher and demand somewhat reduced. However, for a long time now, recharge has not always been sufficient to maintain the level of the water table. During the last 150 years overextraction has lowered it by 60 metres in Central London. Saline water has also begun to seep in from undersea aquifers, polluting this precious resource.

India experiences a monsoon climate, meaning that rain only falls in the wet season, equivalent to the northern hemisphere's

Case study

Australia's Great Artesian Basin

The Great Artesian Basin, the largest in the world, is big enough to underlie one fifth of the Australian landmass. It includes parts of four states – Queensland, the Northern Territory, South Australia and New South Wales – and stretches from the Great Dividing Range in the east to the Lake Eyre Depression in the west. Much of this region is either semi-arid or arid on the surface, which obviously increases the potential value of this resource. The Basin's total 1,711,000 square kilometres is estimated to store 8700 million megalitres of water (1 megalitre = 1000 litres).

This artesian basin contains several aquifers. Sandstone beds contain the water, which are surrounded by impermeable layers of mudstones and siltstones that confine the water, preventing it from seeping away. This particular sedimentary rock structure was formed 250–100 million years ago. Some of the rock beds also contain oil and natural gas reserves.

The water stored in this rock formation falls as rain on the Great Dividing Range, several hundred kilometres to the east and north-east. Today, it is utilised in several ways and it makes a significant difference to the economy of this environmentally difficult region. Water for irrigation systems enables improved quality pasture to be provided, allowing higher stocking densities in this outback zone. The oil and gas reserves are also exploited. Even a small tourist industry is emerging, as at Julia Creek, which claims to be the centre of the Great Artesian Basin.

The leisure facilities at Julia Creek are largely based around water. There are pools and picnic areas, shaded by woodland that could not possibly grow without the underground water source. Information centres let visitors find out more about the geological background to the area and the ways in which the artesian water is exploited. The advertising material refers to the area as an 'oasis' – a comfortable base for an outback holiday.

summer months. When it does rain it is truly torrential, so there is plenty of water, but for the other eight months of the year a dry climate prevails. The River Ganges remains a year-round water source and so attracts dense settlement. However, population can spread out over the entire region because it is underlain by an artesian basin from which well water is drawn to feed irrigation systems in the dry season.

Some richer countries, though not often as arid, also benefit from such a rock structure. North America from Saskatchewan to Kansas is underlain by a huge artesian basin. Here, the resource supports large-scale grain production and cattle ranching in states that would otherwise be much less productive.

WEBSITES

Julia Creek (Centre of the Great Artesian Basin):

www.juliacreek.org/artesian.htm

Tas Walker's website:

www.uq.net.au/~zztbwalk/gab.html

The US Geological Survey:

http://capp.water.usgs.gov/GIP/gw_gip/how_c.html

SEE ALSO

water supply

ATMOSPHERE

'All I need is the air that I breathe', sang the Hollies in 1974. All life *does* depend upon the thin layer of air surrounding the Earth – the atmosphere.

The atmosphere is a layer of gases and aerosols that surround and are held to the Earth by gravity. Composed primarily of nitrogen (78.8%) and oxygen (20.95%), the atmosphere also contains lesser amounts of other gases. Although present in small amounts, some of these gases, especially water vapour, carbon dioxide and OZONE are very important to the environment. Water is needed for life and energy transport, carbon dioxide for its beneficial GREENHOUSE EFFECT and ozone because it absorbs dangerous ultraviolet wavelengths of sunlight. The atmosphere also contains small non-gaseous solid and

Earth's atmosphere as seen from the space shuttle Columbia. NASA

liquid aerosol particles, including beneficial ones that make up **CLOUDS** and precipitation and others which serve as nucleating agents for cloud and precipitation formation. Non-beneficial aerosols include pollutants.

Because air is a compressible gas and gravity is always pulling it towards the Earth, the number of air molecules decreases in a regular fashion with altitude. As a result, air pressure always decreases regularly as one goes higher in altitude. Atmospheric composition also varies depending on altitude. In the heterosphere, below 80 kilometres, the gases are well mixed. Overall proportions of gases are fairly constant, except for some, such as water vapour and carbon dioxide, that vary regionally and seasonally. In the homosphere, above 80 kilometres, the gases are not mixed by vertical and horizontal motion and separate somewhat by mass with the lighter elements preferentially occupying the highest locations. At the very top of the atmosphere, individual atoms of some of the lightest gases, like hydrogen, drift off into space, making it difficult to determine where the atmosphere ends and space begins.

The atmosphere may also be subdivided according to the temperature characteristics found at different altitudes. The lowest layer, the troposphere, which is about 16 kilometres thick at the equator and 6 kilometres thick at the poles, contains most of the water vapour and has most of the weather. The troposphere's main characteristic is that temperatures typically fall as one goes higher into it. The temperature profile will differ depending on location, but averages a 6.5°C decrease per

1000 metres. Have you ever been in an aeroplane and were amazed to discover how frigid the outside temperature was on the digital displays? At the top of the troposphere, temperatures range from –60°C to –70°C.

Above the troposphere, up to about 50 kilometres, is the stratosphere. This layer is defined by a halt in the decline of temperatures that occurs in the troposphere, with constant or isothermal temperatures at its base and then steadily increasing temperatures as one goes further up. This heat is caused by absorption of solar ultraviolet radiation by ozone in this layer. The fact that the temperature no longer gets colder within this layer puts a stability 'cap' on the troposphere below. Some aeroplanes fly in the lower stratosphere because it is quite stable.

From about 50 kilometres to 80 kilometres the temperature once again declines through what is called the mesosphere. From between 80 and 120 kilometres temperatures in the thermosphere once again increase due to the intensity of solar radiation at this, the very top of the atmosphere. At 120 kilometres the atmosphere is so thin that it is virtually indistinguishable from the vacuum of space.

Between each layer of the atmosphere is a fairly abrupt transition zone. The tropopause separates the troposphere and the stratosphere. By moving across this boundary we leave the zone of weather. The stratopause is between the stratosphere and the mesosphere, and the mesopause is found between the mesosphere and the thermosphere.

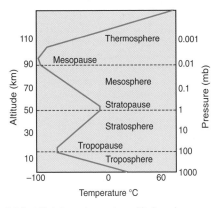

Relationship between temperature, altitude, and pressure. Shodor Education Foundation

Pressure always decreases with height because there are fewer air molecules as one goes up into the atmosphere. The air gets progressively thinner because most of the compressible atmospheric gases are held near the surface, pulled by gravity. Because temperature depends on collisions of air molecules it is usually cooler as one goes up in the lower parts of the atmosphere.

WEBSITES

NASA Atmosphere:

http://liftoff.msfc.nasa.gov/academy/space/atmosphere.html

UCAR's Windows to the Universe:

www.windows.ucar.edu/tour/link=/earth/Atmosphere/layers.html&edu=elem

Layers of the Atmosphere:

http://wings.avkids.com/Book/Atmosphere/instructor/layers-01.html

BATTERY FARMING

Battery farms are a type of factory farming, characterised by highly intensive production, usually associated with eggs and poultry.

Battery hens are placed into cages at 18 weeks old, when they begin to lay eggs. The cages do not allow the hens free movement – they cannot flap their wings and in some cases it is difficult for them to stand upright. The hens may lose their feathers as a result of the stress they suffer or from being pecked by other birds. In most battery farms the hens feed from a conveyor belt that passes the outside of the cage. Each hen produces about six eggs a week for a year. Having finished egg production it is slaughtered and used in the processed food industry. This type of farming is highly regulated and has been much maligned in recent years as being cruel to the birds. Organically farmed produce is promoted as being more desirable.

WEBSITES

SAFE, an animal welfare organisation in New Zealand, produces an interesting factsheet:

www.safe.org.nz/information/issues/print/batteryhensp.php

Vancouver Humane Society:

www.vancouverhumanesociety.bc.ca/issues_battery_hens.html

Australia's Animal Liberation, with material on battery farming:

www.animalliberation.com.au

SEE ALSO

intensive farming

Case study

UK egg production

Today about 20 per cent of eggs bought in the UK are free-range eggs, but why is this percentage not higher? It is probably because of cost. Battery farming is highly efficient and a cheap method of production. Alternative production systems would add about a third more to the price of eggs and cost-conscious consumers currently seem unwilling to pay the extra for the free-range commodity.

Recent legislation in the UK now restricts the number of birds to four per cage (formerly it was five). This will be more comfortable for the hens but will also affect the price of eggs. A farmer who had 20,000 hens in 4000 cages can now only house 16,000 hens. The farmer's income falls by 20 per cent but the costs remain about the same. In addition, a new system of labelling is to be introduced. Instead of the egg boxes being dated, every individual egg must be dated – adding further to the egg farmer's costs. Does the farmer accept the cut in income or put up the price of eggs? Recent press coverage in the UK has warned of a likely shortage of eggs as the new legislation comes into force. So is the likely higher cost of eggs and a potential shortage, at least in the short term, worth it?

The Government is now considering an outright ban on keeping hens in battery cages – but will consumers pay for the alternative free-range eggs? *continued*

According to Mark Williams, Chief Executive of the British Egg Industry Council:

. . . A total ban will remove consumer choice . . . We produce eggs in all different systems from cages, from barns and from free-range and organic . . . [and] figures show that 70 per cent of British consumers are still purchasing caged eggs. At the moment a pack of six eggs is 38p in the supermarket, the similar free-range eggs would be about 68p . . . reflecting the additional cost of production. If we ban cages ahead of time, all that will happen is that we will have eggs from continental competitors produced in cages . . . We have to be practical about this.

But Jacky Turner from Compassion in World Farming believes such a move is long overdue:

. . . There are people in the industry who are really unwilling to get rid of cages . . . Even in enriched cages the hens are still squashed into a cage and it's completely unacceptable. We want to see hens in a barn, not in a cage. Germany has banned enriched cages, so there's a good chance they will go. Cheap imports are a serious worry . . . Countries that have welfare laws must be allowed to refuse imports – we need to keep out eggs that don't meet the standards, keep out battery eggs. Our job here is to defend the interests of farm animals and people are prepared to pay more. For the minority there is a cost implication . . . it might mean some sort of government subsidy.

Case study

This little piggy went to market

Pigs can live up to 20 years and are believed to be highly intelligent, sociable animals. But 95 per cent of them are raised in confined buildings, spending their entire, short lives indoors, and being slaughtered at six months of age. In some cases the female pigs used for breeding are kept in crates where they are unable to walk or even turn around. They live on cold concrete floors for the four months of their pregnancy. Piglets can freely access the mother's teats but the sow cannot move away, leading to lacerations. The piglets are taken away from the mother after just two to three weeks, in comparison to the 15 weeks usually, and crowded into small nursery pens. Appalling conditions and great suffering are described in graphic detail by some commentators – castration, teeth and tail clipping (all without anaesthesia), high levels of disease including pneumonia, high death rates, signs of madness and stereotyped behaviours – all to increase productivity. As the *National Hog Farmer* journal put it 'Crowding pigs pays'.

In 1997 a veterinary committee in the EU reported serious health and welfare problems that led to the banning of sow stalls in the UK and other EU countries. However, they are still legal in the USA and used by almost all producers.

BIODIVERSITY

More species of ants may be climbing a single tree in Amazonian Peru than exist in all of the UK. This is a statement about biodiversity. Biodiversity is the full variety of the biosphere – all the species of living organisms, including plants, animals and microorganisms. Biodiversity also includes the genetic variability within species, and the biotic communities in which they interact.

Whether or not we realise it, humans depend on biodiversity for our survival and quality of life. The food that we eat, medicines that we use, energy we use and structures that we live in – everything depends upon the amazing variety of species found on the Earth. Biodiversity underpins the processes that make life possible. Healthy ecosystems are necessary for maintaining and regulating atmospheric quality, climate, fresh water,

Jesus Christ Lizard, so named because it can walk on water, also known as Common Basilisk, *Basiliscus basiliscus,* in Costa Rica. Photograph by Joseph Kerski

marine productivity, soil formation, cycling of nutrients and waste disposal. Biodiversity is intrinsic to values such as beauty and tranquillity. Underpinning biodiversity is the concept of environmental ethics – that no species, and no generation, has the right to sequester the Earth's resources solely for its own benefit.

A different but related and sometimes confounded concept is ecological diversity. Ecological diversity considers the number of species in a given area (species richness) and the degree to which individuals are distributed across the given species (species evenness).

The extraordinary extent of biodiversity around the globe has led ecologists to observe that biodiversity is not only more complex than we know, but also more complex than we can know. Even the simple number of species times their numbers of interactions in the Earth's biotic communities probably exceeds the number of potential connections in the human brain. In the face of the Earth's prodigious biodiversity, it is not boastfulness but humility that has caused some ecologists to muse that 'Ecology is not rocket science. It's more difficult.' Thus the most significant impediment to the conservation and management of biodiversity is our lack of knowledge about it and the effects of human population and activities on it.

Even the simplest of questions about biodiversity, such as 'How many species are there?', cannot be answered even within an order of magnitude. The number of species described to date is roughly 1.5 million. About one-fifth of these are plants, and most of the rest are animals. Only about 5000 kinds of bacteria have been cultured and named to date, but some ecologists have speculated that the number of 'species' of microbes may exceed the number of 'higher' organisms.

'Higher' organisms – animals and plants – are certainly better known, but our ignorance is still so large that it is not possible to estimate the number of higher organisms with any real confidence. Dr Terry Erwin of the Smithsonian Institution has estimated the total number of species on the Earth at between 3 to 30 million. He based this estimate on the known to unknown species in samples of insects from a variety of tropical environments, and on some assumptions about degrees of specialisation and whether species are native to a particular place (known as 'endemism'). Therefore, the task of cataloguing the Earth's biota is somewhere between only 50 and 5 per cent finished. Using either estimate, there is still much work to be done.

How many species have there ever been? We don't know that either. To come up with a conservative answer, suppose that there really was a 'Cambrian explosion' – a tremendous increase in the number of species during the Cambrian Period, 500 years ago. Further suppose that the number of species was low prior to about 600 million years ago. Then suppose that the average lifetime of a species is 1 to 10 million years. This is a reasonable enough number, based on the history of the biota of the past 50 to 100 million years as known from the fossil record. Therefore, since the Cambrian Period, there has been enough time for 60 to 600 'turnovers' of the biota. That suggests that the biota at any point in time – such as the present – is no more than about 1 per cent of the total number of species over evolutionary time, and it may be a only a very small fraction of 1 per cent. And how many species is not the question. Biodiversity is not merely an accounting problem – a list of species. Rather, it is the number of species times the genetic diversity that they encompass times the interactions between and among species in the biosphere:

Biodiversity = number of species × genetic diversity × interactions between and among species.

WEBSITES

World Resources Institute:
www.wri.org/wri/biodiv
United Nations Environmental Programme, Convention on Biological Diversity:
www.biodiv.org/default.aspx
'The Tree of Life' – a global, collaborative effort documenting biodiversity, compiling all available information on the pattern of evolution of life on Earth:
http://tolweb.org/tree/phylogeny.html
US National Biological Information Infrastructure, consortium of governmental and non-governmental organisations providing information about biodiversity:
www.nbii.gov/issues/biodiversity/
NatureServe, a collaborative between the Nature Conservancy and the H. John Heinz III Centre for Science, Economics, and the Environment:
www.natureserve.org
The Smithsonian Institution's booklet *Biodiversity: Connecting with the Tapestry of Life*:
http://nationalzoo.si.edu/Conservationand Science/MAB/publications/default.cfm
Species 2000 project, with a goal to enumerate all known species of organisms on the Earth (animals, plants, fungi and microbes), serving as the baseline dataset for studies of global biodiversity, at the University of Reading, UK:
www.sp2000.org/sp2000org.html

BIOFUELS

Humans have utilised the energy from organic matter for thousands of years – ever since the first cave dweller started burning wood to cook food or just to keep warm. This organic matter used for energy is referred to as a biofuel.

The search for renewable sources of energy in the past few years has resulted in the development of biofuels as alternative forms of energy. These are crops or organic wastes that can be used for fuel, harnessing the chemical energy produced via photosynthesis and stored in green plants. The plants can be burnt to produce steam, roasted into charcoal or converted into ethanol or biogas through anaerobic (without oxygen) reactions. These are the same conditions that were necessary for **COAL** to be formed.

Biofuels also include energy crops such as willow, alder and other fast-growing, high carbon-content plants that are grown specifically for energy generation. Research has shown that the burning of 40 hectares of willow could provide enough electricity for 5000 houses.

Schemes have already been introduced at the local scale that have had a number of positive impacts. For example, at Weobley Primary School in Herefordshire, UK, locally grown willow and poplar is used successfully to heat the school. Once harvested, the wood is dried on-site before being chipped and then transported to the school. There the chips are burnt to produce hot gases, which pass through the water boiler. The ash produced after combustion is then used as fertiliser on the school garden.

There are many misconceptions about the environmental and sustainable credentials of bioenergy. Some people mistakenly believe that it is about cutting down ancient woodland or tropical rainforest. Instead, it is about making better use of current biomass resources as well as planting new woodland and energy crops that can provide truly sustainable sources of renewable carbon fuel.

A well-managed woodland or energy crop also works in harmony with nature, and in particular follows the natural carbon cycle. On balance, wood is never removed faster than it is added to by new growth. This means that the carbon dioxide released when the wood-fuel is burnt is never more than the carbon dioxide being taken up by new growth.

As far as electricity generation is concerned, bioenergy has life-cycle emissions of 20 to 80 grams of carbon dioxide per kilowatt hour (g/kWh). This compares very favourably to fossil fuels such as coal with its life-cycle emissions of 955 g/kWh and gas with emissions of 446 g/kWh, while other renewables

Case study

Bagasse: a successful biofuel in Australia

Bagasse, or sugar cane fibre, is one of the sources of biomass energy that has been exploited in the past few years by Australian scientists. With support from the country's government, they have found that bagasse can be fed into boilers to generate steam for heat and electricity production.

In 1999 the Rocky Point Green Energy Corporation was awarded a $3 million grant to develop a biofuel plant at its Beenleigh Mill in Queensland. The plant will use bagasse during the 20-week cane crushing season, while at other times wood waste and other crop residues will be utilised.

By using bagasse, it is forecast that Australia's carbon dioxide emissions will be reduced by over 220,000 tonnes per year.

such as wind power and PV solar cells have emissions of 9 g/kWh and between 80 to 160 g/kWh respectively.

In addition to these environmental advantages, bioenergy developments also offer farmers and foresters a growing and sustainable market, and one with long-term price stability. In addition, whereas the money spent on fossil fuels goes to multinationals and shareholders, the money raised by bioenergy stays in the local economy.

There are other ways that bioenergy enhances the rural environment, as it can help bring unmanaged areas of woodland back into production. Organisations such as Friends of the Earth have shown how the cultivation of energy crops has improved wildlife habitats as well as helped to conserve and promote biodiversity.

WEBSITES

Details on energy crops:
www.britishbiogen.co.uk/bioenergy/
energycrops/encrops.htm
Planet Energy, containing further information on biofuels and suggesting other sites to visit:
www.dti.gov.uk/energy/renewables/
ed_pack/crops.html

BIOGEOCHEMICAL CYCLES

Sub-cellular organelles in a blade of grass remove carbon dioxide (CO_2) from the atmosphere and combine it with hydrogen and water to make carbohydrate. A grasshopper eats the leaf and breaks the carbohydrate down into carbon dioxide and water. Both the grass and the grasshopper are participating in a biogeochemical cycle. Biogeochemical cycles are exchanges of chemical elements between life and the physical Earth. The raw materials of life are the chemicals of the Earth's crust – oxygen, hydrogen, carbon and so on. Biogeochemical cycles are movements of elements or compounds through living organisms and the non-living environment (the Earth). Metabolism of organisms produces waste products that return to the physical Earth. When organisms die, they decompose, and this process of disassembly returns the atoms that compose the organisms to the Earth's crust. Small-scale biogeochemical cycles are important to the concept of ECOSYSTEMS. Biogeochemical cycles define the notion of the ecosphere – the global sum of ecosystems, the sum total of mutual interactions between the BIOSPHERE and the Earth's crust.

Some 24 chemical elements are known to be essential to life. Of these essential elements, 21 (87.5 per cent) are located in the first third of the periodic table of the elements. These simpler, lightweight elements can cycle between the Earth and life as gases; others flow readily through ecosystems in the form of salts dissolved in water. Of the essential elements, only molybdenum, tin and iodine lie in the more complex second third of the table. Of the non-essential elements in the first third of the periodic table, three (helium, neon, argon) are inert gases, virtually incapable of forming chemical compounds, one (beryllium) is toxic, and scandium and titanium appear not to be used by living things. Aluminum and nickel are under study

Anthrobiogeochemical cycles

Burning fossil fuels has released enough carbon dioxide (CO_2) to increase atmospheric concentration by 10 to 20 per cent in the past half century. Natural weathering of rock and decomposition of organic matter release about 20 million tonnes of phosphorus per year. Industrial mining of phosphate rock adds about 20 million tonnes of phosphate to the ecosphere each year. Biological nitrogen fixation represents 90 to 140 million tonnes per year, and industrial nitrogen fixation is about 140 million tonnes per year.

It is obvious that the nitrogen cycle is not just a biogeochemical cycle. It is an anthrobiogeochemical cycle – an inadvertent conspiracy of humankind, life in general and Earth. Similarly, the phosphorus cycle is not merely a biogeochemical cycle; it is an anthrobiogeochemical cycle. In addition, humans are involved in the global carbon cycle on a geological scale. Indeed, it is possible to make similar statements about any biogeochemical cycle. It seems likely that never – at least not since the age of primordial slime, 3.5 to 4 billion years ago – has a single species had so much influence on global biogeochemical processes as humans have. Further, our ascendancy has been geologically instantaneous, an eye-blink in the history of the biosphere. Industrial human culture is barely 200 years old, working its will on a biosphere that is 3.5 to 4 billion years old. Modern human time is 0.000005 per cent of the age of the biosphere.

Of course, human involvement in biogeochemical cycles is unavoidable. We are homeostatic individuals, which means that we necessarily participate in the chemistry of ecosystems and, by extension, the ecosphere. Further, there are over 6.4 billion of us, so in the aggregate we make a big difference. Also, despite differences between nations and across socio-economic groups within nations, our per capita demands on the ecosphere are large. We humans take perhaps 30 to 40 per cent of the annual production of the biosphere for human purposes: food, fibre, fuel and general sloppiness. We have a large negative influence on the ecosphere on a geological scale, and much of the impact is simply to obtain and use fossil fuels, a finite resource.

So what? What difference does all this make? Unfortunately, the response can only be that most unsatisfying of answers: 'It depends'. It depends upon how we value the Earth and the ecosphere, other species, other people and the future.

as possibly essential; boron is known to be essential to some plants and may prove to be essential to animals as well.

Of the 24 chemical elements known to be essential to life, six are important to building the molecules of the living cell: carbohydrates, fats, proteins and nucleic acids. These elements are carbon, hydrogen, nitrogen, oxygen, phosphorus and sulphur – C, H, N, O, P and S. These elements comprise 99 per cent of the atoms in the biosphere. Thus a large portion of what is essential to life is contained in an amazingly small number of common elements.

Some elements are essential in electrochemical reactions, such as nervous conduction, four as positively charged cations (Na^+, K^+, Ca^{++}, Mg^{++}) and one as a negatively charged anion (Cl^-). Sulphate (SO_4^{--}), and phosphate (PO_4^{--}) are other important anions. The remaining essential elements are trace elements, mainly needed in small amounts, mostly as requisite pieces of enzyme systems.

A biogeochemical cycle can be described for each of the essential elements of living things. The **CARBON CYCLE** provides a familiar and appropriate example. Because it is a cycle, we can begin at any point. Via **PHOTOSYNTHESIS**, green plants take six molecules of carbon dioxide from the atmosphere and combine them with the hydrogen atoms from six molecules of water to make a molecule of glucose, composed of carbon, hydrogen and oxygen ($C_6H_{12}O_6$). The energy to break water into two hydrogen molecules

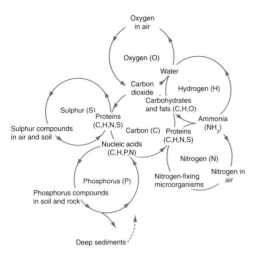

Oxygen
in air

Oxygen (O)

Water

Carbon
dioxide   Hydrogen (H)

Carbohydrates
and fats (C,H,O)

Sulphur (S)   Proteins
(C,H,N,S)   Ammonia
(NH₃)

Sulphur compounds   Carbon (C)   Proteins
in air and soil   (C,H,N,S)

Nucleic acids
(C,H,P,N)

Nitrogen (N)

Phosphorus (P)   Nitrogen-fixing   Nitrogen in
microorganisms   air

Phosphorus compounds
in soil and rock

Deep sediments

Biogeochemical cycles are linked with one another, as
this simple two-dimensional diagram hints. The real
situation in the ecosphere is much more complex, mostly
run at the level of microbes and sub-cellular organelles.
Source: Ehrlich et al, *Biology and Society*, 1976,
McGraw-Hill, with permission of the McGraw-Hill
Companies

Oxygen enters the atmosphere as a by-product
of photosynthesis and re-enters the biosphere
in the process of aerobic respiration. Indeed, it
probably is true that all biogeochemical cycles
are linked, as shown in the figure.

WEBSITES:

Information on the human impact on
nitrogen cycle from the Ecological Society of
America
www.esa.org/sbi/sbi_issues/issues_text/issue
1.htm
Biogeochemical cycles from Environmental
Literacy Organization:
www.enviroliteracy.org/subcategory.php/
198.html
An example of biogeochemical cycles in
wetlands, with implications for other
ecosystems:
www.na.fs.fed.us/spfo/pubs/
n_resource/wetlands/
wetlands7_biogeochemical%20cycles.htm
Biogeochemical cycles information:
www.marietta.edu/~biol/102/ecosystem.html

SEE ALSO

carbon cycle, ecosystems

BIOMASS

Biomass is simply the mass ('weight') of living
material in a given area. Usually biomass is
measured as grams per square metre (g/m^2).
The amounts of biomass vary widely across the

$(2H^+)$ and one oxygen molecule (O_2^-) comes
from the Sun. The 'know-how' to get the job
done resides in the chlorophyll molecule. In
the process, six molecules of oxygen gas are
released to the atmosphere as a 'waste product'. That 'waste' is, of course, essential to life,
in the air that all humans breathe.

Eventually, a molecule of glucose will be dissected in the process of respiration, or breathing. The glucose molecule's energy is captured
to power the activities of living cells, the cells of
the plant itself, or the cells of some consumer
organism (an animal, fungus or bacterium, for
example). Glucose combines with oxygen to
produce carbon dioxide and water. While it
does so, it releases energy to do 'biological
work', which includes growth, maintenance,
movement, reproduction and so forth.

In summary, through the process of photosynthesis, carbon enters the biosphere from
the atmosphere. Through the process of respiration, carbon exits the biosphere to the
atmosphere – in short, a biogeochemical cycle.

The cycle of carbon is not independent of
other biogeochemical cycles. The carbon cycle
is linked to the cycle of oxygen, for example.

The high degree of biomass of the lowland tropical
rainforest biome is evident in Selva Verde, Heredia, Costa
Rica. Photograph by Joseph Kerski

Earth's biomes. Tropical rainforests, which occupy about 11 per cent of the planet's land area, boast about 42 per cent of the Earth's terrestrial biomass. The tree canopy is high, and such biomes support a high degree of bio-diversity and plant density. By contrast, desert and semi-desert shrublands, which occupy about the same amount of the Earth's surface as rainforests, support less than 1 per cent of global biomass. Biomass sometimes is called 'standing crop', even when non-agricultural ecosystems are being studied.

The 'litter mass' – organisms that are dead but not yet decomposed to humus or returned to the atmosphere and soil – is not included in estimates of biomass, but is a related concept. Litter mass also varies widely between biotic communities. Surprisingly, it is not the tropical rainforests that contain the highest amounts of litter mass, since the highest values for litter mass do not correspond with highest values for biomass. Rather, litter mass increases as the activities of decomposer organisms decrease. Hence highest values for litter mass tend to be in boreal coniferous forests (taiga) where the producer organisms are woody plants and where cold weather slows rates of microbial activity. Therefore, taiga, covering about 8 per cent of the land surface, contains 43 per cent of the Earth's terrestrial litter mass. The litter mass here is comprised of dead trees, needles and the like. Tropical rainforests – where temperatures and humidity are high and therefore microbial activity is high – have, by contrast, just 3 per cent of the litter mass.

WEBSITES

Biomass and other biological terms explained at Marietta College, Ohio: www.marietta.edu/~biol/102/ecosystem.html
Biomass Energy Research Association: www.bera1.org/about.html

Biomass as renewable energy

With increasing concern about the supply of petroleum-based energy and its effect on the environment, biomass is increasingly being considered as a renewable energy source. The only known naturally occurring, energy-containing carbon resource that is large enough to be used as a substitute for fossil fuels is biomass. Since biomass is organic material, it contains stored sunlight in the form of chemical energy. When burned, the chemical energy is released as heat. Biomass fuels include wood, wood waste, straw, manure, sugar cane, municipal solid waste (refuse), sewage and by-products from a variety of agricultural processes. Ironically, despite new interest in biomass as renewable energy in the more developed countries, bio-mass in the form of wood and manure was the chief source of heating homes and other build-ings for thousands of years. In fact, biomass continues to be a major source of energy in much of the developing world.

Sugar cane is an excellent example of a biomass crop, grown in many southern states in the USA and in the Caribbean. The chief commercial product, sugar, is extracted from the cane by removing the juice. The remainder of the plant, called 'bagasse', still contains the chemical energy of the sun. As with any biomass, bagasse produces heat when burned.

Biofuel is the term used to describe the direct generation of energy from burning natural bio-mass. When burned, a hydrolysis reaction takes place. This reaction produces ethanol in the majority of cases, but there is also the production of turpentine from pine trees, and a potential for hydrogen production from algal blooms. Waste combustion can be used for direct heating or elec-tricity production. Two and a half tonnes of household waste can contain as much energy content as one tonne of coal. Refuse-derived fuel also helps reduce the amount of waste sent to landfills. Bacterial decomposition of waste typically emits methane gas and carbon dioxide. Methane is a combustible gas, which, along with the small amounts of oxygen given off, can act as a direct fuel source derived from a waste tip. Biogas has less energy value than natural gas but is still able to produce electricity and heat water, and is purely a by-product of the bacterial decomposition of waste. Unlike fossil fuels, biomass is renewable in the sense that only a short period of time is needed to replace what is used as an energy resource.   *continued over*

US Department of Energy Biomass Program: www.eere.energy.gov/biomass

Biomass from the US Department of Energy: www.eia.doe.gov/cneaf/solar.renewables/page/biomass/biomass.html

European Biomass Association: www.ecop.ucl.ac.be/aebiom

SEE ALSO

biofuels

BIOMES

Biomes are the broadest, most comprehensive communities recognised by ecologists. Biomes are so large that they are even recognisable from space – the tans and browns of deserts, the deep green of the boreal forest, the bright green of the tropical rainforest, the winter brown and summer green of the Earth's grasslands, for example. Biomes are groups of ecosystems that are similar in structure and function. Wherever a particular biome is found, climate and soils tend to be similar and organisms tend to have a similar range of life forms. For example, open country is dominated by grasslike plants, fed upon by grazing animals.

Biomes are usually recognised by the structure of the vegetation. In part, this is a matter of convenience. Because it is relatively stationary, vegetation is more easily observed than the animals of a biome. However, animal life (as well as fungal and microbial communities) is structured in characteristic patterns unique to the biomes in which they are found. All living things and non-living characteristics must be considered. According to biogeographer Susan Woodward, these considerations include the global distribution pattern – where each biome is found and how each varies geographically. A given biome may be composed of different species on different continents. Continent-specific associations of species within a given biome are known as formations and often are known by different local names. For example, the temperate grassland biome is called prairie, steppe, pampa or veldt, depending on where it occurs (North America, Eurasia, South America and southern Africa, respectively).

To understand biomes, it is also important to consider the characteristics of the regional

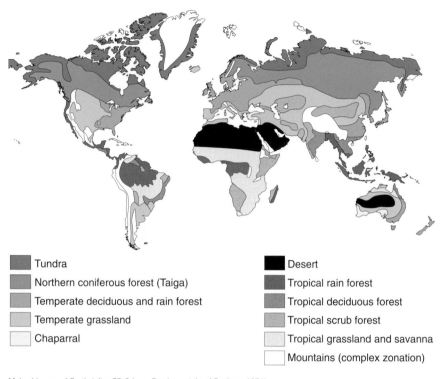

Tundra

Northern coniferous forest (Taiga)

Temperate deciduous and rain forest

Temperate grassland

Chaparral

Desert

Tropical rain forest

Tropical deciduous forest

Tropical scrub forest

Tropical grassland and savanna

Mountains (complex zonation)

Major biomes of Earth (after EP Odum, *Fundamentals of Ecology*, 1971)

climate and the limitations or requirements imposed upon life by specific temperature and precipitation patterns. The physical environment may exert a stronger influence than climate in determining common plant forms. Usually these factors are conditions of the surface itself, for example, whether it is waterlogged, dry or nutrient-poor, or if it has been disturbed by periodic flooding or burning. Biomes also include certain types of soils. Plant life is important to understanding biomes, including leaf shape, size and special adaptations of the vegetation, for example, peculiar life histories or reproductive strategies, dispersal mechanisms and root structure. The types of animals characteristic of the biome and their typical physiological and behavioural adaptations to the environment also contribute to an understanding of a biome.

The distribution of the major terrestrial biomes on the Earth is shown on the map. Perhaps the most fascinating thing about biomes is that they can be mapped at all. Biomes are not geographically continuous, which is

an important part of the definition. **GRASSLANDS**, for example, occur in disjointed areas around the world – North American prairie, South American pampas, Eurasian steppe – but all represent the same biome. South Africa and California are 19,000 kilometres apart, and it is futher still to Western Australia or South Africa, yet each of those places has areas subject to summer drought, winter rains and periodic fire. Such areas are clothed with shrub vegetation called **CHAPARRAL** in California, fynbos in South Africa, macchia in Italy and maquis in France. If one were to arrive unknowingly in any of these places, only a widely experienced botanist might be able to identify which place it was. Close up, the species are different, but they appear similar. They contain similar plants with leaves of similar shape and texture, because they have faced similar environmental challenges in their evolutionary history. Their similarity is the product of **CONVERGENT EVOLUTION**.

At first, the idea of biomes seems simplicity itself – the broadest patterns of the ecosphere.

Major biomes of the Earth

Terrestrial biome	Mean annual Temperature (°C)	Mean annual Precipitation (cm)	Producers	Typical primary Consumers	Net Production g/m²/ year	Other Characteristics
Temperate forest	−3–20	6–250	Deciduous trees, shrubs, annual and perennial herbs	Browsers, frugivores	1200	Highly seasonal, cold winters, hot summers
Temperate grassland	−5–20	50–130	Grasses	Grazers, seed-eaters	600	Fire-, drought-, wind-influenced, extremes, burrowers
Tropical savanna and scrub	17–30	50–150	Scattered trees over grassland	Grazers, browsers	900	Fire-prone
Taiga (boreal coniferous forest)	−5–5	5–200	Needle-leaved trees	Browsers, seed-eaters	800	Minimal under-story, high litter mass
Tundra	−15–5	10–100	Sedges, grasses	Grazers (lemmings, musk ox)	140	Permafrost, many breeding shorebirds
Chaparral	−5–20	50–75	Hard-leafed scrub	Browsers, seed and fruit predators	700	Summer drought, winter rain, fire-prone
Desert	−5–30	0–50	Stem- or leaf-succulents	Browsers, seed predators	90	Water-rich tissues protected by spines, etc.
Tropical rainforest	17–30	25–450	Tall buttressed trees, vines, epiphytes	Specialised herbivores, tree animals	2200	Little seasonality

Major biomes of the Earth continued

Aquatic/ marine biome	Mean annual Temperature (°C)	Mean annual Precipitation (cm)	Producers	Typical primary Consumers	Net Production g/m²/ year	Other Characteristics
Lotic (flowing) waters			Rooted aquatic plants	Invertebrates, fish	250	Well-aerated
Lentic (lakes, ponds)			Floating plants	Invertebrates, fish phytoplankton	250	Poorly aerated
Swamps – wetland with trees			Rooted plants, trees and graminoids	Invertebrates, fish, some mammals	<2000	High productivity
Marshes			Herbaceous (graminoids) wetlands	Invertebrates, fish, specialised rodents	<2000	
Estuaries			Rooted plants, phytoplankton	Invertebrates, fish	<1500	High productivity
Mangrove swamp			Mangroves (various species)	Invertebrates, fish	1100	Littoral or estuarine
Littoral (shoreline)			Phytoplankton	Invertebrates		Moderate productivity
Neritic (continental shelf)			Phytoplankton	Invertebrates, fish		Moderate productivity
Reefs			Algae symbiotic with corals	Fish, corals and other invertebrates,	<2500	Tropical, high species richness, high productivity
Pelagic (open ocean)			Phytoplankton	Invertebrates, fish	125	Low productivity

However, the idea contains an important insight into the evolutionary process. Biomes represent convergent ecosystems – a similar biotic response to similar challenges from such abiotic factors as the climate and natural hazards. In addition, native ecosystems of a place can tell us about its potential use by humans, such as what crops the place can support. It is hardly surprising, then, that convergence has occurred in cultural evolu-tion as well as in biological evolution, so that there are 'ecologically equivalent' patterns of culture and cultivation around the Earth.

Sometimes the biome concept is limited to terrestrial environments. However, organisms of freshwater and marine environments exhib-it convergent evolution and ecological equiva-lence as well, and major aquatic biomes are included in the table.

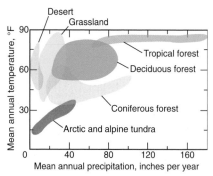

Relationship between mean annual temperature, mean annual precipitation, and some terrestrial biomes (after Whittaker, 1974). The size of the polygons in the diagram indicate how versatile biomes really are. Most can occur over a fairly wide range in temperatures and precipitation.

Biomes versus ecoregions

Ecoregions are **ecosystems** that are distinctive across entire regions – the name comes from 'ecosystem' and 'region'. Ecoregions result from the large-scale interaction of solar radiation (energy flow) and moisture that creates unique regional biotic communities. An ecoregion is a substantial and continuous area of the Earth, variable in size but smaller than a continent, and usually larger than a province or state. Ecoregions are characterised by a combination of climate and vegetation, reflected in their names such as the Valdivian Temperate Rainforest in Chile and Argentina, the Eastern Siberian Taiga, and the East African Acacia Savanna. Organisms other than plants, such as animals, fungi and microbes, are also included.

The term ecoregion is not used consistently by ecologists. Ecoregions are roughly equivalent to **biomes** or subdivisions of biomes. However, the term ecoregion should be reserved to mean regional ecosystems, including both the biota and the physical environment. These ecoregions have become the focus of conservation planning. The term bioregion should be reserved for regional biotic communities; the term is not intended to include the non-living (abiotic) environment such as landforms and soils.

WEBSITES

Biomes from Radford University's biogeographer Susan Woodward:
www.runet.edu/~swoodwar/CLASSES/
GEOG235/biomes/main.html
Images of biomes of the world from Up-Island Regional School District, Massachusetts:
www.blueplanetbiomes.org/index.htm
'Exploring biomes' from the Missouri Botanical Garden:
http://mbgnet.mobot.org
NASA site on the Earth's biomes:
http://earthobservatory.nasa.gov/
Laboratory/Biome
World biomes from the University of California-Berkeley:
www.ucmp.berkeley.edu/glossary/gloss5/
biome
The Sierra Club's descriptions of North American ecoregions with a focus on conservation issues:
www.sierraclub.org/ecoregions
Global 200 planning project from WWF, an ecoregional approach to the conservation of global biodiversity:
www.panda.org/about_wwf/where_we_work/
ecoregions/index.cfm

SEE ALSO

chaparral, deserts, forests and forestry

BIOSPHERE

The biosphere is the global sum of all living (biotic) communities and therefore represents the totality of life on Earth. A biotic community is the living component of an **ECOSYSTEM** and comprises all the organisms present at a particular place at a particular time. The biosphere is the living component of the ecosphere – the global sum of ecosystems.

The term biosphere was coined by Russian scientist Vladimir Vernadsky in 1929. The biosphere distinguishes our planet from all others in the solar system. The chemical reactions of life, such as photosynthesis and

Case study

Biosphere II

Biosphere II was built during the early 1990s to test if and how people could live and study in an environment closed to the outside world. The inhabitants carried out scientific experiments, explored the possible use of closed biospheres in space colonisation, and were able to manipulate and study a biosphere without harming Earth's. Biosphere II was one attempt to see if humans could recreate the balanced ecosystem we require to survive. To test Biosphere II's ability to maintain a self-sustaining ecosystem in total isolation, an eight-member crew was chosen to spend two years sealed in the facility. The four men and four women would produce their own food in the agricultural area, while at the same time monitoring their environment and carrying out a variety of scientific experiments.

Biosphere II is an artificial biosphere in Oracle, Arizona built by Edward P. Bass, Space Biosphere Ventures, and others. The name came from the idea that it was modelled on Biosphere I – the Earth.

Photograph from http://www.bio2.com/index.html

As an attempt to create a balanced and self-sustaining replica of the Earth's ecosystems, Biosphere II was a miserable and expensive failure. Numerous problems plagued the crew almost from the very beginning. Of these, a mysterious loss of oxygen and widespread extinction were the most notable. The reason was that Biosphere II's soil was unusually rich in organic material. Microbes metabolised this material at an abnormally high rate, and, in the process, used up much oxygen and produced much carbon dioxide. The plants in Biosphere II should have been able to use this excess carbon dioxide to replace the oxygen through photosynthesis, except that another chemical reaction was also taking place. A vast majority of Biosphere II was built out of concrete, which contains calcium hydroxide. Instead of being consumed by the plants to produce more oxygen, the excess carbon dioxide was reacting with calcium hydroxide in the concrete walls to form calcium carbonate and water.

Biosphere II included a carefully chosen variety of plant, animal and insect species in desert, intensive agriculture, marsh, thorn scrub, ocean, savanna and rainforest ecosystems. The eventual extinction rate was much higher than expected, with only 6 of the 25 small vertebrates surviving to the mission's end. Almost all the insect species went extinct, and since this included those that pollinated plants, the plants could no longer propagate. Some species absolutely thrived in this manmade environment such as ants, cockroaches, katydids and certain vines such as morning glories that threatened to choke out every other kind of plant. The crew members were forced to put vast amounts of energy into simply maintaining their food crops. Biosphere II could not sustain a balanced ecosystem, and therefore failed to fulfil its goals.

Biosphere II's water systems became polluted with too many nutrients. The crew had to clean their water by running it over mats of algae, which they later dried and stored. Also, as a symptom of further atmospheric imbalances, the level of dinitrogen oxide became dangerously high. At these levels, there was a risk of brain damage due to a reduction in the synthesis of vitamin B12. In addition, inevitable disputes arose among the crew, as well as among those running the project from the outside.

Lessons learned throughout this mission may help scientists in further attempts to create self-sustaining ecosystems. More importantly, according to the Biosphere II project, it 'serves as a chilling reminder that humankind is currently incapable of surviving apart from the well-balanced ecosystems that already exist on Earth. First and foremost, we must strive to preserve Biosphere I, because we have nowhere else to go'.

Columbia University now owns the facility, which it uses to conduct public tours and a variety of studies, particularly ones related to the carbon dioxide cycle.

respiration, have influenced the chemical composition of the atmosphere, transforming the atmosphere to an oxidising environment with free oxygen. The biosphere is structured into a hierarchy known as the food chain whereby all life is dependent upon the first tier – mainly the primary producers that are capable of photosynthesis, such as plants and algae. Energy and mass is transferred from one level of the food chain to the next, with an efficiency of about 10 per cent. For example, if a deer eats 10 kilograms of shrubbery, 1 kilogram of deer tissue is produced. The rest of the food energy is expended on movement, cellular processes, maintenance of a relatively high constant body temperature, and the other processes of being alive. Only energy stored for growth or reproduction is available for transfer down the food chain to carnivores. All organisms are intrinsically linked to their physical environment. The relationship between an organism and its environment is the study of **ECOLOGY**.

The biosphere can be thought of in a similar way to the atmosphere, lithosphere and hydrosphere, which describe the Earth's crust in terms of a state of matter: gas, solid and liquid. The biosphere integrates these different 'spheres' of the physical Earth, combining bits of hydrosphere (water) and atmosphere (carbon dioxide) to make carbohydrate. Other molecules from the lithosphere (major and trace nutrients – see **BIOGEOCHEMICAL CYCLES**) are incorporated as needed. Some scientists like to think of a fourth sphere – the anthrosphere (humans) – because the impact of humans on the rest of the biosphere far exceeds that of any other animal.

At their molecular 'cores', plants have carbohydrate, following the 'blueprint' encoded in their genetic material. Plants and photo- or chemosynthetic microbes (producers) synthesise the range of biomolecules (fats, proteins, and nucleic acids) that we recognise as molecular **BIODIVERSITY**.

The basic resources that organisms need to make a living are air, water and minerals. The biosphere is most diverse and productive where the lithosphere, hydrosphere and atmosphere meet, and where temperatures are optimal for the biochemical processes of life. Biotic communities that fit this description include tropical lowland forests and swamps, tropical estuaries, and tropical algal and coral reefs.

WEBSITES

The Atlas of the Biosphere from the Center for Sustainability and the Global Environment (SAGE) at the University of Wisconsin: www.sage.wisc.edu/atlas/index.php
The UN Educational, Scientific and Cultural Organization's Man and the Biosphere Programme: www.unesco.org/mab/wnbr.htm
Biosphere II Project: www.bio2.edu
International Geosphere-Biosphere Programme: www.igbp.kva.se/cgi-bin/php/frameset.php
European Environment Agency, with detailed information on the components of the biosphere: www.eea.eu.int/main_html

SEE ALSO

ecosystems

CARBON CYCLE

Life is possible because of the chemistry of carbon. An individual organism is built from over 100,000 different chemical compounds. Well over 99 per cent of those compounds contain carbon. The chemistry of carbon is so closely linked with life that carbon chemistry is called organic chemistry – the chemistry of organisms. Until Friedrich Wöhler first prepared the organic compound urea in the laboratory from ammonia and carbon dioxide in 1828, chemists believed that carbon compounds could only be made inside organisms. Many people believed that some mystical 'vital' principle in living things made the complexity and variety of carbon compounds possible.

Today, hundreds of thousands of different compounds of carbon are prepared routinely in academic and industrial laboratories. There is nothing mystical about this. Rather, the wondrous diversity of organic chemistry is a function of the structure of the carbon

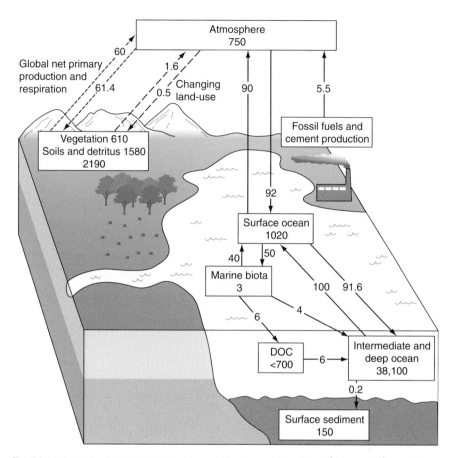

The global carbon cycle – flow units are gigatons/year; stock units are gigatons (Gt = 10^9 tonnes = 10^{15} grams) (from Woods Hole Research Center http://www.whrc.org/science/carbon/)

atom. Because of the arrangement of its electrons, carbon readily forms bonds with other carbon atoms, and with atoms of a number of other elements such as oxygen, hydrogen, nitrogen and phosphorus. These are called covalent bonds, formed when electrons are shared between atoms. Carbon can bond to itself in chains of any length – branched, unbranched, or turned upon themselves as rings. These carbon compounds are responsible for the diversity of life.

Carbon is second only to oxygen in its proportional contribution to biomass. Because of its atomic structure and chemical behaviour, carbon is responsible for the very possibility of life. Silicon, just below carbon in the periodic table of the elements, can also form long chain-like molecules, and it behaves enough like

carbon to inspire science fiction writers to fantasise about silicon-based 'life'. However, real scientists searching for signs of life elsewhere in the Universe look for carbon chemistry.

The principal reservoirs of carbon are the carbon dioxide of the atmosphere, carbonate in rocks and carbonate in the shells of animals like snails and clams. The carbon cycle is driven by the complementary processes of **PHOTOSYNTHESIS** and respiration (the biological equivalent of combustion). The figure shows how the linked carbon cycles of land and water are interrelated. The carbon cycles of **ECOSYSTEMS** are rather small-scale, short-term, 'high-speed' affairs when considered from a global, geologic perspective. This is because most of the carbon on the Earth is not carbon dioxide in the atmosphere or dissolved in the oceans. Nor is it in living or recently living organisms on land or in the sea. Rather, the bulk of the Earth's carbon is in 'dead storage', fairly well sequestered from the day-to-day commerce of the ecosphere. Thus carbonate rocks such as limestone, dolomite and chalk contain 20 million billion (2×10^{16}) tonnes of carbon, and coal and oil contain an additional 10,000 billion (10^{13}) tonnes. In other words, the amount of carbon actively being cycled through the ecosphere amounts to only 0.02 per cent of the Earth's carbon supply. However, the carbonate rocks were partly laid down by organisms – they are the shells of long-since departed clams and ancient coral reefs – and fossil fuels also have a biological origin. Therefore, the chemistry of the Earth's carbon horde is partly a product of biological activity.

The atmospheric reservoir of carbon is remarkably small but also remarkably dynamic. Currently, the atmosphere averages about 0.034 per cent carbon dioxide, or 340 parts per million (ppm). The figure is given as an average because the local proportion varies daily and seasonally, depending upon the activity of the producer organisms – more precisely, the balance between photosynthesis and respiration. In the immediate vicinity of the forest canopy at noon on a long spring day, carbon dioxide concentration of the atmosphere may drop to 10–15 ppm. At night, when photosynthesis slows and respiration predominates, it may rise to 400 ppm or more.

These figures also apply only to current rates. Over the past century the average carbon dioxide content of the atmosphere has risen from about 290 to 340 ppm. The source of this carbon dioxide is easy to identify: it is the product of burning fossil fuels. Combustion – like respiration – breaks carbohydrates and hydrocarbons into carbon dioxide and water. The water released is impressive locally, clearly evident in the white 'smoke' from a fossil-fuel-fired power plant on a crisp winter day. However, this local effect is largely irrelevant because it is a minute fraction of the Earth's water budget. The carbon dioxide released *is* highly relevant, however, because it amounts each year to nearly 1 per cent of the atmospheric reservoir, and has much to do with concerns about global climate change and the GREENHOUSE EFFECT. Therefore, humans have much to do with the carbon cycle, and thus can have a great effect on the environment, which is based on carbon.

WEBSITES

Animation of carbon cycle from the US Environmental Protection Agency: www.epa.gov/globalwarming/kids/carbon_cycle_version2.html

Carbon cycle diagram and information: www.physicalgeography.net/fundamentals/9r.html

US Global Change Research Program carbon cycle information and links: www.usgcrp.gov/usgcrp/links/carbonlinks.htm

Human interaction with the carbon cycle, from National Academies Press: http://books.nap.edu/books/0309084202/html/R1.html#pagetop

SEE ALSO

ecosystems

CARBON SINKS

There is one thing that links every person on the planet – we are all carbon-based forms of life.

Carbon, the fourth most abundant element, circulates through the oceans, the lithosphere (underneath the Earth's surface) and the biosphere (plants and animals). It forms a global cycle, with sinks or accumulations forming at points where large quantities of this essential building block of life are stored for varying lengths of time. It is stored, mainly in the form of carbon dioxide, over a range of time scales, from days and months to millions of years. This carbon cycle has two distinct components – biological and geological – each with its own sink where carbon is stored.

Throughout the Earth's history, carbon dioxide has been released from the geological sink by volcanic eruptions and by the weathering of carbon-rich rocks. However, since the Industrial Revolution began, people have been releasing carbon dioxide from both of these sinks. The geological sink has been tapped into by the extraction of oil, gas and coal, while in the case of the biological sink, people have been releasing carbon dioxide

The two types of carbon sink: biological and geological
The biological sink results from the conversion of carbon held in the air, water and soil into a biological form, that is, into plants. Animals consume plants and plants die. The carbon animals and plants both contain is released back into either the atmosphere or the ground, so the world's vegetation forms one of the two great holding areas of global carbon.

The geological sink occurs in the oceans, where a large amount of organic carbon – from plants and animals – settles on the ocean floor and is buried in the sediment that eventually turns into solid rock. This carbon may take tens of million years to return to the atmosphere, if it does at all. It includes carbon-rich deposits of coal and oil, as well as carbonate rocks such as chalk and limestone.

through slash-and-burn agriculture; clearing land for permanent pasture, cropland and settlements; the development of infrastructure such as roads and dams; accidental and intentional forest burning; and unsustainable logging and fuelwood collection.

As a result of human use of the carbon sinks the levels of carbon dioxide in the atmosphere have increased considerably. Levels have climbed from about 275 ppm (parts per million) in 1750 to 365 ppm today – a 30 per cent increase. Scientists agree that carbon dioxide is the main greenhouse gas and that its increasing presence has already enhanced the **GREENHOUSE EFFECT**. This, in turn, is expected to lead to corresponding changes in the Earth's climate, including global warming.

Some countries, notably the USA, have suggested that restoring the biological sink, by afforestation and reducing the clearance of vegetation, is one way of meeting their obligations under the Kyoto Protocol on Climate Change. Researchers have shown that natural ecosystems can store between 20 to 100 times more carbon dioxide than types of agricultural land use.

However, most scientists argue that relying on trees and vegetation to absorb carbon

dioxide will do little to tackle global warming. They say that the amount of carbon the biological sink can store is significantly less than the larger quantities released by burning fossil fuels. Emissions from the burning of fossil fuels account for about 65 per cent of the additional carbon dioxide currently in the Earth's atmosphere.

The stark truth is that there is no viable alternative to reducing the consumption of fossil fuels and protecting the geological carbon sink. So the upshot for the future population of the planet is that the global carbon cycle has been permanently altered – some scientists would even go further and say irreversibly knocked out of balance.

WEBSITES

World Resources Institute:
www.wri.org/climate/sinks.html
Manaaki Whenua Landcare Research:
www.landcareresearch.co.nz/research/
greenhouse
Physical Geography.net's carbon cycle information:
www.physicalgeography.net/fundamentals/
9r.html
Woods Hole Research Center's carbon cycle:
www.whrc.org/science/carbon/carbon.htm
NASA's earth Observatory:
http://Earthobservatory.nasa.gov/Library/
CarbonCycle

SEE ALSO

carbon cycle, climate change

CATALYTIC CONVERTERS

At this moment, there are over 600 million cars being driven on roadways all across the world. Each one is potentially a source of much air pollution, except if it is fitted with a catalytic converter. These are ingenious devices that treat the exhaust before it leaves the car, thereby removing much of the pollution and allowing people to get from A to B as quickly as possible.

In simple terms, a catalytic converter transforms the noxious hydrocarbons and **NITROGEN OXIDES** in **VEHICLE EXHAUSTS** into water, carbon dioxide and nitrogen gas. Most modern cars are equipped with a three-way converter. Three-way refers to the three regulated emissions that the device helps to reduce – carbon monoxide, volatile organic compounds and nitrous oxide. The converter uses two different types of catalysts – a reduction catalyst and an oxidisation catalyst – and consists of a ceramic structure coated with a metal catalyst, usually platinum, in order to create a structure that exposes the maximum surface area of catalyst to the stream of exhaust fumes.

How do catalytic converters work?

There are three stages to the process. The reduction catalyst is the first, with a nitrogen oxide (NO) or nitrogen dioxide (NO_2) molecule contacting the catalyst. The catalyst 'rips' the nitrogen atom out of the molecule and then holds on to it, freeing the oxygen in the form of O_2. The nitrogen atoms bond with other nitrogen atoms that are also stuck to the catalyst, forming N_2, or nitrogen gas, the same non-toxic gas that makes up three-quarters of the air that we breathe.

The oxidation catalyst is the second stage of the process. It reduces the unburned hydrocarbons and carbon monoxide (CO) by burning (oxidising) them over a catalyst. This catalyst aids the reaction of the carbon monoxide and hydrocarbons with the remaining oxygen in the exhaust gas.

The third and final stage is a control system that monitors the exhaust stream and uses the information to control the fuel injection system. There is usually an oxygen sensor mounted upstream of the catalytic converter, which tells the engine computer how much oxygen is in the exhaust. The engine computer can increase or decrease the amount of oxygen in the exhaust, simply by adjusting the air-to-fuel ratio. This reduces the emissions of harmful polluting gases.

However, not all environmentalists hail the catalytic converters as a total success. This is because the converters rearrange the nitrogen–oxygen compounds to form nitrous oxide, which increases the amount of a potent greenhouse gas – one that is more than 300 times more potent than carbon dioxide in trapping heat.

In 1998 it was estimated that nitrous oxide comprised about 7.2 per cent of the gases that cause global warming, with cars and trucks, fitted with catalytic converters, producing around half of that total. As one scientist for the Environmental Protection Agency said: 'You've got people trying to solve one problem, and, as is not uncommon, they've gone and created another'.

WEBSITES

Howstuffworks:
auto.howstuffworks.com/
catalytic-converter.htm
National Synchrotron Light Source:
www.nsls.bnl.gov/about/everyday/
catalytic.html
Junk Science.com:
www.junkscience.com/news2/catalyt.htm

SEE ALSO

nitrogen oxides, vehicle exhausts

CAVES AND KARST

Caves are among the environment's most fragile places. Caves provide irreplaceable habitats for rare plants and animals, some of which spend their entire lives in complete darkness. Troglobites – cave-dwelling animals – include insects, crustaceans and fish, many without skin pigment or eyes. Biologists have recently discovered cave-dwelling extremophiles whose food web is based on chemosynthetic, or mineral-eating, bacteria. These organisms provide clues about the earliest forms of life on Earth, and NASA scientists study them to learn about the potential for life on Mars. The organisms also show promise as a source for new antibiotics and other medications.

Trogloxenes are animals that live in caves but return to the surface to feed. These include the amazing flying mammal – the bat, as well as bears, pack rats, snakes, raccoons, swallows, moths, foxes and even people. A single small bat can eat 1200 mosquito-sized insects each hour. A colony of 150 big brown bats can protect farms from up to 33 million rootworms each summer. Other bats feed on

Delicate helictite formations growing in a Colorado cave, USA. The growth of helictites is still not well understood.
Photograph by Richard Dieter

fruit and nectar, and pollinate and disperse seeds. Bat droppings are an important natural fertiliser source. However, during the twentieth century the use of pesticides reduced the food source for bats, and their numbers began declining.

Some caves form as lava tubes in volcanic regions or along coasts as a result of coastal erosion, but most caves are holes that begin as cracks in the sedimentary rock called limestone. Limestone was formed from the calcium coral and shell remains of sea life. As rain falls through the atmosphere, it absorbs a small amount of carbon dioxide; it gathers additional carbon dioxide as it moves through the soil. When water is mixed with carbon dioxide it forms a weak carbonic acid solution. Carbonic acid in rainwater reacts to dissolve away limestone along the cracks, which widen still further into underground streams, rivers and even lakes. When the water drains away, the waterways turn into open cave tunnels, passages and caverns.

It may require 10,000 to 100,000 years to form a cave big enough for people and animals to move around inside. Many caves are still living – where water is still dripping and doing its work. The drips dissolve limestone minerals in one part of the cave. As water dries up, the calcium-rich minerals are deposited to build up in other places as stalagmites, stalactites, columns and other rock and crystal formations. These rock formations change dark limestone caves into wonderlands of fantastic beauty.

Not only are the caves unique, but also the landscapes in which caves develop – called karst – are environmentally unique places. The word karst is derived from the Serbo-Croat word *krs* and the Slovenian word *kras* meaning stony bare ground. The Kras is a limestone region, now a part of Slovenia and Croatia, in which the distinctive karst landforms are exceptionally well developed. Karst landscapes include caves, sinkholes, underground streams and other features formed when bedrock is dissolved by water. Typically, karst regions lack rivers and other surface waters because the rain is swallowed up by fissures, depressions and conduits in the rock, and it then flows underground. Eventually, the waters return to the land surface, often as large springs.

Karst regions have always been important to human societies. Karstified limestone covers approximately 10 per cent of the land surface of the Earth. About 25 per cent of the world's population lives in these regions – for example, southern China, large areas of central and southern Europe, and much of Central America. About 20 per cent of the USA is underlain by karst; nearly every state contains caves. Caves have served as homes, burial grounds and sites for religious practices. In prehistoric times karst regions were used as refuges and are thus very important for archaeological remains – the cave paintings of Lascaux in France and of Altimira in Spain being famous examples. Thus caves offer clues to both geologic events and the pre-historic past.

Karst areas are among the world's most diverse and resource-rich areas, containing the largest springs and most productive groundwater supplies on the Earth. However, karst landscapes are vulnerable to environmental impacts. Constant temperatures and conditions help caves preserve and protect formations and life forms, but these same conditions mean that caves can never fully recover from direct or indirect damage of any kind. At best, even a small number of human visitors to caves changes the air quality, and brings in mud and dust. At worst, cave formations and wildlife are inadvertently or purposefully damaged. When caves are gated to regulate the number of visitors, the gates may change the amount of air that flows through the cave opening. This 'breathing hole' of the cave is important in regulating cave temperature, humidity and life forms.

Caves are damaged not only by human activities occurring within them, but also from human activities above ground, often many kilometres distant. In highly permeable karst areas, surface water flows into caves quickly with little filtration, and if polluted by agricultural, mining or urban activity on the surface, caves and cave life are damaged. Furthermore, this contaminated water may travel great distances underground, polluting wells and aquifers far away. The most common form of groundwater pollution is human waste (sewage), usually from poorly planned

Case study

Horse Cave, Kentucky, USA: from stinking pit to success story

The town of Horse Cave was built around the giant sinkhole that was the entrance to Hidden River Cave. Behind a row of Main Street businesses, a three-storey drop off leads down to the sinkhole and the yawning entrance to the cave. Cool temperatures in and near the sinkhole made it the centre of town activity in the early days. Tennis courts were created with safety nets to keep the balls from rolling down into the cave. Dr G. A. Thomas began showing Hidden River Cave around 1900. His son, Dr Harry Thomas, installed the country's first electric lights in the cave in 1916. From 1927 until 1943 a total of five different tours, showing a variety of cave features unequalled anywhere, were available in the three caves that he owned.

Due to industrial and residential growth, large amounts of household and other wastes were dumped into sinkholes upstream from the cave, and the town let its raw sewage run down the river, which eventually entered Hidden River Cave. As the underground streams were depleted of oxygen, terrible odours began to emit from the cave entrance, and the cave's eyeless fish population began to decline. By 1943 the cave became such a mess that it was closed and allowed to fester. According to local townspeople, 'The smell was so bad, the jewellery store across the street would have to close by midday'.

In 1989 a new municipal sewage treatment centre was built, which ended the dumping of waste into Hidden River Cave. The odours have disappeared, the cave fish are returning, and the cave is part of the American Cave and Karst Center. The centre serves as a living exhibit to teach people about the impact of humans in karst environments.

or faulty septic field systems. Other pollution is caused by industrial and hazardous waste. Each person in the USA generates about 1.6 kilograms of solid waste per day. Much of this waste is deposited in landfills. In karst areas, substantial volumes of water commonly move through these landfills, creating a liquid called landfill leachate. These leachates routinely contain large concentrations of heavy metals, chemicals and toxic materials.

Karst regions often attract large numbers of visitors to their spectacular scenery and fascinating caves. The most famous cave regions include the Greek islands, the Dordogne, Vercors and Tarn areas of France, the pinnacle karst of the Guilin area in southern China, the Mammoth Cave area of Kentucky in the USA, and the Postojna Caves in Slovenia. However, this influx of visitors is putting pressure on many states and countries to examine their policies regarding the protection of caves and karst.

For many years, caves were used as rubbish dumps by communities or were otherwise neglected. During the second half of the twentieth century people began to realise how important caves were for local and regional water quality, as well as for their unique environment, archaeology and habitat. In 1988 the US Congress passed the Federal Cave Resources Protection Act. It was an important step forward, but the protection only covered caves on federal lands. Sound management of karst requires the wholehearted participation of citizens and land-use decision makers.

WEBSITES

National Speleological Society:
www.caves.org
National Caves Association:
www.cavern.com
Bat Conservation International:
www.batcon.org
National Park Service Caves:
www2.nature.nps.gov/grd/tour/caves.htm
National Cave and Karst Mangement Symposium:
www.nckms.org
American Cave Conservation Association:
www.cavern.org/CAVE/ACCA_index.htm

The Virtual Cave:
www.goodearthgraphics.com/virtcave.html
Hidden River Cave in Horse Cave, Kentucky:
www.cavern.org/ACCA/ACCA/index.htm
Caving UK:
www.caving.uk.com

CFCs

For many years, chlorofluorocarbons, or CFCs, were only studied by those with an interest in chemistry, and for the average person they were nothing to worry about. But from the mid 1970s CFCs became public enemy number one, as an increasing body of evidence was assembled by environmental researchers that these substances were having an extremely adverse effect on global climate.

CFCs include several artificial compounds composed of carbon, fluorine, chlorine and hydrogen. Developed during the 1930s, CFCs found wide application after the Second World War, most notably trichlorofluoromethane (CFC-11) and dichlorodifluoromethane (CFC-12), as aerosol-spray propellants, refrigerants, solvents and foam-blowing agents. The CFC compounds were ideal for these and other applications because they are non-toxic and non-flammable and can be readily converted from a liquid to a gas and vice versa. However, when released into the atmosphere, they attack the ozone in the stratosphere, as well as acting as greenhouse gases, trapping heat in the lower atmosphere.

The atmospheric concentration of CFCs ranges from 2 to 484 parts per trillion by volume, but from the mid 1970s they were shown to be increasing at a very rapid rate, of between 4 and 15 per cent per annum. This was the time when environmentalists discovered that CFCs were one of the greatest agents in ozone destruction, and a leading contributor to the enhanced greenhouse effect.

The discovery by Joe Farman of the ozone hole, backed by further scientific research in the mid 1980s, showed that global controls were needed to reduce the emissions of CFCs, and in 1987 27 nations signed a global environmental treaty, called the Montreal Protocol to Reduce Substances that Deplete the Ozone Layer. Under the Montreal Protocol, these countries agreed to reduce 1986 production levels of these compounds by

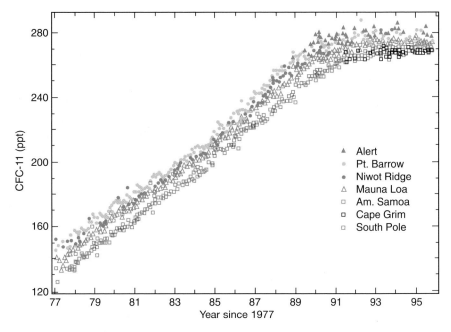

The increase between 1977 and 1996 in the amount of chlorofluorocarbon (CFC-11) in the atmosphere, based on data from Port Barrow, Alaska; Mauna Loa, Hawaii; Cape Matatula, American Samoa; the NOAA station at the South Pole; Alert, Northwest Territories, Canada; Niwot Ridge, Colorado; and Cape Grim Baseline Air Pollution Station, Tasmania, Australia. Source NOAA/CMDL

50 per cent before the year 2000. An amendment approved in London in 1990 was more forceful and called for the elimination of production by the year 2000. The Montreal Protocol also included enforcement provisions by applying economic and trade penalties should a signatory country subsequently trade in, or produce, these banned chemicals. A total of 148 countries have now signed the Montreal Protocol.

Companies have also developed two classes of CFC substitutes – hydro chlorofluorocarbons (HCFCs) and hydro fluorocarbons (HFCs). HCFCs include hydrogen atoms in

Joe Farman: the man who found the ozone hole

In 1985 the scientific journal *Nature* published a paper by Joe Farman and his colleagues working at the British Antarctic Survey in Cambridge. The paper revealed a dramatic decline in stratospheric ozone in perhaps one of the most unlikely regions of the planet – over Antarctica.

Farman's groundbreaking discovery was based on measurements of the total amount of ozone in a column of atmosphere above the British Antarctic Survey's base at Halley Bay, dating back to 1957. In particular, Farman showed that a dramatic decline in ozone concentrations had taken place during the spring, starting in the late 1970s, and reaching around 30 per cent by 1985.

Farman and his colleagues suggested that the so-called 'ozone hole' was due to gases emitted into the atmosphere by human activity, and subsequent research unequivocally linked the ozone decline to CFCs.

addition to chlorine, fluorine and carbon atoms. The advantage of using HCFCs is that the hydrogen reacts with tropospheric hydroxyl (OH), resulting in a shorter atmospheric lifetime. However, HCFCs still contain chlorine and therefore they can still destroy ozone, and the amendments to the Montreal Protocol, agreed upon at Copenhagen in 1992 and Vienna in 1995, called for their production to be eliminated by the year 2030. In contrast, HFCs do not contain chlorine, and are therefore considered one of the best substitutes for reducing the loss of stratospheric ozone.

Despite all these regulations and new substances, there is still quite a concentration of CFCs in the atmosphere, largely because the substances do not just disappear overnight. This so-called 'residence time' in the atmosphere is in some cases up to 400 years. So, although modern society is now pumping smaller amounts of these chemicals into the atmosphere, their effects will only end in several generations' time.

Vegetation and rocky hills of the chaparral biome in San Diego County, California, USA. Nearby, large areas of hills like this one were devastated in wildfires during October 2003. Photograph by Joseph Kerski

WEBSITES

Climate Monitoring and Diagnostics Laboratory:
www.cmdl.noaa.gov/noah/publictn/elkins/cfcs.html
NASA:
http://daac.gsfc.nasa.gov/CAMPAIGN_DOCS/ATM_CHEM/ozone_atmosphere.html
Queens' College, Cambridge University:
www.quns.cam.ac.uk/Queens/Record/1998/Academic/ozone.html
Our Planet:
www.ourplanet.com/imgversn/104/farman.html

SEE ALSO

Antarctica, ozone

CHAPARRAL

The chaparral **BIOME** occurs in the Earth's temperate zone, between about 30° and 40° north and south of the Equator. This biome is clothed with shrub vegetation of the same name – chaparral – but only in California. The vegetation is named fynbos in South Africa,

macchia in Italy and maquis in France. This biome is most commonly found on the western and southern coasts of continents – California, Spain, Portugal, Italy, Greece, Western Australia, South Australia, South Africa and central coastal Chile. Its presence around the Mediterranean gives it its alternative name, the Mediterranean scrub biome. Chaparral biomes usually receive more rain than deserts and grasslands but less than forested areas: between 200 and 1000 millimetres annually. There is a noticeable dry season and wet season and the precipitation is highly variable. Chaparral is often grouped in a shrublands biome, which includes not only chaparral, but some woodlands and savanna. These regions are usually found surrounding deserts and grasslands. True chaparral occupies less than 1 per cent of the Earth's land surface, but it is a biome of economic importance, because the moderate climate appeals both to people and to wine grapes.

Chaparral biomes are covered with shrubs that thrive on steep, rocky slopes, since not enough rain falls to support tall trees. Chaparral biomes are usually fairly open, so grasses and other short plants grow between the shrubs. In the areas with little rainfall, plants have adapted to drought-like conditions. Many plants have small, needle-like leaves that help to conserve water. Some have leaves with waxy coatings and leaves that reflect the sunlight. The woody vegetation is harvested by browsers (which mainly eat leaves of trees, shrubs and ferns) and by seed and fruit predators.

Case study

San Diego fires of 2003

Chaparral is a fire-prone biome. Every year, news from California, USA, carries images and text of large, high-intensity fires that sweep the chaparral landscape, threatening lives and homes. Ecologists have long known that chaparral biomes burn extensively and often, and that much of the dominant vegetation in these biomes is highly adapted to a fire-prone environment. Dry winds called the Santa Ana winds are the drivers for many of these fires. Many native plants have seeds that require fire to germinate, or need the kind of disturbed habitat fires leave behind to grow.

The worst fires in the history of California ravaged San Diego County during the week of 26 October 2003. The wildfires killed 16 people and burned down 2427 homes and businesses. The fires rekindled controversies about whether humans should practise controlled burns or practise fire suppression in rural chaparral areas. It was long thought that fire suppression played the same role in chaparral as it has in forests – that suppressing fires builds up fuels that can eventually lead to more destructive fires. However, some new findings cast doubt. According to USGS fire researcher Dr Jon Keeley:

Past fire suppression is not to blame for causing large shrubland wildfires, nor has it proven effective in halting them. Under Santa Ana conditions, fires carry through all chaparral regardless of age class. Therefore, prescribed burning programmes over large areas to remove old stands and maintain young growth as bands of firebreaks resistant to ignition are futile at stopping these wildfires.

One thing is certain – with 5 million people already in San Diego County, and growing, and foothills encroached upon with more and more homes, fires will continue to be a concern for the residents there. Fighting fires is economically expensive and traumatising to homeowners and the entire community. Homeowners are required to plant water-retaining shrubs near their homes, but even this causes environmental disturbance. First, these shrubs are sometimes not natural to the biome, and alter the plant ecosystem. Second, ground and surface water resources become overused, in order to keep alive the vegetation designed to suppress fires.

Two types of chaparral vegetation exists – hard and soft. Also called coastal scrub or coastal sage scrub, soft chaparral is dominated by small shrubs with 'soft' leaves that have a pliable, thin texture. A wonderful characteristic of this biome is that the shrubs are fragrant – soft chaparral leaves may be heavily scented of sage, turpentine or mint to keep animals from eating them. These fragrant oils also evaporate on hot days to cool leaves and inhibit the growth of competing plants. All these ploys prevent shrubs from losing precious leaves, since it requires energy and water to produce new ones. During prolonged dry summers, soft chaparral shrubs may lose most of their leaves to keep from dehydrating faster than their roots can replenish water from bone-dry soils. Winter rains bring temporary supplies of water, during which leaves are replaced. Soft chaparral is typical of rocky promontories in the fog belt, but components of this same community appear as temporary replacements for hard chaparral shrubs after brush fires. Shrubs include coyote brush, California sagebrush, monkeyflower, black sage, blue witch and poison oak.

Hard chaparral is so named because its component species have stiff, tough, durable leaves that are seldom shed, even in midsummer. Leaf design varies – manzanitas have stiff leaves turned edgewise to avoid the full brunt of sun, and some even have whitish leaves that reflect away excess light and heat. Chamise uses narrow, needlelike leaves clustered together to conserve water by minimising the surface area exposed to sun. Wild lilacs cover their leaves with a thick, waxy coating that makes them shiny. Bush poppy's bluish green leaves are also obliquely angled to reflect away heat and minimise the impact of the fierce summer sun. Hard chaparral replaces soft chaparral in hotter, drier inland areas, usually on steep, rocky slopes. (Shrubs favour the summer heat of south-facing slopes.) From a distance the dense, tall shrubberies of hard chaparral look like a uniform dark green velvet draped over the mountainsides.

In addition to their ingeniously designed leaves, chaparral shrubs have deeply probing roots that hold shrubs in place and find sources of deeply hidden water. Roots may also carry on 'chemical warfare' with neighbouring shrubs to prevent invasion into their own root zone. Chaparral peas (*Pickeringia montana*) have tiny knobs on their roots that house nitrogen-fixing bacteria. As a result, these shrubs can move on to nutrient-poor soils. When they die, they may pave the way for other shrubs to move in by releasing these nitrogenous compounds into the soil.

Chaparral shrubs may grow into nearly impenetrable canopies – from head high to well over 3 metres. The best way to pass through is to crawl beneath the branch canopy, as small mammals do.

WEBSITES

Shrubland biome from NASA:
http://earthobservatory.nasa.gov/
Laboratory/Biome/bioshrubland.html
Chaparral from Blue Planet Biomes:
www.blueplanetbiomes.org/chaparral.htm
Maps, satellite images and data on San Diego fires:
http://map.sdsu.edu/fire2003/
fire2003main.htm
Mediterranean shrublands biome:
www.runet.edu/~swoodwar/CLASSES/GEO
G235/biomes/medit/medit.html

SEE ALSO

biomes, wildfires

CHERNOBYL

Before 1986 Chernobyl (Chornobyl' in Ukrainian) was a city unknown by most of the world. Situated on the Pripiat' River in north-central Ukraine, almost incidentally its name was attached to the V.I. Lenin Nuclear Power Plant located about 25 kilometres upstream. On 26 April 1986 the city's anonymity

vanished forever when, during a test at 1.21 a.m., the number 4 nuclear reactor exploded. This released 30–40 times the radioactivity of the atomic bombs dropped on Hiroshima and Nagasaki in 1945. It took nine days and 5000 tonnes of sand, boron, dolomite, clay and lead dropped from helicopters to put out the graphite fire in the reactor. The radiation was so intense that many of the helicopter pilots died. It was several days before the world first learned of history's worst nuclear accident, when abnormal radiation levels were registered at one of Sweden's nuclear facilities.

Ranking as one of the greatest industrial accidents of all time, the Chernobyl disaster had a far-reaching effect on humans, the environment and the course of Soviet politics. While 31 lives were lost immediately, it will be impossible to predict the exact number of human victims of the genetic impact that will reveal itself over several generations. Following the accident, 116,000 Ukrainians, Russians and Belorussians had to be evacuated and between 1990 and 1995 an additional 210,000 people were resettled. This created a need to build a new town, named Slavutich, for the personnel of the Chernobyl power plant. While estimates vary, it is likely that 3 million people, more than 2 million in Belarus alone, are still living in contaminated areas. The city of Chernobyl is still inhabited by almost 10,000 people. Billions of rubles have been spent, and billions more will be needed to relocate communities and decontaminate the rich farmland.

After the explosion villages had to be decontaminated and major infrastructure had to be rebuilt, for example, water and gas pipelines. The contamination was from radionucleotides (caesium-137, strontium-90 and plutonium-239). It is too early to

Researchers outside the plant, ten years after the accident. Source: Yuri Khotyaintsev

The plant with its sarcophagus as it looked in 2004.

Source: Elena; www.kiddofspeed.com

determine the long-term effects of the radioactive contaminants on all affected habitats, because genetic changes are often not evident in a general population for two or three generations. About 40 per cent of the contaminated area was used for agriculture. The remainder was forest, water bodies and cities. Plants and animals living in the 30-kilometre exclusion zone received the highest level of radiation. As radionucleotides migrate very slowly in soil, the radiation level in this region remains high. In Belarus, 2640 square kilometres of farmland and 1900 square kilometres of forest have been taken out of use by humans forever, according to Igor V. Rolevich, Belarus's first deputy minister for emergencies.

Winds carried the radioactive dust over large parts of north-western Europe and some parts of upland Britain were affected when rainfall washed the dust out of the atmosphere. Sheep grazing was particularly badly affected and farmers had to be compensated for livestock that they could not sell. For many years soils in some areas remained contaminated.

The Chernobyl accident took place during the growing season. It took only two weeks for the conifers to suffer significant damage from exposure. Initially, many trees suffered damage to reproductive tissue. Within three years of the accident, however, the forests began to thrive. Areas within the heart of the exclusion zone now have the largest density of animals as well as the greatest diversity. According to Environmental Chemistry.Com, because people and livestock do not enter the exclusion zone, there is no overgrazing, grasses are thick, and the habitats are doing very well. The area outside the exclusion zone has been severely overgrazed by cattle so that there is little grass and, in addition, most trees have been cut for firewood. The report cautions, 'One must avoid the temptation to become too overly optimistic about the rapid recovery of the contaminated area. Currently there is no standard by which to predict the long-term effects on populations, species or ecosystems.'

Contamination of the soil by radionucleotides is a particular problem for the local residents near the Chernobyl nuclear reactor. These radionucleotides remain in the soil for a long while, are taken up by plants and transferred to the milk and meat products of cattle that graze the area. Human exposure could have been reduced by prohibiting residents from working in the fields, eating fresh vegetables, and preventing their livestock from eating the contaminated forage. Even changing the type of crop planted, because each species has a different rate at which it absorbs specific chemical elements or compounds, or adding chemicals such as lime or potassium fertilisers could have protected the population.

The water supply does not seem to have been as contaminated as the soil. The radionucleotides tend to settle out with time. Aquatic habitats also tend to be more tolerant of radioactive contamination. However, there are many sites where radiation-contaminated equipment was dumped that could be harming the groundwater. While the environment in the exclusion zone seems to have recovered, it still has to deal with long-term effects, such as genetic mutations, which may not surface for a several generations. The enormous 300,000-tonne concrete and steel sarcophagus that was built at Chernobyl to entomb the destroyed reactor still contains uranium fuel. The long-term stability of the sarcophagus continues to cause concern for the environment and people of the area.

The closure of Reactor 4 and the freeze on construction of new reactors reduced the availability of electricity supplies. In addition, the birth rate in the area has decreased, out-migration has resulted in a workforce shortage, and the near ending of agriculture and forestry production has further hampered the economy.

The Chernobyl catastrophe derailed what had been an ambitious nuclear power programme and formed a fledgling environmental movement into a potent political force in Russia, as well as a rallying point for achieving Ukrainian and Belorussian independence in 1991. Chernobyl has become a metaphor not only for the horror of uncontrolled nuclear power but also for a political system of secrecy, disregard for the safety and welfare of workers and their families, and an inability to deliver basic services such as health care and transportation, especially in crisis situations. The Chernobyl accident continues to be a focal point for those who advocate that nuclear power can do great human and environmental harm in the event of an accident, negating the environmental benefit that nuclear power has over the burning of fossil fuels.

WEBSITES

Chernobyl information site:
www.chernobyl.info/en
UK Chernobyl information site:
www.chernobyl.co.uk
A grim motorcycle ride through Chernobyl in 2004:
www.kiddofspeed.com

SEE ALSO
nuclear power

CLEAN AIR ACTS

Fog everywhere. Fog up the river, where it flows among green aits and meadows; fog down the river, where it rolls defiled among the tiers of shipping and the waterside pollutions of a great (and dirty) city. Fog on the Essex marshes, fog on the Kentish heights. Fog creeping into the cabooses of collier-brigs; fog lying out on the yards, and hovering in the rigging of great ships; fog drooping on the gunwales of barges and small boats. Fog in the eyes and throats of ancient Greenwich pensioners, wheezing by the firesides of their wards; fog in the stem and bowl of the afternoon pipe of the wrathful skipper, down in his close cabin; fog cruelly pinch-ing the toes and fingers of his shivering little 'prentice boy on deck. Chance people on the bridges peeping over the parapets into a nether sky of fog, with fog all round them, as if they were up in a balloon, and hanging in the misty clouds.

Charles Dickens, writing in his famous novel *Bleak House* in 1853, was one of the first writers to provide a graphic description of the fogs that engulfed London. The fogs were made worse by the emission of smoke and other pollution from the thousands of chimney stacks that lined the capital's skyline.

In fact, the term clean air is a misnomer – there is no such thing as completely clean air. The atmosphere that forms a swirling envelope of liquids, solids and gases above the Earth's surface has, and will always, contain microscopic particles of dust and other substances. There have, however, been times when the unregulated emission of particulate matter, largely from heavy industry, has resulted in serious health problems and other difficulties for the people living in industrial settlements.

It was not until almost a hundred years after *Bleak House* that the British Government took action to reduce the amount of harmful substances emitted into the troposphere, and then only after the heavy loss of life in the notorious London smogs of the winter of 1952–3. Around 4000 died as a result of the pea-soup fogs that engulfed London, as mortality rates from bronchitis and pneumonia rose dramatically above the average.

Steps to curb the emission of harmful substances began with the City of London (Various Powers) Act of 1954, and then the Clean Air Acts of 1956 and 1968. Both these acts banned emissions of black smoke and decreed that residents of urban areas and operators of factories should convert to smokeless fuels. However, the policies were not immediate and gave people and businesses time to convert to these clean fuels. Therefore, outbreaks of severe air pollution continued and even in 1962 750 Londoners died as a result of another outbreak of smog, which thankfully was not as potent as that ten years before.

The discovery and widespread use of natural gas and oil from the North Sea sped up the

Case study

Clean air acts in the USA

Clean air policies were also introduced in the USA during the 1950s. In 1955, after many state and local governments had passed anti-pollution legislation, the federal government decreed that a national solution was needed, and the Air Pollution Control Act of 1955 was passed.

The passing of the act increased public awareness of the problem. In the following years a number of other acts were introduced, modifying the regulations and tightening up the controls, especially on power plants and steel mills. Many of these acts did not take into account the mobile sources of air pollution from vehicle emissions, which by the end of the 1960s had become the largest source of many dangerous pollutants. The Clean Air Act of 1970 was a major revision and set much more demanding standards, especially on mobile sources, but further modifications in the 1980s did not happen, since President Reagan's administration placed economic goals ahead of environmental ones. However, oxygenated automobile fuels were required by 1990 in many cities to reduce smog problems, such as Denver Colorado.

switch in the UK to cleaner fuels. By the end of the 1960s less coal was burnt, both on domestic fires and in industrial furnaces, while a switch by British Railways from steam-driven locomotives to diesel engines assisted the process.

In recent years the decline of heavy industry and the closure of coalmines have also played a role, but even in this age of smokeless fuels and anti-pollution legislation, we still do not have totally clean air. The air in urban areas still contains other types of pollutants, many of them from vehicle exhausts, including carbon monoxide, nitrogen dioxide, benzene and various aldehydes. They are less visible than the pollutants of yesteryear, but these modern pollutants are equally as dangerous, causing eye irritation, asthma and bronchial complaints.

Although to a large extent the notion of clean air still remains a myth, perhaps the various legislation introduced in the UK and North America needs modification before the next wave of disease and death from air pollution takes place.

WEBSITES

Manchester Metropolitan University:
www.doc.mmu.ac.uk/aric/eae/Air_Quality/Older/Clean_Air_Acts.html
American Meteorological Society:
www.ametsoc.org/sloan/cleanair/cleanairlegisl.html

UK Met Office:
www.meto.gov.uk/education/historic/smog.html

SEE ALSO

nitrogen oxides, pollution, vehicle exhausts

CLEAR FELLING

We live in an interesting age. What will our descendants think of us as they look back? The world's tropical rainforests are not all lost yet but it is difficult to be optimistic about their future. Major parts of the rain forest have been irretrievably lost. I write this in Singapore, and have recently made several visits to the adjacent southern tip of Malay, South Jahor. Rain forest has been virtually eliminated from the landscape . . . the local endemics are now extinct. (T. C. Whitmore, 1998, *An Introduction to Tropical Rain Forests* 2nd edn, Oxford University Press)

The main cause of the forest loss described above has been felling (cutting down), initially for agriculture but later for timber. The primary driving force in each case is profit. Loggers usually use one of two methods to remove trees: clear felling and selective felling.

In clear felling all the trees in an area are felled. In Europe and the USA clear felling has been carried out for centuries to create land for agriculture and settlement. The forest was never able to re-establish itself and over time much of the forest cover across Europe and the USA has disappeared.

Most felling today tends to be for wood as a raw material for furniture, timber and paper. Clear felling does not prevent the resource being renewed, and replanting often leads to the regeneration of the forest cover with little permanent damage to the ECOSYSTEM. Clear felling is also a valuable management technique employed in many woodland areas. However, long rotation times are needed, often over 50 years, to allow the replanted areas to reach maturity.

In the twentieth and twenty-first centuries clear felling has been much more prevalent in the less developed countries and especially in the world's rainforests. The felling has been largely for the timber resource, although some clearances have made way for new settlements and for agriculture, such as the ranches in the Amazon rainforest and plantations in Malaysia. The clear felling may be highly damaging, since it is rarely followed up with any replanting. Unlike temperate forests dominated by one tree species, for example an oak wood, in the rainforests one hectare of rainforest may contain over 500 different species of tree. Some of these rare species are at risk of extinction by clear felling. It is reported that some logging companies will slash and burn half a hectare of rainforest just to obtain one commercially valuable hardwood tree.

The second technique to remove trees is selective felling, in which only the required species of trees are felled in an area. While this sounds good, selective felling can be almost as damaging as clear felling. Other trees still need to be felled to be able to gain access and transport the selected trees out of the forest. In addition, considerable damage occurs accidentally as felled trees fall and crash against neighbouring trees, bringing them down too. In Indonesia it is reported that while only three to ten trees may be targeted in a hectare of forest the damage to the overall stand may be as high as 50 per cent when operations use heavy machinery.

Most logging today is mechanised, whether the technique is clear felling or selective felling. The trees are usually cut down using hand-held chain saws. Other heavy machinery, including trucks, tractors, bulldozers, wheeled skidders (used to drag the logs out of the forest) and loaders, are used for skidding, lifting and transporting the logs, and roads are built to provide access.

Case study

Hazlehead woodland management scheme, Scotland

Hazlehead Woodlands is managed by Aberdeen City Council in Scotland, UK. Plans for its management have been approved by the Forestry Commission and Scottish National Heritage and discussed at two public meetings. The plans involve both selective and clear felling. The selective felling will remove between 25 and 33 per cent of the trees, giving space for those remaining – the strongest trees – to grow to maturity and allow more light to penetrate to ground level. This will encourage a wider range of wildflowers and the germination of tree seedlings. This technique reduces the visual impact of felling and causes less disruption to the wildlife. For example, areas of Sitka spruce will be clear felled where the trees are at risk of being blown down.

The plans involve the use of large timber harvesting machinery that will complete the job more quickly than chain saws. All the harvested timber will be transported to a sawmill and the income generated used to support the project. Replanting will take place in these areas using a variety of species, including rowan, ash, Scots pine and Norway spruce. This will create a varied food source for birds and small mammals, in particular the red squirrel, which is in danger of being ousted by the North American grey squirrel (as has happened in other parts of the UK).

During felling some trails will be closed and others may be damaged. However, once the work is complete it is expected that trails will be upgraded and extended. Access will be given to both walkers and horse riders and the woodland will become sustainable with a greater diversity of species.

Global forest losses

Research carried out by World Wide Fund for Nature suggests that the international timber trade is the primary cause of forest losses rather than shifting cultivation and fuelwood collection. Findings from a selection of countries are shown below:

- Finland: Only 1–2 per cent of old-growth forest remains.
- UK: Illegal felling of broadleaved trees to sell as firewood is on the increase.
- Canada: Boreal forest logging is taking place on a large scale in many areas, particularly in Alberta. In Ontario two thirds of the remaining 1 per cent of old growth forest is slated for commercial felling.
- USA: Logging of old growth forests in the Pacific North West looks likely to increase again in response to government aims to deregulate the industry and overturn environmental legislation.
- Brazil: Illegal logging of mahogany is having a major impact on the ecology and the survival of forests in many areas. Until recently 80 per cent of mahogany exports were of illegally felled trees.
- Argentina: Temperate forests are rapidly being logged by foreign companies, including many from North America
- Cameroon: Numerous transnational companies are operating, including companies from Belgium, France, Germany and Italy. A survey in 1993 identified 100 forest operations, 60 of which were foreign owned. Logging has increased 100 per cent in the last few years.
- Indonesia: The Government intends to replace 2 million hectares of forest with plantations.
- Australia: Logging is the major cause of forest degradation and loss particularly in the South West and Tasmania.

WEBSITES

World Wide Fund for Nature:
www.panda.org
Friends of the Earth:
www.foe.org
Information on all aspects of boreal forests:
www.borealforest.org/world/scan_mgmt.htm
Finland's Ministry of Agriculture and Forestry:
www.mmm.fi/english/forestry/forests
The Royal Botanical Gardens at Kew, UK, produces an informative fact sheet:
www.rbgkew.org.uk/ksheets/pdfs/r3_rainforest.pdf

England's green landscape (above, Elham, Kent, in southeast England) is a result of its mild temperatures and sufficient rainfall. Photograph by Joseph Kerski

SEE ALSO

deforestation, woodland ecosystems

CLIMATE

Climate refers to the long-term, average weather in a particular place or region. While weather events such as hurricanes have an effect on the environments in which they occur, climate impacts on the environment much more than the weather does. Climate influences what types of plants and animals can live in an area, the energy and water balances, and the biodiversity. Climate even influences the types of human activities and population that the land can support, and, therefore, influences the kind and amount of human pressure on the environment.

Given the variety of climates in the world, how can they be understood? In the late 1800s, Vladimir Koppen examined the correspondence between natural vegetation and the climate type around the world. Vegetation adapts to the average climate condition of an area and therefore provides clues to an area's climate. Koppen developed a climate classification system, still in use today, which uses monthly averages of temperature and precipitation throughout the year. In places with incomplete records of temperatures and precipitation, which was quite common during the 1800s, Koppen used vegetation as a surrogate. As an alternative, Thornthwaite used potential evapotranspiration to classify climate, which may be more useful when considering the effects of climate on agriculture. Other specific systems to classify climate have been created to deal with such phenomena as frost and freeze, drought (such as the Palmer Drought Index) and flood.

One of the major reasons why people are concerned about climate is that if it changes it can have a significant effect on the environ-ment. In the twentieth century global mean surface air temperatures increased by between 0.3°C and 0.6°C. This may not seem like a large amount, but when the changes accumulate, the impact could be far-reaching. Effects of climate change range from droughts, floods, coastal erosion, changes in biodiversity, risks of disease, agricultural changes from soil alterations, to changes in the ocean currents, and more. In the UK, temperatures are expected to rise by between 0.1 and 0.3°C every ten years, easily adding up to 1.5°C over the next 50 years.

WEBSITES

US National Climate Data Center:
www.ncdc.noaa.gov/oa/ncdc.html
Climate statistics for specific cities in the USA, from the National Weather Service:
www.wrh.noaa.gov/wrhq/nwspage.html
Environment Canada:
www.climate.weatheroffice.ec.gc.ca/Welcome_e.html
UK Met Office, the UK's weather agency:
www.metoffice.gov.uk

World weather extremes

The highest recorded temperature ever recorded on Earth occurred on an afternoon in El Azizia, Libya, in 1922, when the thermometer read 57.8°C. Interestingly, the hottest recorded temperature on Antarctica was 15°C at Vanda Station in 1974. Vostok's −89.5°C reading in 1983 was the lowest recorded on the planet, and Asia's record of −67.8°C at Oimekon, Russia, was also bitter. However, Australia's lowest reading was only −22.8°C in 1994.

The wettest place on the Earth may be Mawsynram, India, which recorded 1187.2 centimetres of precipitation in one year, or Mount Waialeale in Hawaii, where 1168.4 centimetres of precipitation fell. The official greatest average annual precipitation for South America is 899.2 centimetres at Quibdo, Colombia, but rainfall in the nearby town of Lloro was estimated to be 1329.9 centimetres during a single year. The Atacama Desert in Chile is famous for being what many consider to be the driest place on Earth – Arica recorded 0.1 centimetres of annual precipitation. However, other places are nearly as dry: Wadi Halfa in the Sudan, Africa, recorded less than 0.3 centimetres of precipitation, and Batagues, Mexico, recorded 3 centimetres in one year. The South Pole Station received only 2 centimetres of precipitation during at least one year of measurement.

These extreme weather events can be understood within the context of their climate. For example, Australia's coldest temperature is mild compared to that of Russia because Australia is completely surrounded by oceans, which help moderate the climate (see **ocean currents**). In Russia, some places are thousands of kilometres from the nearest ocean. The South Pole Station's dry years reflect the fact that Antarctica is actually a desert. Although most of its surface is covered by ice, little new precipitation falls – most sleet and snow in the air is blown off from a nearby mountain or icecap.

World Meteorological Organization:
www.wmo.ch/index-en.html

Compendium on climates of the world:
www.ncdc.noaa.gov/oa/documentlibrary/
pdf/climatesoftheworld.pdf

World weather extremes:
www.ncdc.noaa.gov/oa/climate/
globalextremes.html

UK Climate Impacts Programme:
www.ukcip.org.uk

UK Environment Agency's climate resources:
www.environment-agency.gov.uk/kids/youth/
infopoint/climate/?lang=_e

SEE ALSO

Antarctica, climate change

CLIMATE CHANGE

Climate change refers to a large-scale change in one or more basic climate components, such as temperature or precipitation. Such a change would be significant enough to have a profound impact on the natural environment and human society. Climate change is perfectly natural and has taken place many times over the geological record. During the last Ice Age, which ran from 2 million years ago until about 10,000 years ago, temperatures fluctuated considerably. This led to cold periods called glacials, when ice sheets advanced southwards over much of Europe and North America, and warmer inter-glacial periods when temperatures exceeded the values of today.

Most recently, however, the term climate change has become synonymous with 'global warming', as scientists have become increasingly convinced that temperatures worldwide are on the increase. Global mean surface temperatures have increased 0.3–0.6°C since the late nineteenth century. The twentieth century's 10 warmest years all occurred in the last 15 years of the century. Of these, 1998 was the warmest year on record. Snow cover in the northern hemisphere and floating ice in the Arctic Ocean has decreased. Globally, sea level has risen 10 to 20 centimetres over the past century. Worldwide precipitation over land has increased by about 1 per cent. The frequency of extreme rainfall events has increased throughout much of the USA. Temperatures are projected to increase by 0.33°C each decade through the remainder of this century, or a total of just over 3°C.

Most scientists would agree that at least some of the warming is human induced, primarily through increased discharges of carbon dioxide and other greenhouse gases, such as methane, into the atmosphere, stratospheric ozone destruction, and land use changes. Since the beginning of the Industrial Revolution, atmospheric concentrations of carbon dioxide have increased nearly 30 per cent, methane concentrations have more than doubled, and nitrous oxide concentrations have risen by 15 per cent. These increases have enhanced the heat-trapping capability of the Earth's atmosphere. The combustion of fossil fuels and other human activities are the primary reason for the increased concentration of carbon dioxide. Plant respiration and the decomposition of organic matter release more than ten times the carbon dioxide released by human activities, but these releases have generally been in balance during the centuries leading up to the Industrial Revolution, with carbon dioxide absorbed by terrestrial vegetation and the oceans.

Despite the general agreement amongst scientists that temperatures are increasing, it is important to take into account the fact that reliable statistics only exist for a matter of a few decades in a few places across the world. Scientists have had to make use of other forms of evidence, such as air trapped in polar ice sheets or fossil pollen data, to access longer-term changes in climate. The period upon which data is collected is very important. Furthermore, because the composition of the Earth's atmosphere has changed over time, and because the radiation balance, global winds and ocean currents are affected by plate tectonics and how much land is mountainous, the task of putting recent measured short-term change into its proper perspective becomes all the more daunting.

Typically, the recent measured upward temperature trends are not very large when compared to inferred changes that took place after the last Ice Age. When these values are introduced into a global climate model (a version of global general circulation models) and predicted increases in greenhouse gases are examined, future temperature trends give

Case study

The National Ice Core Laboratory

Even when it is hot outside, Geoffrey Hargreaves, Curator at the National Ice Core Laboratory (NICL) in Denver, Colorado, USA, dons fleece-lined boots, fur-lined gloves and a parka to go to work. Temperature inside the laboratory is −36°C, day in and day out, where ice core staff assist scientists from around the world to uncover the mysteries of past climate history locked inside ice. This ice has been painstakingly brought to the laboratory in Colorado from Antarctica and Greenland. Unlike tree rings or layers of sediment, ice contains samples of the ancient atmosphere, including 'greenhouse gases' trapped in bubbles dating back hundreds of thousands of years. Determining the age of samples within cores is done partly by counting layers of ice, which represent annual snowfall, much like dating tree wood by counting annual rings. Snow falling on the continents forms layers, and the snow of summers, with their sun and dust, has a different appearance from the cleaner snow that falls in sunless winters, making the layer counting easier.

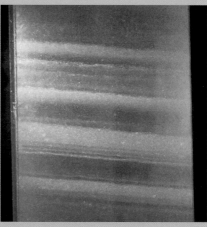

A piece of ice from a core at about 1850 metres below the surface. White layers are summer layers and dark (clear, really) layers are winter layers.

Photograph: National Ice Core Laboratory

Information available from analysis of the ice record can include temperature, from measurements of the isotopes of the elements hydrogen and oxygen in the ice. Information also includes rates of ancient precipitation, chemistry and gas composition of the lower atmosphere, volcanic eruptions, solar variability, sea-surface productivity and other climate and atmospheric indicators. This data makes ice cores a powerful tool in researching ancient climates and the composition of the atmosphere, as far back in time as 800,000 years or more.

Ice comes to the laboratory in the form of 'core' – cylinders of ice drilled from the polar ice sheets using drills that are like large hollow screws. Ice near the surface represents periods immediately before the present, whereas deeper ice, up to 3 kilometres down, is older – to nearly 1 million years, depending on location. Acquiring each ice core from a remote region of the world and transporting it back to the National Ice Core Laboratory safely can require years of planning and execution. Drilling the core is the responsibility of the Ice Coring Drilling Services. Safely transporting ice cores from the drilling site back to Denver requires the diligence and cooperation of several organisations, including the National Science Foundation, Raytheon Polar Services and the New York Air National Guard. Several other countries have ice core laboratories and field drilling programmes, including sites in high mountain glaciers (which may give past climate information about temperate latitudes) as well as polar regions.

significant cause for concern. These projections are of a magnitude that would significantly affect life on Earth, including human life.

Significant changes are possible for the environment locally and regionally, and globally with global warming. Deltas and coastal wetlands are already in rapid retreat in many areas, due to higher water levels and stronger sea storm surges. Coral reef ecosystems would be impaired by a warmer sea temperature. Agriculture and plant distribution would be altered throughout the world. Precipitation might be too much in some areas, but not

Once the new ice has come to thermal equilibrium with its new surroundings, it is carefully unpacked, organized, racked and inspected. Photograph: National Ice Core Laboratory

enough in others. Melting polar ice caps would inundate numerous coastal cities and endanger many other coastal areas. More heat waves and droughts would result in more demand and conflicts for water resources. More extreme weather events could produce floods and destroy property and human life. There would be a greater potential for heat-related illnesses and deaths, and a wider spread of infectious diseases carried by insects, rodents and people.

WEBSITES

The National Ice Core Laboratory:
http://nicl.usgs.gov/index.html
US Global Change Research Program:
www.usgcrp.gov
US EPA global warming site:
http://yosemite.epa.gov/oar/globalwarming.nsf/content/index.html
US global climate change policy:
www.state.gov/g/oes/climate/

Manchester Metropolitan University's global climate change student guide:
www.ace.mmu.ac.uk/Resources/gcc
International Energy Agency's global climate change resource:
www.ieagreen.org.uk/ghgs.htm
British Antarctic Survey:
www.antarctica.ac.uk
UK Met Office:
www.met-office.gov.uk

SEE ALSO

greenhouse effect

CLOUDS

The Earth is not the only planet in the solar system with clouds. Modern astronomy and probes to Venus, Mars and several of the gas giants, including Jupiter and Saturn, have shown that clouds are more the rule than the exception in our solar system. These planets have clouds of carbon dioxide, methane and other gases, and sometimes huge dust storms appear to be clouds through telescopes.

Here on Earth, clouds are made up of tiny water droplets or ice crystals. These liquid and solid particles are suspended as aerosols in the atmosphere. Clouds are ephemeral in most regions of the world, meaning that they come and go. They are affected by changes in temperature and humidity, constantly being formed through condensation and deposition, and dissipating through evaporation and sublimation.

Clouds are extremely important to the environment. Without clouds, the Earth would not support life. Not only do clouds supply much needed rainfall, but they also help to maintain the Earth's heat balance. Moreover, they affect vegetation: different plants grow in sunny locations compared with shaded locations.

Clouds are classified into four main categories: low clouds, middle clouds, high clouds and vertical clouds, which build through tens of thousands of metres of the atmosphere. Clouds are further categorised according to appearance, with the layered, horizontal clouds (stratus), the puffy convective clouds (cumulus) and the wispy high ice clouds (cirrus) being among the most common.

Clouds are formed when air is cooled to the dew point temperature, at which point it is saturated. The process of condensation then begins turning water vapour into water droplets. It is these water droplets that form clouds. When condensation occurs, latent heat is released. Releasing heat makes air even more buoyant, and, if enough is released, the cloud will rise and chill further, and more condensation will result. This, in turn, releases more latent heat and further lifting occurs. This lifting, cooling, condensing and further lifting becomes almost self-perpetuating, resulting in high, towering cumulonimbus clouds.

Low-level, 'puffy' fair-weather cumulus clouds are typically formed when solar heating warms a patch of earth. The land then warms the air immediately above it (by conduction), making it buoyant. The buoyancy provides lift, and the lift provides cooling. The moisture in the air condenses into tiny cloud droplets that reflect the white light of the Sun.

Cumulus clouds building in the atmosphere. Photo credit: National Oceanic and Atmospheric Administration/U.S. Department of Commerce

Low stratus clouds depend on the slow, gentle lifting and cooling of large parcels of air that are hundreds or even thousands of square kilometres in extent. Clouds can be formed from convection resulting from the heating of the Earth's surface, from air flowing up mountains or hills, air forced up along the edges of frontal boundaries and upper air conditions such as jet streams.

Some regions are almost perpetually covered by clouds. In these places, such as Central and South American rainforests, detailed mapping of the surface was not possible until the development of remote sensing technologies such as side-looking radar imaging. Elsewhere, such as in the Atacama Desert of Chile, persistent high pressure with descending and diverging air, prevents cloud formation, with month after month of clear sky being the result. Whether clouds are present to block and reflect incoming solar radiation is an important component in determining the Earth's energy balance. For example, if warmer global temperatures cause greater evaporation, resulting in greater condensation and cloud cover, it might be argued that further warming might be forestalled by the clouds that would then shade more of the surface. Modellers of global climate are working hard to include the role of cloud cover in the global general circulation models that are used to predict future climate changes.

Altostratus lenticularis clouds at sunset. These clouds typically form as air flows over the leeward sides of mountains, as here in Colorado USA. Photo: Joseph Kerski

WEBSITES

Cloud types from the University of Illinois: http://ww2010.atmos.uiuc.edu/(Gh)/ guides/mtr/cld/cldtyp/home.rxml
University Corporation on Atmospheric Research Cloud Resources: www.windows.ucar.edu/tour/link=/earth/ Atmosphere/clouds/cloud_types.html
Thompson's weather photographs: www.inclouds.com/gallery.html
Cloud types from Michigan Technology University: www.geo.mtu.edu/department/classes/ ge406/tjbrabec/cloud.html

SEE ALSO

climate change, precipitation, thunderstorms

COAL

Coal ranges from a soft substance to a hard rock, formed by the decomposition and bacterial decay under anaerobic (oxygen-free) conditions of plant remains. It forms in environments that in the geological past were warm and humid, thereby promoting both rapid vegetation growth and the accumulation of plant debris (for example swamps, bogs and marshes).

The word coal is something of an umbrella term, as there are many types of coal, each representing different stages in the coal formation process (coalification), and therefore varying in their physical and chemical properties. The four main types of coal are peat, brown coal or lignite, bituminous coal and anthracite. The latter has the greatest carbon content, and therefore the smallest amount of other volatile substances that could lead to pollution.

The primary use of coal is as a fuel. At the local scale it is burned on domestic fires, while on a larger scale it fires turbines in enormous power plants to provide electricity for whole regions. Despite being widely distributed across the world and an efficient provider of heat, its burning has been linked with environmental issues such as ACID RAIN, GLOBAL WARMING and smog.

The USA has more coal reserves than any other nation on the Earth – estimated at around 250 years at current production rates. The USA has 25 per cent of the world's coal reserves, and it contributes around 28 per cent of world coal production.

Global coal reserves (million tonnes – data for 2002)	
USA	249 994
Russia	157 010
China	114 500
India	84 396
Australia	82 090

Coal production by state in the USA (thousand tonnes – data for 1993)	
Wyoming	210 129
Kentucky	156 299
West Virginia	130 525
Pennsylvania	59 700
Texas	54 567

The story of coal in other industrialised nations is a very different one, where the cycle of boom has turned into one of bust, as coal seams have become exhausted. The nineteeth-century boomtowns of the German Ruhr and South Wales became problem regions, with high levels of unemployment.

Coal was the first fuel of the Industrial Revolution and its exploitation from the late eighteenth and into the nineteenth century spawned vast industrial regions in Western Europe. 'Coal was king' in Britain from the 1780s onwards, and the growth of heavy industry – steelmaking, shipbuilding, and others – saw British coalfields develop into thriving economic regions. Hundreds of deep shaft and drift mines were established across the UK, as the demand for coal grew at home and abroad.

The second half of the twentieth century saw a decline in coal output in the UK, and restructuring took place during the 1980s in the face of competition from cheaper coal from other European countries, especially Poland, as well as from the opencast mines in Australia and Canada. Some mechanisation had taken place in British mines after the Second World War in an attempt to reduce costs, but British coal was still relatively expensive, and as seams became deeper and more dangerous to exploit, questions started to be raised about the economic viability of mining the coal.

Many seams had already become exhausted, and as the table below shows, the number of mines in the UK had dropped dramatically by the mid 1990s.

Coal production in the UK

	Number of collieries	Output (million tonnes)	Workforce (thousands)
1955	858	211	700
1965	535	186	500
1975	245	117	220
1985	171	105	171
1994	17	80	18

WEBSITES

Kentucky Geological Survey, with an overview on coal in the USA:

www.uky.edu/KGS/coal/webcoal/pages/coal3.htm

National Coal Mining Museum UK, with a virtual underground tour:

www.ncm.org.uk

Further information about Tower Colliery:

www.welshcoalmines.co.uk/GlamEast/Tower.htm

SEE ALSO

mining reclamation, mining waste

Case study

Tower Colliery: the last deep mine in South Wales

There had been hundreds of coal mines in South Wales in the 1890s, yet by the 1990s there was just one left at Tower Colliery, near Hirwaun. The colliery was opened in 1878, and by the Second World War over 1000 men worked at the mine.

Tower was one of the mines that British Coal wanted to close, and on 22 April 1994 it ceased production on the grounds that it would be uneconomic to continue. However, the local workers believed otherwise and £2 million was raised by 239 miners, with each pledging £8000 from their redundancy payouts. They successfully bought the mine, resumed production and in 1995 produced 450,000 tonnes of high-grade anthracite, making over £4 million in pre-tax profit.

Tower Colliery is therefore an example of how privatisation has successfully taken over from the previously nationalised mining operations, allowing energy demands to be met from domestic supplies.

COMBINED HEAT AND POWER

The pressure on energy companies to cut costs and reduce harmful emissions of pollutants and greenhouse gases has resulted in the owners and operators of these facilities looking at ways of using energy more efficiently. One option is combined heat and power, or CHP – a highly efficient method of generating both electricity and heat at a single power plant. Known as 'cogeneration' and 'total energy' in the USA and in EU nations, it utilises heat from electricity generation and avoids transmission losses because electricity is generated on site. It is also highly efficient, producing a given amount of electric power and heat with 10–30 per cent less fuel than it takes to produce the electricity and heat separately.

Although the basic CHP concept has been understood for over a century, the development of new technologies during the 1990s greatly increased the attractiveness of CHP.

Many successful cogeneration schemes exist in the USA. For example, the University of Florida has a 42-megawatt gas turbine cogeneration plant, built in partnership with the Florida Power Corporation. The State University of New York at Stony Brook has a 45-megawatt facility, while in Washington, DC, there is a cogeneration plant that produces chilled water for cooling and electricity to power eight Smithsonian Institution facilities.

The US Government has recognised that cogeneration has significant potential to increase the overall efficiency of energy use, as well as reduce total greenhouse gas emissions. The Bush Administration's national energy policy proposal, released in May 2001, included recommendations to encourage the growth of CHP, including investment tax credits for cogeneration projects.

CHP has also increased in popularity in the UK, with an annual average growth rate of 7 per cent per annum since the 1990s. A number of community heating schemes have seen the adoption of CHP principles. An example is the Manchester Energy Company (MECo) in South Manchester – a partnership between Manchester City Council and Powerminster – that has upgraded and replaced heating systems in 417 flats on two separate sites.

MECo replaced gas warm air and electric storage heating systems with a radiator system and community heating scheme involving CHP, each with a capacity of 150 kilowatts. It is

Case study

CHP at the Massachusetts Institute of Technology, USA

Some cogeneration schemes in the USA had to face a number of legal hurdles and bureaucratic obstacles before being introduced. An example occurred at the Massachusetts Institute of Technology (MIT) which in 1985 began to consider generating its own electricity for a variety of very laudable reasons. With its students increasingly using PCs, stereos, hair dryers and toaster ovens, the university faced soaring electricity costs from the local utility, Cambridge Electric Company (CelCo). There were also concerns that many of MIT's world-class research projects could be ruined by service interruptions, while its 1950s steam-powered heating and cooling system was a major source of local air pollution.

The university selected a 22-megawatt natural gas-fired CHP system, which would cut annual energy bills by 40 per cent and allow MIT to recoup its investment within seven years. MIT's first major hurdle was in obtaining the environmental permit needed before construction could begin. Although the scheme would reduce annual pollutant emissions by 45 per cent, the stumbling block lay with the amount of ammonia needed by the system. It contravened the state's nitrogen oxide guidelines, ironically designed for power stations more than ten times larger than MIT's CHP generator. After a period of legal wrangling, MIT received the go-ahead and completed its scheme in September 1995.

Some 20 years after MIT's initial decision, the view of CHP has changed for the better, and more CHP schemes are now being given the green light. The experience of MIT remains as a salutary reminder that it was arguments based on economics and profits, rather than those based on the environment and pollutants, that frequently won the day in earlier energy debates.

estimated that the scheme has saved residents around £15 per month, giving annual energy savings of £75,000 for both sites.

WEBSITES

Information about CHP schemes in the UK: www.chpa.co.uk
MidWest Cogeneration, with useful graphics about how CHP works and lists of other sites to visit:
www.cogeneration.org

SEE ALSO

energy conservation

COMMON AGRICULTURAL POLICY (CAP)

After the Second World War agriculture in Europe was in a dire state. Food shortages continued and many items were rationed – people made their cakes using dried eggs and children queued to spend their sweet vouchers. World markets were topsy-turvy and so were the farmers' incomes and food prices. Western Europe was not producing enough food to feed its people. Although the farmers were a powerful lobby, with their votes especially important in France and Germany, their economic position was very fragile and their futures uncertain.

When the EEC, the forerunner of the EU, was established in 1957, one of the key driving forces was the state of agriculture within the member countries. The Common Agricultural Policy (CAP) was established in the 1960s to solve the problems of food shortages, to stabilise prices and to give farmers a minimum standard of living.

Prior to 1992, the main policies of CAP involved price support policies, subsidies and restricting imports from outside the EU. A target price was set for all produce and an intervention price. If the market value of the product fell to the intervention price then the EU bought up the produce to ensure a reasonable income for the farmers. There was also a minimum price for imports that stopped EU farmers being undercut by imported goods. In addition, subsidies were put in place for certain products in which the EU was not self-sufficient, to encourage more production.

Overproduction in the EU

The system of subsidies led to overproduction of many products – the press had a field day with wine 'lakes' and butter, beef and wheat 'mountains'. These surplus products were then bought up by the EU and put into expensive storage. These surpluses led to quotas, such as for milk and potatoes, the policy of 'set aside' and the removal of some subsidies. The farmers were asked to produce less – and were paid to do so.

After 1992 farmers had quotas that controlled how much they were allowed to produce. Surplus milk was simply poured away down the drain. They were also required to designate at least 20 per cent of their land as 'set aside' for five years. This land could be left fallow, planted with trees or used for non-agricultural purposes. In 1995 UK farmers were paid £253 per hectare for fallow land.

The CAP remains the largest common policy in the EU and claims to have brought advantages of cheaper food prices, improved food quality and higher standards of animal welfare and environmental protection. However, it has many critics who claim the policies have been a failure. Certainly, farm production has been reduced and farmers' incomes lowered, although other factors include the outbreaks of BSE and foot and mouth disease. The policy is very expensive and requires direct income support for farmers.

The CAP costs about £30 billion a year, almost half the EU's total budget. It is paid for by value-added tax (VAT) charged on most purchases and services in the member states. Some production still generates surpluses that are being dumped cheaply in developing countries, undercutting local farmers. The policy is also blamed for encouraging environmentally damaging intensive farming, such as hedgerow removal and the use of chemicals. Rural depopulation has continued and concerns exist about who is going to farm in the future. The policies have also upset many of the traditional trading partners who imported foodstuffs into the EU. Given these problems, it was clear that the CAP needed to be reformed.

In June 2003 a new agreement was reached to alter the CAP radically. The changes implemented in January 2005 remove the link between production and payment. Quotas were abolished, except for milk, and compensation payments will be made where subsidies are removed. An annual payment will be made to farmers based upon income in previous years regardless of the yields. It is hoped that farmers will farm for the market rather than the subsidies available. The new policies also allow farmers to be paid to maintain the rural landscapes.

WEBSITES

Information about the CAP from the Department for Environment, Food, and Rural Affairs:

www.defra.gov.uk/farm/capreform

University of Nottingham:

http://agrifor.ac.uk/whatsnew/detail/5042394.html

European Commission:

www.cec.org.uk

COMPETITION, PREDATION AND PARASITISM

Competition, predation and parasitism are forms of **SYMBIOSIS**. Competition might be thought of as 'lose–lose', and predation and parasitism as 'win–lose.'

Competition

Farmers and grasshoppers compete with each other. If there were no grasshoppers, there would be more wheat for the farmer. If the farmer did not harvest the wheat, there would be more for the grasshoppers. Competition is a contest between two or more individuals (or populations of individuals) for a common resource. In order for competition to occur, that resource must be limiting; that is, it must be in short supply. Unlike in competitive sports, for example, where there are winners and losers, in the environment, competition is reciprocal, a 'lose–lose', proposition (see the table in **SYMBIOSIS**).

As a form of symbiosis, competition is *between* species (interspecific). Competition can also occur between individuals of the same species (intraspecific). However, intraspecific competition is not a form of symbiosis. To say that it is not symbiosis is not to say that it is unimportant. Darwin noted in *The Origin of the Species* that the 'struggle for existence' would be more intense if organisms were more closely related, because the resource needs of similar organisms are similar. Indeed, the most intense competition ought to be among members of the same species, which have roughly the same genetic programme and, hence, roughly the same resource needs.

The famous experimental results of Russian ecologist G.J. Gause demonstrated interspecific competition. Gause grew populations of two species of the organism, *Paramecium*, in laboratory cultures with a constant food supply. The figure below indicates the growth curve of each species living separately and in a mixed culture. After a brief initial lead, *Paramecium caudatum* declined to extinction when grown in mixed cultures.

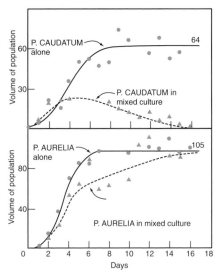

A classic demonstration of competition, populations of *Paramecium caudatum* (above) and *Paramecium aurelia* (below) grown separately and in mixed culture. Source: EP Odum, *Fundamentals of Ecology*, 1953, with permission from Elsevier

Over time, competition may foster evolutionary responses to avoid competition. Indeed, examples of species driving each other to local extinction, like Gause's *paramecia*, rarely are observed in nature. Such competitive exclusion is a transitory process. We are much more likely to observe the product of competitive exclusion (habitat segregation or non-overlapping geographic ranges) than the actual process of exclusion. Humans, however, have inadvertently or purposefully driven many species to extinction in the past, and our actions continue to do so today.

Predation and parasitism

Predation and parasitism are different forms of symbiosis from competition. They are variants on a common theme: they are a benefit to one kind of organism and a detriment to another. The terms usually refer to feeding relationships. We sometimes read that if the diner is larger than the 'dinee', then it is predation, and if the diner is smaller than the 'dinee', it is parasitism. This simple quantitative distinction misses the ecological point. Predators kill their prey; parasites usually do not. Predators usually kill and eat prey quickly, and must eat many prey. Parasites may spend their entire life in or on a single host, and may not necessarily kill the host.

The cheetah is a predator, chasing down and killing small antelope. Inside the average antelope are a variety of parasites – muscle worms, lung worms, liver flukes, bloodworms – and outside there are fleas and mites. The cheetah hosts a comparable menagerie of its own. These parasites all normally occur at populations low enough that they do not materially influence the health of the host. A healthy antelope, alert and fast enough to escape the cheetah, is a wonderful provider, turning grass into food fit for its diverse throng of parasitic residents. If the symbionts make too many demands, however, the antelope will be slowed or weakened and the parasites will end up as so much inadvertent garnish on the cheetah's lunch.

Parasitic disease in humans or domestic organisms results from an imbalance between resources and demands. These diseases can lead to crippling deformity or even death in humans. But this does not mean that parasitism is somehow 'evil'. We all are parasites as well as predators. We are predators when we take the lives of organisms; eating a cheeseburger involves predation on a steer, and on some wheat plants, and on an onion. In the ecological sense, eating the cheeseburger also involves parasitism on a dairy cow from which the cheese came.

Parasites control populations such as predators, they mediate competition and they can be precursors of **MUTUALISMS**. They have tremendous economic importance; affecting agriculture and fishing. They are also used to control pests.

WEBSITES

McGill Institute on Parasitology's detailed notes on parasitism and symbiosis: http://martin.parasitology.mcgill.ca/jimspage/biopage/parasit.htm
Discussion on parasitism and predation: www.modares.ac.ir/elearning/Dalimi/Proto/Lectures/week1.htm

CONSERVATION BIOLOGY

Conservation biology is an emerging scientific discipline, important to the understanding of and protection of the environment. Conservation biology applies principles of ecology, population genetics, biogeography, economics, sociology, anthropology, philosophy, geography and other fields to the task of maintaining **BIODIVERSITY** in the face of a growing human population with growing expectations and demands on resources.

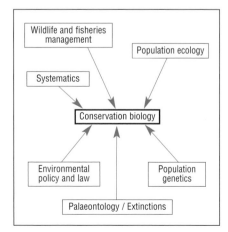

Some of the antecedents and connections of conservation biology.

Case study

The Endangered Species Act of 1973 and the Sixth Extinction

Important to conservation biology in the USA is the Endangered Species Act of 1973 (ESA), subsequently amended. The ESA noted that '. . . fish, wildlife and plants are of aesthetic, ecological, educational, historical, recreational value to the Nation and its people'. The purpose of the Act was '. . . safeguarding, for the benefit of all citizens, the nation's heritage of fish, wildlife, and plants.' In both of these statements, a focus on fish and wildlife and their habitats is evident. Conservation is clearly directed to meeting human needs and wants, rather than to the welfare of the biota *per se*. Despite this bias, however, ESA has been applied fairly broadly, to protect and restore a diversity of organisms, from 'lowly' snails and insects to 'charismatic' mammals and birds such as the black-footed ferret, grizzly bear, bald eagle, peregrine falcon and California condor.

We may identify three broad stages in the evolution of conservation attention and concern, based on the experience in the USA:

1. Conservation of stocks, herds, populations (for the sake of human exploitation – evidenced by game, fish and forestry laws, etc.).
2. Conservation of species (for human needs and for their own sake).
3. Conservation of habitats and biotic communities (for the conservation of species).

Some would urge that a further step is needed. Why not transform the Endangered Species Act into an Endangered Landscapes and Ecosystems Act? Or better still, the Endangered Ecosphere Act or the Endangered Evolutionary and Ecological Processes Act? A grand scale of thinking and action might actually allow us to reverse some of the effects of the ongoing 'Sixth Extinction'.

In 1993, Harvard biologist E.O. Wilson stated that Earth is currently losing around 30,000 species per year, which breaks down to the even more daunting statistic of some three species per hour. The previous mass extinctions were due to natural causes. The first major extinction, about 440 million years ago, was due to the relatively severe and sudden global cooling that caused pronounced change in marine life (little or no life existed on land at that time). Approximately 25 per cent of families were lost (a family may consist of a few to thousands of species). The second major extinction, about 370 million years ago, near the end of the Devonian Period, may or may not have been the result of global climate change, with 19 per cent of families lost. The third major extinction, about 245 million years ago, with 54 per cent of families lost, at the end of the Permian Period may have been from climate change rooted in plate tectonic movements or from meteor impact. The fourth major extinction, about 210 million years ago, signalled the end of the Triassic Period, shortly after dinosaurs and mammals had first evolved, with 23 per cent of families lost. The fifth major extinction, about 65 million years ago, was the most famous event, ending the Cretaceous Period with 17 per cent of families lost. It wiped out the remaining terrestrial dinosaurs and marine ammonites, as well as many other species. Many believe that this event was caused by one or more collisions between Earth and an extraterrestrial object, probably a comet, although alternative volcanic theories also exist.

The current (sixth) mass extinction, however, is being caused by humans, through transformation of the landscape, overexploitation of species, pollution and the introduction of alien species. Because *Homo sapiens* is also a species of animal, the Sixth Extinction would seem to be the first recorded global extinction event that has a biotic, rather than a physical, cause.

Biodiversity has three components – all life forms, individual organisms and their genetic material; groups of similar organisms, such as populations and species; and groups of species in communities, ecosystems and landscapes (groups of adjacent ecosystems), and all of the interactions among the forms of life and their levels of organisation, including **COMPETITION, PREDATION AND PARASITISM**.

The underlying philosophy behind conservation biology is that biodiversity has utilitarian value because it benefits people directly and maintains interactions between the living and non-living parts of the environment. For example, biodiversity provides plants for crops that feed billions of people, as well as decomposing organisms (such as bacteria and fungi) that release nutrients from organic material into soil and water. Biodiversity also has inherent value. In other words, it has worth simply because it exists – beyond the goods and services it provides humans and ecosystems.

Despite recent attention, in a sense, conservation biology is a fairly old science. For hundreds of years wildlife managers have practised the rudiments of conservation biology by managing populations and habitats to allow a sustainable harvest of game animals. Therefore, some of the roots of conservation biology lie in wildlife management.

Conservation ecology is a major part of conservation biology. Some trace conservation biology to the 1930s and 1940s and the work of Aldo Leopold, the visionary founder of modern wildlife management in the USA. Leopold looked well beyond the practicalities of producing deer for harvest to consider the 'right relationship' between people and the ecosphere of which we are a part.

The founding of a scientific discipline is often dated from the initiation of professional societies and professional journals. By that criterion, conservation biology was born with the founding of the journal *Biological Conservation* in the UK in 1968, and the founding of the Society for Conservation Biology together with the launch of its flagship journal, *Conservation Biology*, in 1987.

WEBSITES

The Society for Conservation Biology has an extensive website and a virtual library on biodiversity and its conservation:

http://conbio.net/vl/

Example of applied conservation biology for the study of insects:

www.ento.csiro.au/conservation/conserv_story.html

Detail on the Endangered Species Act of 1973:

http://endangered.fws.gov/esa.html

Information on the Sixth Extinction:

www.actionbioscience.org/newfrontiers/eldredge2.html

SEE ALSO

endangered species, trade in, overfishing, overgrazing

CONTINENTAL SHELF

The continental shelf is a unique part of the ocean, termed the 'neritic **BIOME**'. This biome occupies about 7.5 per cent of the marine realm and ocean depths descend to about 200 metres. Unlike in the deep oceans, light can penetrate so productivity is moderate, about $350g/m^2/yr$, comparable to that of a typical wheat field. In this 'photic', or 'light' zone, producers are algae (phytoplankton) and consumers include invertebrates (zooplankton), macroinvertebrates (crustaceans, molluscs) and fish. Most of the consumers are ectotherms ('cold-blooded') and food chains may be long – six or even seven links to a top carnivore such as an orca (killer whale).

The shelf typically ends at a very steep slope called the shelf break. After moving over the shelf break, one descends the continental slope, which has a much steeper gradient than the shelf. The continental shelf ends at the ocean floor. The width of the continental shelf varies significantly, but averages about 80 kilometres. Some areas have virtually no shelf at all, while the largest shelf, the Siberian shelf in the Arctic Ocean, extends 1,500 kilometres in width. The depth of the shelf varies from 30 to 600 metres.

The continental shelf is by far the best understood part of the **OCEANS** on account of its relative accessibility. Virtually all commercial exploitation from the sea, such as oil and gas extraction and fishing, takes place on the continental shelf.

Not surprisingly, considering the relatively high primary and secondary productivity relative to the open sea, most commercial fisheries exploit the continental shelf. In fact, some of the world's most important commercial fisheries are on the continental shelves, including the North and Barents Seas of western Europe, the Grand Banks and other shallow waters of northeastern North America, the Gulf of Mexico and in the Pacific off the coast of Chile. Fish are an excellent source of protein for humans and fish products such as oils and bones are used in industry to produce livestock feed, fertilisers, glues and drugs. Most of the world's catch comes from the

Case study

Kelp: The hidden forest

Kelp forests, a unique and fascinating part of some continental shelves, can reach 20 to 30 metres high. They are unique because they act like roots, have a stem or trunk, a floating 'top' structure and have blades or fronds, like leaves. Like forests on land, kelp has a canopy and an understory. They take up nutrients directly from seawater and rely on constant water movement to bring nutrients to them. Kelp can grow rapidly, up to 50 cm per day! Net annual productivity of kelp beds, at approximately 800 to 2000g/C/m^2 can be several times the production of phytoplankton. Their life cycle represents an alternation of generations between an asexual sporophyte (macroscopic kelp) and the sexual gametophyte (microscopic kelp, which produces the male and female gametes).

Kelp provides a home for a diversity of marine life. Surf grass, algae, dolphins, sea lions, sea urchins, sea stars, garibaldi and jellies are just a few of the species of plants and animals found in kelp. Sea otters have been known to wrap themselves in kelp so they won't float away while they sleep! Kelp grows in both the Northern and Southern hemispheres. In the Southern hemisphere, kelp occurs off Argentina, through the Straits of Magellan to Chile, off South Africa,

Gazing upward in a kelp forest. Source: © Royalty-Free/Corbis

Australia, New Zealand and many sub-Antarctic islands. In the Northern Hemisphere, kelp grows from central Baja California to Sitka, Alaska. The kelp forests off the California coast are the most well developed. Kelp is harvested commercially for food and for other substances, like alginates, which are used as thickeners in ice cream, beer and paint. Kelp as a food source is an important part of many island cultures. In some parts of the world, it is grown on huge frames suspended in the water so that it is easier to harvest.

What harms kelp? Certainly not herbivores; only about 10 per cent is eaten by them; the rest decays. Instead, high waves, nutrient depletion and warmer water temperatures are the culprits. In the advent of global warming, warmer water temperatures are a very real concern, just as they are for the survival of CORAL REEFS. When sewage from cities reaches kelp beds, the number of urchins rises. Urchins chew the kelp streamers free from their holdfasts. Unless the populations of sea otters and other predators rise as well, the numbers of urchins can grow too great and large portions of kelp beds can be destroyed.

oceans. In 1995, the world's total fish catch was estimated at about 100 million tonnes a year.

The world fish catch increased by an average of 7 per cent each year from 1950 to 1970. Refrigerated factory ships allowed filleting and processing to be done at sea and Japan evolved new techniques for locating shoals by sonar and radar and catching them with electrical charges and chemical baits. By the 1970s, OVERFISHING had led to a serious depletion of stocks and heated confrontations arose between countries using the same fishing grounds. A partial solution was the extension of fishing limits to 320 km. Countries bordering the North Sea have experimented with artificial breeding of fish eggs and release of small fry into the sea. In 1988, overfishing of the northeastern Atlantic led to hundreds of thousands of seals starving on the north coast of Norway. A United Nations (UN) resolution was passed in 1989 to end drift-net fishing by 1992. Marine **POLLUTION** is blamed for the increasing number (up to 30 per cent) of diseased fish in the North Sea.

As traditional fisheries in the continental shelves became depleted, interest grew in deepwater fisheries. However, deepwater fish are slow growing and do not reproduce rapidly enough. Stocks of several deepwater fish have been decreasing since the 1970s because of the global boom in deepwater fishing. For example, stocks of New Zealand's Orange Roughy, one of the first deepwater fish to be exploited, fell by 90 per cent during the 1970s and there are no signs of recovery. This may have a permanent effect on deepwater **ECOSYSTEMS**, with the disappearance of slow-growing species in favour of faster-reproducing fish.

WEBSITES

Continental shelf information:
www.slider.com/Enc/O/Oa/ocean/
Continental_Shelves,_Slopes,_and_Rises.htm
The underwater world – landforms of the oceans:
http://pao.cnmoc.navy.mil/pao/Educate/
OceanTalk2/indexunderwater.htm
Kelp information and photographs:
http://life.bio.sunysb.edu/marinebio/
kelpforest.html
Kelp forests under threat article:

www.abc.net.au/science/news/enviro/
EnviroRepublish_987401.htm

CONTOUR PLOUGHING

Contour ploughing involves ploughing around a hillside or along the contours, rather than ploughing up and down the slope. It is a method of tilling the land that promotes soil conservation. In contrast, ploughing up and down the hillside encourages water to collect in the furrows and run down hillsides. Water carries with it particles of soil as well as any chemicals that have been applied to the land, and this can lead to the pollution and silting of waterways, as well as local flooding. The resulting soil erosion has damaging consequences for agriculture, such as decreasing yields, as the fertile topsoil is washed away. In response to soil erosion, contour ploughing is a frequent management technique employed particularly in the developing world to improve food production.

Contour ploughing has been a feature of farm practices for many years. In the UK, hill farmers used contour ploughing on sloping land. However, it was potentially very dangerous as the tractors often turned over; it also was time-consuming and required considerable skill. As a result in the UK today ploughed fields on hilly land can be seen with furrows going straight up and down rather than across the slopes. There is no attempt to hold back the rainwater and the furrows encourage the water to run off even faster, creating gullies and leading to soil erosion.

Contour ploughing is best used on gently sloping hillsides, with terracing being the better solution where the land slopes more steeply (although terraces can hinder the use of machinery).

Case study

Claveria in the Philippines

Claveria is an area in the Philippines with acidic soils on sloping land that has suffered from severe soil erosion. Local and international agencies have worked with the farmers to develop contour ploughing and the practice of 'ridge tillage', whereby a variety of perennial plants and trees are grown on the ridges to provide extra protection from wind and water erosion. Plants used include fruits, coconuts, mulberry and other fast growing trees – useful additions to the local people's diet and fuel wood supplies. In areas where these improvements have taken place maize yields have increased by 15–25 per cent and land values by 35–50 per cent.

WEBSITES

General farming information, including details about contour ploughing, from the UN's Food and Agriculture Organisation:

www.fao.org

Specific information, including photographs:

www.fao.org/ag/ags/AGSe/7mo/69/chap10_1.pdf

SEE ALSO

soil erosion and conservation

CONVERGENT EVOLUTION

In many instances, animals and plants that live in similar habitats resemble each other in outward appearance. These similar looking animals and plants may, however, have quite different evolutionary origins. This is known as the theory of convergent evolution. Convergence is to meet, or approach the same point from different directions. Convergent evolution is the development of similar traits in organisms that have independent ancestry. The concept is related to natural selection, which means that an organism that has the most effective way of dealing with the environment is the one most likely to survive. Convergent evolution results from natural selection operating in environments that make similar demands. For example, deserts are dry. To survive in deserts, successful plants have often evolved mechanisms to help retain water. Waxy stems help keep them from drying out. Often, stems become the plant's photosynthetic tissue, and leaves are reduced in number or are absent entirely. Storing water in the stem, and thus producing a fleshy, or succulent, plant body, is another good solution to the problem of water conservation.

Of course, a juicy plant in the desert will attract thirsty animals. Selection will favour plants with protection from such marauders. Where do the plants get such protection? In the evolutionary process, the usual source of a new structure is to recycle something old. Water-wasting leaves can be turned into sharp, protective spines. Thus spiny, leafless, stem-succulent plants have evolved from different ancestors in hot **DESERTS** around the world.

The result of such convergent evolution is ecological equivalence, a situation in which unrelated organisms have similar roles with similar adaptations in physically similar ecosystems. For example, the cacti of American deserts are the ecological equivalent of spiny, stem-succulent milkweeds of Africa. Plant and animal structures that are the result of convergent evolution are called homoplasies. Structures that have a common origin are called homologous structures.

Sea otters with thicker fur will survive cold northern waters better than those with thinner fur. The sea otters with thicker fur will survive to pass this trait on to their offspring, and, over time, all sea otters end up with thicker fur. This adaptation is a characteristic that an animal has that allows it to survive in its environment. Sometimes these adaptations are very similar among different animals, but sometimes entirely different species can have the same type of adaptation to survive in similar environments. Wings, fins, armour, horns, large ears, large eyes, long necks, spines, stripes and tusks are all good examples. To be considered an example of convergent evolution, animals or plants must be ancestrally unrelated, have structurally similar characteristics, and be using these characteristics for the same function.

According to Dr Rudolf A. Raff, an evolutionary developmental biologist at the Molecular Biology Institute of Indiana University in Bloomington, 'Convergences keep happening because organisms keep wanting to do similar things, and there are only so many ways of doing them, as dictated by physical laws'. For example, in tropical and subtropical savannas around the world, ants and termites provide a rich protein resource for mammals. Several groups of mammals have evolved adaptations to exploit that resource. It takes several things to be a first-class anteater: a probing snout, a long, sticky tongue and strong claws, for example. Teeth are not particularly useful for handling such tiny prey; ant-eating mammals tend to dispense with them.

In another example, swifts, swallows and martins all hunt for insects while they fly. They have streamlined bodies with long wings. Hummingbirds and sunbirds feed on nectar from flowers. These birds have long bills to reach the nectar at the base of flowers. Based

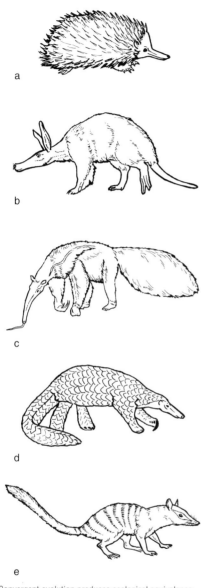

Convergent evolution produces ecological equivalence among anteaters from various continents. A = Australian echidna (an egg-laying monotreme); B = African aardvark; C = South American giant anteater; D = South Asian or African pangolin; E = Australian numbat (a marsupial) (after FH Pough et al., Vertebrate Life, 2002).

on appearance only, we would conclude that sunbirds are related to hummingbirds and that swifts are related to swallows and martins.

In reality, genetic techniques have shown that swifts are related to hummingbirds, while sunbirds are related to swallows and martins. Since they are not closely related but have evolved similar traits as they adapted to similar environments, these are more examples of convergent evolution.

Antarctic and Arctic fish have a sort of antifreeze composed of molecules called glycoproteins that circulate in their blood. The glycoproteins surround tiny ice crystals and keep them from growing. This lowers the temperature at which their body fluids would otherwise freeze and kill the fish. The populations of fish in the Arctic and Antarctic split a long time before they developed these genes and proteins. Researchers have found that the genes that produce the antifreeze proteins are quite different for Arctic versus Antarctic fish. Therefore, quite separate, independent episodes of molecular evolution occurred, with the same functional results.

WEBSITES

Science Fair Encyclopedia:
www.all-science-fair-projects.com/
science_fair_projects_encyclopedia/
Convergent_evolution
Convergent evolution history and discussion from Singularity Watch:
www.singularitywatch.com/
convergentevolution.html
Discussion on convergent evolution:
www.thegreatstory.org/convergence.html
BBC article about convergent evolution, with examples from tuna and sharks:
http://news.bbc.co.uk/2/hi/science/
nature/3683885.stm
New York Times article on convergent evolution:
http://dml.cmnh.org/1998Dec/
msg00381.html

COPPICING

The term coppice derives from the French word *couper* – to cut. A coppice is a thicket or dense growth of small trees and bushes, sometimes also called a copse. Coppicing is the technique used to create the coppice. The trees and bushes are regularly trimmed back to stumps so that a continual supply of small shoots is generated. Trees and shrubs that are cut down this way can produce shoots that grow over 30 centimetres in a week and a coppiced tree can live many times longer than if the tree had not been cut down at all.

The technique of coppicing generates sustainable fuel wood without causing deforestation and can also provide materials for industries such as basket making, broom making and dressed sticks, famous among the hill sheep farmers. The main trees and bushes that coppice well in temperate environments are willow and hazel, which have vigorous regrowth. Coppice woodlands are cut on rotation, normally from between 6 to 25 years and usually one part of the wood, called a coupe, is harvested each year. The coppice trees respond to cutting by sending up multiple stems from the cut stump, which is called a stool. In many coppices, some trees are left uncut to grow as standards – tall and single-stemmed.

Coppicing can be traced back to Neolithic times (c. 4000 BC) and through the Bronze, Roman, Saxon and medieval periods. It was the most common form of woodland management in Britain until the mid 1800s. Because of this long history, ancient coppice woodlands are often considered to be direct descendants of the original forest that covered most of Britain after the last Ice Age. Coppicing declined greatly in the twentieth century and especially after the Second World War. Remains of the old coppices are now a tangled mass where little of the original fauna and flora has been able to survive in the dark undergrowth.

As a sustainable woodland management technique, coppicing is still relevant today. A crop of wood is obtained annually, yet no trees are removed – only cut and allowed to regrow. In ancient coppices much of the wildlife has come to rely on the periodic cutting and regrowth of the stools. The current expansion of coppicing, especially in the UK, is partly a result of the strategy to encourage wildlife and in particular to increase the population of dormice. In addition, continued cutting ensures that a woodland with a healthy coppice structure is passed on to the next generation, as it has been by countless woodcutters through history.

Coppicing is also increasingly being adopted in many developing countries as a renewable energy source that is also sustainable. It provides a local supply of fuelwood and limits the deforestation that has occurred and the consequent long journeys women in these societies need to make to collect fuelwood.

WEBSITES

Forestry Commission:
www.forestry.gov.uk/forestry/infd-52jbr7
Coppice UK, an organisation promoting awareness of coppicing:
www.coppice.org.uk/background.htm
Quoditch Moor Nature Reserve, Devon, UK:
www.rdg.eurobell.co.uk/
quoditchcoppicing.html

SEE ALSO

deforestation

CORAL REEFS

Coral reefs are the most diverse marine **ECOSYSTEMS**. A hectare of coral reef off Southeast Asia may support some 2000 species of fishes, more than the number of species of birds in the whole of North America. Coral reefs are built by coral animals, powered by stable algal-animal **SYMBIOSES**. In these symbioses, the animal 'host' has living within its tissues or sometimes even within its cells, algal or microbial cells. The term 'alga' (plural = algae) is a vernacular term with no real scientific meaning. Because of this, a large, diverse and unrelated variety of organisms are lumped together under that name. About the only thing most algae have in common is that they are organisms capable of photosynthesis and are similar in some biochemical regards to plants.

A reef is a carbonate structure of biological origin, in a sense, similar to limestone above ground. Reefs are sufficiently raised above the seabed that they influence local sedimentation and ocean currents. Some living reefs are millions of years old. Coral reefs are restricted to sea temperatures of about 20–30°C – and are found, at present, between latitudes of about 30° north and 30° south of the equator. The great coral reefs of the world are located in the Caribbean, the Indian Ocean and the

Case study

Coral reefs – beautiful, yet fragile ecosystems

Coral reefs have long been known as fragile ecosystems and a number of human activities can harm them. These include mechanical disturbance by recreational divers, outflow of sediment and sewage from human settlement and fishing with dynamite and cyanide. Now, more subtle perils are affecting reefs. EUTROPHICATION, where high nutrient concentrations stimulate blooms of algae, is particularly harmful. Algae cloud the water and block sunlight, causing some underwater plants to die and reducing the process of PHOTOSYNTHESIS. Also, when algae die and decompose, oxygen is used up. As it turns out, coral reefs are essentially self-contained systems, their nutrient cycles only minimally coupled with those in the deeper waters in their neighbourhood. Infusion of nutrients from outside the system quickly leads to population explosions in non-symbiotic algae. These then out-compete the zooxanthellae, leading to weakening or death of the living reef.

Coral reefs are also remarkably sensitive to temperature increase. For reasons that remain a mystery, above 30°C, coral polyps begin to jettison their zooxanthellae. With the algal symbionts reduced or absent, the moribund reef turns from colourful to chalky white, as the carbonate of the mineral reef shows through the transparent coral. This phenomenon is known as 'coral bleaching'. It seems to be associated with EL NIÑO episodes, but it may also be an early and ominous warning of the effects of general global warming on marine ecosystems.

tropical Pacific, including the Great Barrier Reef of Australia. Reefs form either on shallow areas of continental shelf or as fringing reefs around oceanic islands.

Corals are cnidarians (formerly known as coelenterates), related to sea anemones, jellyfish and the common freshwater hydra. Coral reefs represent a tight symbiosis between coral animals and specialised green algae known as zooxanthellae. The algae actually live within the coral, in the gelatinous region between the endoderm and ectoderm. The pigmented algae are responsible for the vibrant and diverse colours of the coral reef. Algae power the symbiosis by transducing solar energy into chemical bond energy of carbohydrates. The coral animals (polyps) provide structure (by secreting the carbonate that builds the reef) and protection (by means of the stinging, harpoon-like nematocysts – 'hair cells' – typical of corals and their relatives). Corals are both sexual and asexual and the coral 'heads' familiar to human scuba divers usually represent a clone of individuals derived by budding from an original coral colonist established perhaps thousands of years previously.

Although they occupy only about 0.1 per cent of the area of saltwater realm, coral reefs are by far the most productive of marine ecosystems. With production of upwards of $2500g/m^2/year$, they are in the same league as

tropical rainforests, so they contribute around 3 per cent of total marine production.

Ecological diversity (especially species richness) for coral reefs is also high. The inventory is far from complete, but some scientists suspect that coral reefs may harbour between one and three million species, approaching levels suspected in tropical rainforests. Coral reefs also are the basis for major tourism. For example, Australia's Great Barrier Reef supports a US$1 billion tourist industry, five times the value of the reef's commercial fisheries. And yet fisheries can be important too; in developing tropical nations, some 25 per cent of fish harvested for human consumption comes from coral reefs.

One characteristic of coral serves to illustrate its mystery and diversity. The algal symbionts found in coral reef animals are often from an algal group called the Dinoflagellata, or as they are commonly called, dinoflagellates. Recent work on the structure of the dinoflagellate genome indicates that dinoflagellates are unique among all living organisms. They are no more closely related to other algae than they are to animals or fungi. It has been suggested that this group of strange organisms be given their own biological kingdom equal in the taxonomic hierarchy to the animal or plant kingdoms.

SEE ALSO

oceans

CROP ROTATION

Growing the same crop year after year on a plot of land can lead to serious problems. The crop will remove the same nutrients every year, leading to losses in soil fertility and the need to apply large quantities of expensive and potentially environmentally damaging fertilisers. Weeds, pests and diseases peculiar to the particular crop can also build up to epidemic proportions. The inevitable consequence is that yields will decline and crops may even fail. One solution to this problem is crop rotation.

The earliest known example of crop rotation is the Norfolk four-course rotation of crops in the UK. It developed in the late eighteenth century and allowed cultivation on the dry, less fertile chalk downs in this part of East Anglia. The land had formerly only been used for sheep rearing. The system of rotation introduced turnips that were grazed by the sheep while at the same time the manure fertilised the soil. In the second year, the turnip field was sowed with wheat and then, in subsequent years, with barley and grass and clover. As the diagram shows, there is a rotation in a clockwise direction. The turnips and the clover add nitrogen to the soil (nitrogen 'fixers'), which is added to by the manure from the animals, while the wheat and barley are heavy nitrogen users. This led to the expression, 'what "corn" took out of the soil, "horn" put back'. The four-course rotation was an efficient system of mixed farming that maintained soil fertility. In Norfolk its use built up the fertility of otherwise poor soils.

Over time, the practice of crop rotation on commercial farms in the developed world has largely disappeared. Due to the pressures for more intensive agriculture, farmers have specialised in the production of just one or two crops. The intensification and the loss of crop rotation has led to the heavy use of chemical fertilisers and pesticides, as well as the loss of hedgerows to accommodate the increasing size of machinery. However, the use of crop rotation has remained an important part of the practices of small-scale vegetable growers in back gardens or allotments, who understand and value the advantages of crop rotation for soil fertility and pest control.

As more sustainable farming practices are sought today there is a strong movement

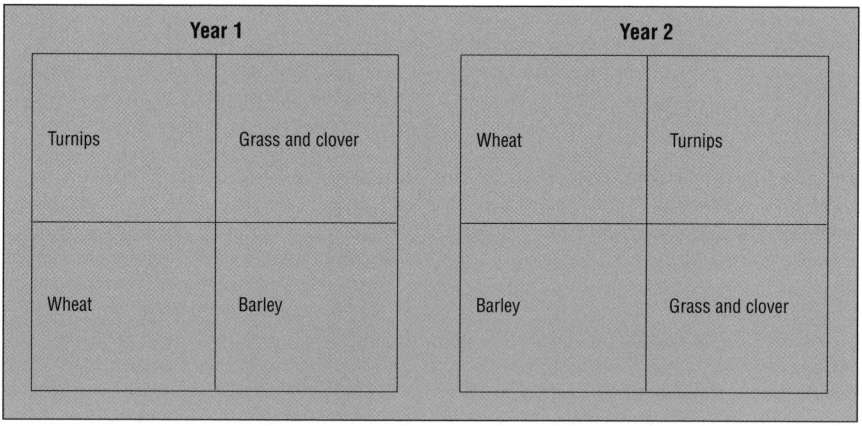

Year 1		Year 2	
Turnips	Grass and clover	Wheat	Turnips
Wheat	Barley	Barley	Grass and clover

The Norfolk four-course rotation of crops

towards organic farming in which limited, if any, use is made of chemicals. As a result there is also a move back towards crop rotation as a natural way to maintain soil fertility. This growth in popularity is helped by the need to reduce farm production to avoid surpluses (see **COMMON AGRICULTURAL POLICY**).

In contrast, farmers in countries in the developing world have always practised crop rotation to maintain soil fertility, control weeds and to maintain a vegetation cover on land prone to erosion by wind and water. In general, farmers in these countries have neither the access to, or the resources to fund, expensive fertilisers and other chemicals, so that organic farming practices have remained the only option. In some areas crop rotation is a key technique in projects designed to improve agriculture in the developed world and research abounds into its impact on crop yields.

WEBSITES

Information on the Agricultural Revolution from the BBC:
www.bbc.co.uk/history/society_culture/industrialisation/agricultural_revolution_03.shtml
Core4, at Purdue University, with information on crop rotation:
www.ctic.purdue.edu/Core4/CT/Choices/Choice6.html
George Washington's method of crop rotation:
www.mountvernon.org/pioneer/farms/rotation.html

SEE ALSO
agriculture

CYCLONES (DEPRESSIONS)

Cyclones (also called depressions or 'lows') may be defined as areas where atmospheric pressure (barometric pressure) is lower than in the area surrounding it. On a weather map, cyclones are shown with an L symbol surrounded bull's-eye fashion by a series of concentric lines. These lines are called isobars, and represent lines of equal pressure.

Where pressure is lower than the surroundings, air flows in from the surrounding area to the centre, where the pressure is lowest. The lower the pressure, the faster the air will rush in, and the greater the wind speed. In the northern hemisphere, movement into the low will be counterclockwise. In the southern hemisphere a clockwise in-spiral will result. Air flowing towards a particular place at a given time finds that place in a different location due to the Earth's rotation. This apparent shift to the right in the northern hemisphere and left in the southern hemisphere is known as the Coriolis Effect.

When the air reaches the centre of the low, it must rise since there is no other place to go. If less air rises than comes in, the pressure rises as the low 'fills up' and the pressure differences from inside to the outside of the low become less. Eventually the low may cease to exist if the pressure equalises. If more air rises than comes in (which may happen if there is significant heating of the surface under the low or if there is a pull from aloft, perhaps from the jet stream) then air pressure will fall and the low will become deeper and stronger.

Lows, or cyclones, are among the most significant dynamic weather phenomena on the Earth, having a great impact on the environment. The rapidly rising air cools readily and clouds form. Cyclones often bring periods of heavy and prolonged rainfall, particularly when they form over warm oceans such as the South Atlantic, where vast amounts of water have been evaporated into the air. Lows are cloudy and stormy weather systems because air converging into the low is forced to rise. Air that rises cools because it is being forced into the upper reaches of the atmosphere where there is less pressure. Molecules of air move away from each other and collide less frequently than in the warmer lower atmosphere. The air may cool sufficiently to drop to its dew point temperature where condensation occurs. Dew point temperature is a measure of humidity. If air has a high relative humidity, the dewpoint temperature will be close to the measured regular temperature. If the air is dry, the dewpoint temperature will be far lower than the measured temperature. If the measured temperature falls to the dewpoint temperature, condensation results and a cloud forms. If this happens at the surface, dew is deposited and surfaces are wet to the touch.

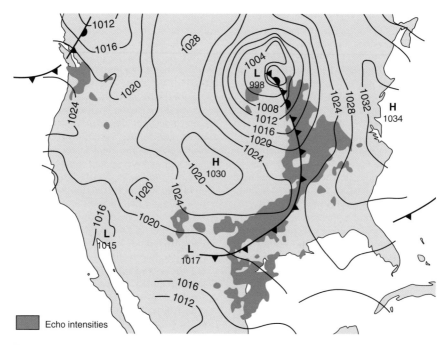

This map shows the typical pattern of isobars associated with a cyclone. This low was responsible for widespread rain (as shown by radar echo intensities). The 998 in southern Canada refers to the barometric pressure in millibars associated with the centre of the low

Latent heat is released whenever clouds form, making the air even more buoyant and likely to rise further, chill further and condense (or deposit) more water. Each time this happens, additional latent heat is released. The lifting, cooling, condensing and further lifting process becomes almost self-perpetuating. Lows 'attract' air masses from their surroundings – air that may have been filled with water vapour from evaporating water surfaces. This means that the fuel for this process, which is moisture with its latent heat of vapourisation, is readily available and constantly replenished. This is the opposite of the high or anticyclone, where air is descending and diverging, pushing air masses that may contain evaporated moisture away from the low-pressure centre and depriving it of energy.

With plenty of lift and a source of energy to continue the lifting, lows can last a long time, creating thick layers of clouds and plenty of precipitation. Some lows retain their identity for several days or as long as a week.

It is possible to identify several different types of cyclone. The midlatitude cyclone is formed when cold, dry air from cooler polar regions meets warm tropical air moving poleward from the low latitudes. The big difference in the temperatures of these air masses prevents an orderly, symmetrical air movement into the centre of the cyclone. Frontal boundaries are formed where these air masses meet. It is along these fronts that most of the cloud and rainfall occurs, as the warmer air is forced to rise up and over the colder air.

Thermal lows develop over very hot regions. If the regions are dry, the thermal low is shallow. Little cloud formation results because the air is generally dry. With the lack of condensation there is no additional release of latent heat to promote further rising and cloud formation. In moister areas, the generalised low pressure will initiate daily showers or thundershowers.

If a cyclone develops over a warm ocean surface, a tropical cyclone may develop. These cyclones trap uniformly warm and moist air from over a vast area of warm water and have a huge supply of latent heat fuel for lift to con-

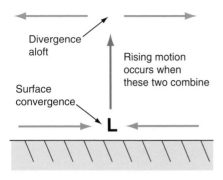

Converging air at the Earth's surface, with diverging air aloft, commonly associated with a low pressure system.

tinue within the centre of the storm. This causes the extremely low pressures and resulting high winds and waves that we associate with tropical storms and hurricanes. Rapid, massive condensation creates thick convective clouds that produce massive amounts of rain.

WEBSITES

Current tropical cyclones map:
www.hawaii.edu/News/storm.tracks.html
Tropical cyclones information:
http://cimss.ssec.wisc.edu/tropic/tropic.html
Midlatitude cyclone resources:
http://ww2010.atmos.uiuc.edu/(Gh)/
guides/mtr/cyc/home.rxml

This image shows the typical cloud pattern of a cyclone north-west of Ireland. The comma shaped clouds form from lifting associated with the inward and upward movement of air into the low. The sharp cloud boundary off the coast of the Iberian Peninsula is associated with the cold front. Source: NASA

Australia Bureau of Meteorology cyclones information:
www.bom.gov.au/info/cyclone

DAMS

The construction of dams is one of the most controversial engineering schemes of recent times. Throughout the world, from the Colorado in the USA, to the Nile in Egypt and the Yangtse in China, major rivers have been dammed to provide water for industry and agriculture, hydroelectric power for cities and to regulate river flows to prevent flooding. Yet,

in their wake, huge areas of land have been inundated, forcing people off their land. Some would say that the huge sums of money spent on these massive prestigious projects have only benefited rich urban dwellers, leaving many in the countryside landless and homeless. One of the most controversial schemes under construction in the early part of this century is the Three Gorges Dam in China (see **RESERVOIRS**).

Large-scale dams were the preferred solution from the 1930s to the 1970s, but often for the wrong reasons. Due to their scale, governments perceived them as symbols of economic and political prestige. The world's financiers, who were needed to put up the money, liked them because of the profits they made from interest. Aid donors favoured them because there was something large to see for their money. Many countries remain in serious debt today as a result of choosing this route to development decades ago. However, from a more practical point of view they could be seen as a tool towards greater economic success. They represent dreams for a better standard of living for a nation's poor through greater local food production and access to electricity supplies, bringing domestic and industrial benefits, as well as potential employment.

Not all dams are large-scale concrete constructions. They can be small scale and made of other materials such as earth and stones, built for local improvement. The Band Aid charity of the 1980s emphasised such small-scale projects in Ethiopia to improve

Case study

The impact of dams

Dams on the Rio Grande

The Rio Grande forms part of the border between the USA and Mexico; it is, in a sense, one river cut and transformed into two – a vigorous and unregulated snow-fed river in northern New Mexico, and a highly regulated river with substantial flow where it reaches the Gulf of Mexico. Dams were constructed to retain all flow on the Rio Grande, releasing water only for irrigation purposes. As a result of this restricted flow, sediment loads, which normally are mixed into the water and transported downstream during high flows, have accumulated on the riverbed. The river channel has been invaded by tamarisk, a non-native, salt-tolerant bushy tree, for hundreds of kilometres below the dams. The channel has diminished in size since the dams were constructed and can no longer contain floods as large as it once could. As a consequence, relatively minor floods have caused significant damage to riverfront properties and structures downstream from the dam.

Dams on the River Nile

The Aswan High Dam on the River Nile in Egypt was completed in 1970. In several ways it illustrates the benefits of big dam construction. The hydroelectricity produced by the dam is cheap, plentiful and ultimately sustainable. It saves on the purchase of fossil fuels and indirectly reduces greenhouse gas emissions. The erratic flow of the Nile prior to the dam construction has now been controlled, benefiting many communities along its shores who were previously negatively affected by its unpredictability. Navigation is easier and this has encouraged the growth of tourism. Nearly half a million hectares of land have been brought into production by irrigation made possible by the construction of the dam.

Following the success of the Aswan High Dam, a new dam is being proposed on the Nile. Uganda is one of the poorest countries in the world: less than 5 per cent of its people have access to electricity supplies. The project to construct an enormous new dam near Bujalagi Falls should provide a major energy resource.

Like all dams this offers advantages and disadvantages. It should improve standards of living, so the World Bank is likely to fund it. Industry is likely to benefit and many new jobs should be created. However, flooding of land to create the reservoir will displace many people in what is now a tourist area that brings in about $4 million per annum. One village of 820 people will be forced to move and a further 6000 will lose their farmland through flooding. Moreover, local habitats will be destroyed. Critics argue that the scheme will only benefit the rich and that the poor local people will lose out.

The Vaiont Dam disaster, 1963

On 9 October 1963, 2043 people died when a huge slab of rock slid 400 metres into the Vaiont Reservoir in the Italian Alps north of Venice. The landslide caused a giant wave some 100 metres high to surge over the dam and into the valley below, where it tore through several villages. Amazingly, the dam itself did not fail.

This was one of the worst dam disasters of all time and was largely the result of human error – the dam should never have been built there in the first place. The site of the dam was very poorly chosen. The steep slopes had failed before, as recently as 1960 when the reservoir began to fill, and there was evidence of prehistoric slip surfaces within the rocks. As the lake continued to fill, water pressure increased within the slopes as the water table rose, making them even more unstable. Heavy rainfall finally triggered the landslide on the south side of the reservoir.

living standards after the famines. Village irrigation systems are still operated from water stored behind earth dams constructed by the local people at that time. Only low-level technology is applied, but it has been sufficient to make a noticeable difference to people's lives.

Dams affect the environment. Biological communities form in response to interactions between species, as well as to their chemical and physical environments. Loss of vegetation along streams (riparian vegetation), channel straightening, increased siltation, reduced sand bars, altered temperature, increased contaminants and the loss of floodplain backwaters have reduced habitats, increased siltation and changed water quality. Even changes in light penetration in streams affects algal and macrophyte (a large aquatic plant) growth.

Further reading

Griffiths, I. L., 1993, *An Atlas of African Affairs*, Routledge.

Middleton, N., 1988, *Atlas of Environmental Affairs*, Oxford University Press.

WEBSITES

International Rivers Network:

www.irn.org

US Army Corps of Engineers North West Division:

www.nwd.usace.army.mil/ps/colrvbsn.htm

Netherlands Friends of the Earth (Milieu Defensie):

www.milieudefensie.nl/earthalarm/eng71.htm

USGS Circular 1126, 'Dams and rivers: primer on the downstream effects of dams', by Michael Collier, Robert H. Webb and John C. Schmidt

http://onlinepubs.er.usgs.gov/lizardtech/iserv/browse?cat=CIR&item=/circ_1126.djvu &style=simple/view-dhtml.xsl&wid= 750&hei=600&props=img(Name,Description) &page=1

SEE ALSO

hydroelectric   power,   reservoirs,   river restoration

DEFORESTATION

Deforestation is the process of clearing of forests, most commonly by burning or cutting down trees, either by clear felling or selective felling. The impact of deforestation has been dramatic – about 40 per cent of the world's forests have been removed and replaced either by smaller and less diverse vegetation or by artificial surfaces such as roads and buildings. Deforestation has occurred for centuries across the globe in order to provide land for settlement and agriculture, and wood for fuel, building materials, furniture and paper. Until the beginning of the twentieth century the greatest impact of deforestation had been in temperate lands, the Mediterranean and monsoon Asia. Today, rates of deforestation are highest in tropical rainforests and the northern boreal or coniferous forests in parts of northern Canada and Scandinavia. Large areas of forest have also been lost in order to build dams and reservoirs and to set up large-scale mining operations.

The impact of the deforestation is both negative and positive. It goes well beyond the national boundaries of the country in which the deforestation is taking place, making it a global issue. Deforestation increases soil erosion and the flood threat because there are no longer any roots to bind the soil or forest canopy to intercept the rainfall. Soil fertility declines due to erosion and the lack of trees available to return nutrients to the soil by falling leaves. As a result, deforestation may lead to desertification. The loss of the forests leads to the loss of habitats for wildlife and the loss of homes and the traditional way of life for local tribes. Some tribes people have lost their lives as they try to resist the loggers or fall prey to western diseases to which they have no immunity. On the global scale deforestation contributes to global warming as less carbon dioxide is being taken in by plants and converted to oxygen.

However, logging companies and governments keen to develop the rainforests see many advantages in promoting deforestation. They wax lyrical about the increased employment and incomes resulting from logging and the additional land created for settlement and agriculture. Governments can reduce debts by earnings from exports of timber and timber products, and developed countries gain valuable foodstuffs, timber and medicinal drugs. The governments of these countries see the logging as a chance to develop and to provide a higher standard of living for their people. The disadvantages are real, yet it is difficult to persuade governments, logging companies and even some companies in the developed world, which profit from rainforest products, to change the pattern of exploitation.

Many of the most outspoken opponents to deforestation come from the developed countries, where just such exploitation took place in the name of development, albeit several centuries ago. Yet it is the developed world that fuels much of the deforestation with its insatiable appetite for rainforest products such as mahogany, special drugs and rubber, along with the products from other plantations and the extensive ranches producing cheap beef for the hamburger market.

The question is to what extent these developing countries have the right to exploit their natural resources in order to move along the path of development and try to provide an acceptable standard of living for their people. The alternative is to look for a middle ground,

Case study

The human impact of deforestation

Air pollution caused by deforestation in Malaysia

Forest fires in Borneo and Indonesia in 1997, resulting from deliberate burning, produced a thick blanket of smoke that spread across neighbouring Singapore and Malaysia. The smoke and other pollutants trapped at ground level presented a real health risk to people. People, especially the elderly and ill, were recommended to stay indoors and everyone was advised to wear special masks. Up to 20 million people in Indonesia were affected by throat and respiratory diseases. At the height of the fires schools and factories were closed, aircraft grounded and the worst affected settlements were evacuated. The air pollution problem lasted for almost three months until the monsoon period began in November.

The Kayapo battle against the world's largest dams

The price Native Americans paid as their land was encroached by Americans of European descent in the USA in the 1800s is well documented and known. Today Indian tribes of the tropical rainforests face similar problems and some are fighting to preserve their ancient cultures and forests.

The Kayapo have fought for centuries to defend their rainforest territory along the Xingu River in the Amazon. The Kayapo are a tribe of about 3000 people living in about 13 villages scattered through the rainforest. They are over 6 feet (1.8 metres) tall and are adorned with beads; the women wear large feathered earrings. This is the tribe famous for stretching their lower lips over large wooden disks. One of the Kayapo chiefs, Paulinho Paiakan, described their way of life and their recent fight:

We are not nomadic but live in villages, we plant narrow bands of crops that stretch for miles along trails. We have a unique knowledge of rainforest plants and use them for food and medicines. We live on a designated reserve but the boundaries are not well marked or permanent. Electronorte, the Brazilian power company, with money from the World Bank, recently tried to alter the boundaries to deforest a massive area to build 47 hydroelectric dams and flood large areas of the Kayapo land, including some villages.

We decided to fight the project – we made videos of other schemes to convince the tribe we had to fight; we told the world press and environmental groups what we were doing. Five hundred of us went to Altamira in full ceremonial dress with war-clubs to confront the Electronorte company. It was a dramatic but peaceful demonstration – it was broadcast around the world. We also went to New York and lobbied the World Bank officials and spoke to members of the US Congress. Thankfully the World Bank withheld the initial funding of $500 million and the project was cancelled.

a way in which sustainable development of the forests can be achieved. In some countries sustainable strategies are now being developed, including conservation of areas by creating national parks, the use of replanting strategies, the development of eco-tourism and increased prices for rainforest products.

WEBSITES

Earth Observatory:

http://earthobservatory.nasa.gov/Library/ Deforestation

Basic Science and Remote Sensing Initiative: www.bsrsi.msu.edu/rfrc/home.html

Greenpeace: www.greenpeace.org.uk

SEE ALSO

forests and forestry, national parks

DEMOGRAPHIC TRANSITION

Environments support populations and are also affected by populations of plant, insect and animal species. Of all of these populations, humans have the greatest ability to negatively impact the environment. This is because of their ability to manufacture harmful substances and objects, and because of their large and expanding numbers. Environmental concern heightened in the twentieth century mainly because it was during this century that rapid population growth became blatantly apparent. Human population growth depends on the ratios of the birth and death rates. The study of human population is called demography. For many countries, these birth and death rates have followed a pattern over time called the demographic transition.

Many countries' populations experienced high birth and death rates in the past, but now experience low birth and death rates. This transition usually occurs when a region moves from a rural, agricultural society to a more urbanised, industrial society with low birth and death rates. In agricultural societies, people have more children to help with farming the land, but they have poorer access to medical care, so the death rate is higher. In urban, industrial or post-industrial societies, people have fewer children and better health, so the death rate and birth rates are lower.

The demographic transition theory, or model, was originally developed to help explain the historical changes that took place in the birth and death rates of Western European countries. Early on, European countries had high birth and death rates. As people were dying at the same rate as they were being born, population stability resulted, with little natural increase or growth. Next, the death rate declined due to advances in medicine and improvements in living conditions and diets. However, because the birth rate was still high, populations grew rapidly. Later, when the birth rate declined due to improvements in standards of living and the introduction of contraceptives, populations grew more slowly, once again approaching stability. This time, instead of stability with high birth and death rates, there is stability with low birth and death rates. This transition from high birth and death rates to low birth and death rates is also associated with a variety of societal changes and linked with the modernisation process. Each stage is marked by environmental impacts, concerns and pressures.

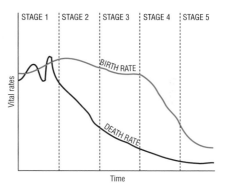
Stages of the demographic transition

The demographic transition model can be divided into five stages. In the first stage, both the birth rate and the death rate are high. This means that there is little natural increase or growth in the population since births and deaths are about equal. This was the situation in pre-industrial Europe and although no major region of the world today is in this stage, several sub-Saharan African countries exhibit these characteristics. However, the environmental concerns are quite different. In pre-industrial Europe, the population base

Case study

China

China, home to 1.3 billion people, does not fit the demographic transition model. Although China is still considered poor by some measures of modernisation, it experienced a rapid decline in births during the last part of the twentieth century. This decline was primarily because of stringent governmental policies related to the one child per family policy, first announced in 1979. In China, the environmental implications of a rapidly growing population were enormous, and were the primary reason for the one child policy. In 1970 the birth rate was 33 per thousand, but this had dropped to 21 per thousand by 1990. The average number of children per woman in 1970 was 5.01 but was only 2.31 by 1990. According to Lily Liu Liqing, a spokesperson for the China Family Planning Association, 'It was a choice we had to make – not one we wanted to make – if we wanted everyone to have a better life in a comparatively short period of time'.

The ideal of a one-child family implied that the majority would probably never meet it. It was hoped that three (or more) child families could be eliminated and that about 30 per cent of couples might agree to forgo a second child. It was argued that the sacrifice of second or third children was necessary for the sake of future generations. People were to be encouraged to have only one child through a package of financial and other incentives, such as preferential access to housing, schools and health services. Discouragement of larger families included financial levies on each additional child and sanctions that ranged from social pressure to curtailed career prospects for those in government jobs. Specific measures varied from province to province, and minorities were excluded from the policy. This policy also favoured males, as many female children were aborted.

In 2000 China began to modify this policy, deemed too harsh by many. The Chinese Government now offers exemptions to children who have no siblings and are now adults. When two of these 'only children' marry, they will be allowed to have two children. The Law on Population and Family Planning in 2002 kept features of the old policy but modified several components. Despite the criticisms, reductions in fertility have eased at least some of the pressures on communities, state and the environment in a country that still is home to one-fifth of the world's people. However, now with increased industrialisation and prosperity, a similar set of environmental concerns loom – the same ones that plague many Western nations. These include how to provide for new wealth and consumption without rapidly degrading land, atmosphere, water, mineral, energy and other resources.

was fairly low, and consequently the pressure on the environment was small. Many sub-Saharan African countries contain millions of people, including Nigeria, with an estimated 137 million in 2004. Pressure on the land leads to soil exhaustion, degradation and desertification.

The second stage of the model represents an area with a population that has a high birth rate but a declining death rate. This means that there is some natural increase or growth in the population because of the disparity between births and deaths, with more births than deaths. Many African nations and some nations in Asia exhibit this type of demographic or population behaviour. Health care improves and agricultural technologies increase the food supply, but pressure on the environment rises with an expanding population.

In the third stage, the gap between births and deaths is the largest. This portrays a society of rapid population change, as in some of the nations in Central America, tropical Africa, and South-West Asia. Countries in the fourth stage have declining birth rates and low death rates, so their overall natural increase is less than those in stage three. This includes countries in temperate Latin America like Chile and Argentina, as well as China and South Korea. Increased industrialisation, wealth and consumption causes environmental pressure on the landscape such as mining, urbanisation, energy production and motor vehicles.

The final or fifth stage of the demographic transition model depicts a population in which both birth and death rates are low. This includes most of the European countries and

Japan. Today, many of the European countries, especially those in Eastern Europe, are actually losing population and have a negative rate of natural increase. Pressure on the environment continues, but the political systems and wealth of some countries permits some environmental expenditures and laws to be enacted that help protect the environment.

Exponents of the demographic transition model believe it is useful in understanding the behaviour of today's populations, showing how there is a causal link between modernisation and a decline in birth and death rates. In other words, as a population becomes more modern, death rates decline due to better health and sanitation conditions, while birth rates decline as people adopt more modern contraceptive practices or want or need fewer children. On the other hand, there are some experts who believe that although the demographic transition model has some value in explaining population changes, it is not applicable to all places or at all times. However, the demographic transition model can help to predict and plan for future population changes. This, in turn, will help environmental planning efforts worldwide.

WEBSITES

Discussion of the demographic transition with excellent graphs:

www.uwmc.uwc.edu/geography/Demotrans/demtran.htm

'China's population: trends and challenges', from the Population Reference Bureau:

www.prb.org/pdf04/59.2ChinasPopNewTrends.pdf

China's one-child policy:

http://geography.about.com/library/weekly/aa092799.htm

New China Child Law:

www.crlp.org/ww_asia_1child.html

SEE ALSO

mortality, population increase

DESALINATION

Desalination (or desalinisation) is the removal of salt from seawater. This provides water for drinking, domestic use and for industry – one tonne of industrial products represent an average use of about 200 tonnes of water. Desalination is an expensive process, so is only undertaken in wealthy regions where demand outstrips supply, for example in Israel and Saudi Arabia. With warm climates, low rainfall and high evaporation rates, alternative sources of water are in short supply. In such coastal areas it seems obvious, therefore, to turn to the Earth's greatest water resource, the sea, which stores 97 per cent of the planet's water.

Although we consider seawater to be 'salty' it only contains 3.5 per cent salt. To serve a community properly, a large volume of water is needed. It takes 0.75 kilowatt/hour (1 kilowatt/1500 litres) for each cubic metre of seawater to be adequately desalinated. The remaining concentrated brine solution is very corrosive and has to be dealt with very carefully. As a result, the desalination plant is both expensive to build and to maintain.

More than 7500 desalination plants operate worldwide, producing several billion litres of water per day. Approximately 57 per cent of these plants operate in the Middle East, and 12 per cent of the world capacity is produced in the Americas, with most of the plants located in the Caribbean and Florida. However, as drought conditions continue and concerns over water availability increase, desalination projects are being proposed at numerous locations.

Coastal desert countries like Kuwait and islands like Guernsey (Channel Islands, UK) have used the 'flash distillation' process since the 1960s. This process involves turning a liquid maintained at high pressure to one kept at a lower pressure. The change in pressure causes part of the liquid to 'flash' (turn quickly) into a distillate, which can then easily have the soluble materials extracted from the basic solvent. In the case of desalination, seawater is 'flashed' to allow the salt compounds to be separated from the basic H_2O. Other methods of desalination include electrodialysis and vacuum freezing. Kuwait uses the former, and Israel the latter. Reverse osmosis is another relatively cheap method and even allows the conversion of sewage into domestic quality water supply. In any system, costs per unit volume do fall with the size of plant.

Active research in these technologies is now continuing. In the future desalinated

Case study

The Palestinian Water Authority

For more than 50 years water has been a source of conflict between Israel and its neighbours, especially the Palestinians. Gaza and the West Bank, including East Jerusalem, are in the midst of an acute water crisis brought about not simply by the area's naturally arid conditions or the current drought, but primarily by the politics of the distribution of water by governments and organisations. The Palestinian Water Authority (PWA) was set up in 1995 to coordinate water resource management in the region.

Two small seawater desalination plants are in the process of being constructed in the Gaza Strip to provide Palestinian communities with good quality water for drinking purposes. This is because some of the current sources are unsafe, with high levels of chloride and nitrate pollution. USAID is also financing a huge-scale desalination plant for Gaza. Its ultimate capacity will be 150,000 cubic metres/day by 2020. The supply will be distributed over the whole Gaza Strip from north to south.

By providing access to safe water, it is hoped that the people of the region will feel better able to engage in political negotiations, with the aim of seeking a lasting peace.

water offers a huge potential, not only in providing much needed fresh drinking water, but also in irrigating dry lands in places like the Middle East and Israel – an estimated 32,000 kilometres of dry coastline remains unexploited worldwide.

WEBSITES

Water Desalination:

www.waterdesalination.com

Desalination Training Centre:

www.medrc.org.om

SEE ALSO

irrigation, water supply

DESERTIFICATION

Desertification describes what happens when once fertile lands become non-productive and barren, or 'like a desert'. As a result of desertification, soil fertility is reduced, fewer desirable plants can be seen growing in the area, the soil becomes more salinised (or salty), topsoil erodes, and the soil becomes compacted and crusted. All these outcomes reduce the value and productivity of the land.

According to the United Nations Convention to Combat Desertification, desertification is happening faster than at any time in human history. Desertification affects about one-sixth of the world's population and 70 per cent of the world's dry lands, amounting to

one-quarter of the planet's total land area. When civilisations damage the resource base that sustains them, they collapse. Well-known examples include Mesopotamia and Easter Island.

Fundamentally, desertification can be traced back to patterns of land abuse. Although natural processes such as droughts can exacerbate desertification, they do not cause it. Depending on the location and land use patterns, a variety of human-induced causes can lead to desertification. For example, in rangelands, overgrazing of livestock and loss of trees that act as windbreaks can be harmful. Native plants can withstand moderate or short-term intensive grazing and still survive. However, intensive, long-term grazing destroys the native plants, leaving the soil friable and easily eroded. When this happens, wind and rain can carry the soil away, causing erosion of the fertile topsoil. Without the topsoil, it becomes harder for new plants to take root and grow successfully.

Desertification of croplands is caused by accelerated water and wind erosion, usually resulting from inappropriate cultivation practices. Native grassland plants, such as grasses, forbs (perennial herbs with broader leaves than grasses) and shrubs have complex root structures that help withstand the pressures of drought and anchor the soil. When native plants are replaced by less hardy agricultural crops, drought becomes a serious concern. As the less hardy plants die off, the soil is left

Case study

The 1930s Dust Bowl, USA

During the 1930s the south central USA was devastated by desertification brought about by a combination of drought, wind erosion and dust storms. This combination and the economic impact it caused became known as the Dust Bowl.

The hardy native grasses and forbs that covered the land had been destroyed by overgrazing of livestock or replaced by agricultural crops like wheat. The new crops, unlike the native plants, could not withstand the severe droughts that took place between 1930 and 1937. The crops died, and without plants anchoring the soil, winds eroded the exposed topsoil. The combination of strong winds and loose topsoil caused hundreds of dust storms between 1932 and 1941, with dust being carried as far away as Washington, DC. In 1935, for example, Amarillo, Texas underwent numerous complete blackouts, with visibility at zero. People suffered respiratory ailments and eye and lung damage. Livestock and wild animals were suffocated or blinded. Many farmers and ranchers were forced to abandon their now desolate land. By 1938, tree planting, better ploughing methods, and other conservation measures reduced the amount of soil blowing around by 65 per cent. The following year, rains came to end the drought period that had exposed the limits of the previous land-use patterns.

A wall of dust approaching a Kansas town. In: "Effect of Dust Storms on Health," U. S. Public Health Service, Reprint No,. 1707 from the Public Health Reports, Vol. 50, no. 40, October 4, 1935. From NOAA

vulnerable. As is common with the effects of overgrazing, drought can wipe out the new crops, leaving the topsoil vulnerable to erosion.

Desertification of irrigated lands is caused by improper water management leading to salinisation. Salinisation (the over-accumulation of salts in the soil) occurs when irrigation water evaporates leaving behind the dissolved salts. Over time, the concentration of salts in the soil becomes so high that plants can no longer grow in the soil and they wither and die.

While these are serious concerns, desertification is avoidable. The erection of fences helps control livestock grazing. Planting of small areas of trees to create a windbreak cuts down on wind erosion. Salinisation can be avoided by providing enough water to leach the salts down through the soil, and by making sure there is suitable drainage. Taking these steps helps to ensure the long-term viability of the soil.

Further reading

John Steinbeck's powerful book *The Grapes of Wrath* tracks the plight of a family who move from Oklahoma to California during the 1930s.

WEBSITES

Desertification of dry lands:
www.ciesin.org/docs/002-193/002-193.html
Food and Agricultural Organization on desertification:
www.fao.org/desertification
PBS – Surviving the Dust Bowl:
www.pbs.org/wgbh/amex/dustbowl

SEE ALSO

desalination, overgrazing, soil erosion and conservation

DESERTS

Deserts are areas with an annual precipitation below 250 millimetres. In most deserts, drought contributes to high daytime temperatures in summer and rapid cooling at night. Other BIOMES are insulated by their humidity – the water vapour in the air. Temperate deciduous forests, for example, may have 80 per cent or more humidity during the day. This water reflects and absorbs sunlight and the energy it brings. At night the water acts like a blanket, trapping heat inside the forest. However, because deserts usually have only between 10 and 20 per cent humidity to trap temperatures and have very few trees and other vegetation to retain heat, they cool down rapidly when the sun sets and heat up quickly after the sun rises. In short, deserts have more extreme climates than most other terrestrial ECOSYSTEMS. Productivity is generally low. Deserts occupy about 12 per cent of EARTH's land area, but contribute only 1.3 per cent of global net primary production.

What most people typically think of when they hear the word 'desert', such as camels walking across a sandy landscape, is in fact fairly rare. Sand covers only about 20 per cent of the Earth's deserts. There are coastal deserts, polar deserts, midlatitude deserts, RAIN SHADOW deserts, and trade wind deserts. Nearly 50 per cent of desert surfaces are plains where wind erosion ('aeolian deflation') has exposed loose pebbles. The remaining surfaces of arid lands are exposed bedrock outcrops, desert soils and water deposits, including alluvial fans, playas, desert lakes and oases. Bedrock outcrops commonly occur as small mountains surrounded by extensive erosional plains.

In addition to vegetated deserts, about 16 per cent of the Earth's land surface is extreme desert, almost or completely without primary production. This includes the cap ice of ANTARCTICA and Greenland as well as the shifting sands of the Sahara. Therefore, deserts are dry, but they don't have to be hot. Even under such extreme conditions, however, life may have a 'toehold' in peculiar, local situations such as oases around a spring among the dunes, or colonies of microbes insulated beneath Antarctic ice. These communities add little to the production of the BIOSPHERE but they can add immensely to our understanding and appreciation of the tenacity of life and its capacity to respond to environmental opportunity, however modest and marginal, under the inexorable force of natural selection.

Some of the larger hot deserts of the world include the Arabian, Australian, Chihuahuan

(Mexico and USA), Kalahari (southwest Africa), Mojave (USA), Monte (Argentina), Sahara, Sonoran (Mexico and USA) and Thar (India and Pakistan). Cool or cold deserts of the world include Chile's Atacama, the Gobi of China and Mongolia, the Great Basin in the western USA, the Iranian, the Namib in Africa, Central Asia's Turkestan and Takla Makan in China ('Takla Makan' means 'place from which there is no return').

The world's driest desert is the Atacama Desert of northern Chile (although some maintain that Antarctica is drier). Although the Tropic of Capricorn passes through the region, the Atacama lies in the rain shadow of Chile's coast range. The desert is completely barren and while most areas only receive moisture from an occasional fog or a shower every few decades, the rain gauge at Calama has *never* recorded any measurable precipitation. The Atacama is a high (most elevations are over 8000 feet) and cold desert, average temperatures range from 0° to 25°C. Most areas have an average of less than .01 centimetres of rain annually, and some places have evidently not had rainfall for over 400 years.

Producers in deserts include small-leafed shrubs, bunchgrasses and stem- or leaf-succulents. **CONVERGENT EVOLUTION** leads to structural similarity; water-rich tissues are often protected by spines, for example. There is convergence even in the characteristic columnar or 'candelabra' shapes of stem succulents. The intense midday sun hits only the top of the column and the rest of the plant is protected by its own shade at the most extreme time of day. Primary consumers include a variety of browsers such as ungulates, rodents and insects, and seed predators, such as birds, rodents and insects. In many deserts of the world, lizards and snakes are important insectivores and other snakes are top predators.

Human impact on desert environments is varied. In some areas, very little impact has occurred mostly because few people live there. In other areas, mining has disrupted landscapes, increased erosion and polluted groundwater. In other areas, irrigation and urbanisation has caused subsidence of the land. Desert fringes are **ECOTONES** that form a gradual transition from a dry to a more humid environment, making it difficult to define the edge of the desert, such as in Africa's Sahel south of the Sahara. These transition zones have very fragile, delicately balanced ecosystems. **DESERTIFICATION** been a major concern for decades. In Africa's Sahel, a drought that began in 1968 was responsible for the deaths of between 100,000 and 250,000 people, the disruption of millions of lives and the collapse of the agricultural bases of five countries.

WEBSITES

USGS publication: Deserts: Geology and resources:

http://pubs.usgs.gov/gip/deserts/contents/

Deserts from About.com's geography resources:

http://geography.about.com/cs/deserts/index.htm

Blue Planet biomes on deserts:

www.blueplanetbiomes.org/desert.htm

US National Park Service Mojave desert resource:

www.nps.gov/moja/mojadena.htm

US long-term ecological research (LTER) programme site on the Sevilleta National Wildlife Refuge in New Mexico. This is a desert grassland system diversified dramatically by mountains and riparian systems:

www.lternet.edu/sites/sev/

The 'hot desert' LTER Site in central Arizona USA, which includes parts of metropolitan Phoenix, and studies human impact on desert ecosystems, including irrigation:

www.lternet.edu/sites/cap/

DISPERSION AND DISTRIBUTION

Every species is spaced differently on the landscape and in the oceans. Dispersion is the pattern of spacing of individual members of a population. There are three broad patterns of dispersion (shown in the figure below). Each represents the predominance of a different kind of ecological interaction.

Uniform dispersion is also called 'hyperdispersed', a pattern where the individuals are more widely spaced than one would predict by chance. Corn plants in a field are hyperdispersed. The farmer spaces the plants uniformly so that each has enough resources to survive

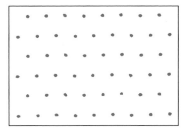

A UNIFORM

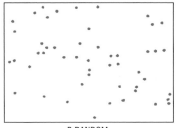

B RANDOM

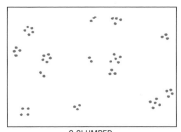

C CLUMPED
(BUT GROUPS RANDOM)

Three kinds of dispersion within local populations:
uniform (hyperdispersed), random, and clumped
(contagious). Each pattern tells an ecological story.
Source: EP Odum, *Fundamentals of Ecology*, 1953, with
permission from Elsevier

and thrive, but there is no 'wasted space'.

Despite the numerous instances of examples such as this one with corn, humans are not responsible for all uniform dispersion that we see in the landscape. Competition in natural populations can lead to hyperdispersion. This is fairly common in desert shrubs, for example, because of competition for water in the root zone. Also, some plants actually produce so-called allelopathic (from the Greek *allo-*, 'other' + *pathos*, 'disease') chemicals that inhibit the germination of seeds or growth of seedlings. The chemical treats the soil beneath the plant's canopy of leaves and assures that no competitors become established.

Random dispersion means that knowing the position of one individual does not allow prediction of the next individual in the population. That is, neighbours may be close together or far apart. It suggests that there is no interaction between individuals, either positive or negative. Not surprisingly, random dispersion of individuals is not particularly common in local populations. However, the rate of migration is limited, such that the dynamics of the group, or 'metapopulation', should be seen as the sum of the dynamics of the individual subpopulations (although the pieces of a metapopulation may be distributed randomly with respect to each other).

The most common pattern of dispersion is *contagious*, or 'clumped'. In many species, if you find one individual, you are likely to find another – a school of fish, a herd of bison, a mother aphid and her daughters, a family of foxes. Between the clumps or concentrations of individuals are areas with few or no individuals.

Dispersion is sometimes confused with *dispersal*. Dispersal is the movement of individuals from one place to another; usually, from place of birth to place of reproduction. Dispersion is a pattern; dispersal is a process.

Any individual has a position in space. Its position may change, but at any moment in time it has a single, particular position. The area over which an individual moves in its daily round of activities is called the *home range*. A portion of the home range that is defended against other members of the species is called the *territory*.

When all of the positions of all of the individuals of the population are considered at once, that constitutes the distribution, or geographic range of the population. Distribution is the extent of a population over space and may be large or small. The chipmunk is very widespread. Its distribution in North America covers perhaps 10 million km^2, from the Yukon to California and eastward to Quebec and Michigan. Palmer's chipmunk, by contrast, occurs only on Charleston Mountain, near Las Vegas, Nevada; the extent of its distributional range is only a few dozen square kilometres.

Distributions change with time. The most dramatic adjustment of distributions over the past few million years occurred during the

Pleistocene ice ages. As glaciers advanced from the poles and high mountains, species' distributions were pushed southward or downslope. As glaciers retreated, species could expand poleward. Sometimes, species of high latitudes were marooned on mountaintops. Many alpine plants represent such glacial artifacts, as do some alpine mammals, such as marmots, in the Rockies, Sierras and Alps, for example.

Phenology is the study of seasonal events, such as the flowering of plants and the arrival or departure of migratory birds. Among the results of phenological studies are insights into the effects of global climatic change.

Perhaps some changes in distributions that are occurring today, as species respond to global climatic change, will eventually be as dramatic as those of the geological past. The distributions of butterflies are very well known because butterflies are studied and collected by lots of enthusiastic naturalists. A recent study of a sample of 35 non-migratory European butterflies showed that the ranges of 63 per cent of species had shifted northward in the twentieth century by 35 to 240 kilometres, whereas only 3 per cent had shifted southward. This could be a result of human-influenced global warming.

WEBSITES

UK Phenology Network:
www.phenology.org.uk/
European Phenology Network at
Wageningen University in the Netherlands:
www.dow.wau.nl/msa/epn/index.asp
Student-led study of monarch butterfly
migration in North America:
http://kancrn.org/monarch/wave/

EARTH

The Earth is our home. The Earth's environment is a result of the location of the Earth in the solar system, and the physical make-up of the planet. To understand the environment, it is important to understand the Earth's physical properties. In addition, if we can understand how the Earth works, we can learn how human activities can cause changes in the delicate balance that allows all elements that make up the Earth to work together.

The Earth is the third planet from the Sun, which is about 149 million kilometres away. Eight other planets orbit the Sun. Only the Earth, at just the right distance from the Sun, supports life as we know it, despite recent evidence of gases relating to life forms being discovered in the Martian atmosphere. The Earth's orbit around the Sun is not a perfect circle, but is slightly elliptical, requiring 365 days, 5 hours, 48 minutes and 46 seconds. As a year is bit longer than 365 days, a day is added to calendars every four years, a leap year, to keep the calendars on the right track.

The Earth has one satellite, the Moon. Even the Moon affects the environment since it exerts a gravitational pull on the Earth, which affects our tides. Tidal pools, wetlands and estuaries host a variety of unique plants that can survive a salty environment, including mangroves, glassworts and lichens. These habitats would simply not exist if there were no tides.

The Earth completes one rotation on its axis each day. This rotation has a profound effect on the environment because of the resulting daily heating and cooling, and also because the Earth's axis is tilted at an angle of 23.5° from vertical. Although the Earth is tilted in the same direction throughout the year, as the Earth orbits, the hemisphere tilted away from the Sun gradually comes to be tilted towards the Sun, and the hemisphere tilted towards the Sun will come to be tilted away from the Sun. The tilt is responsible for the seasons, which have an enormous effect on the environment – climate, soils, plants and animals. Areas that receive sunlight that does not come from straight overhead, but near the horizon, are colder. The environment in these polar regions is harsh, making it difficult to sustain many life forms.

The Earth's **ATMOSPHERE** is composed of nitrogen (78 per cent), oxygen (21 per cent) and other gases (1 per cent). Oxygen is essential to life because animals and plants breathe it, but also because some of the oxygen forms **OZONE**. The ozone layer is important because it filters out the Sun's harmful ultraviolet radiation. Human activity, such as the creation and release of fluorocarbons, threatens the ozone layer and our protection from the Sun.

The Earth is made up of the crust, mantle and core, which since the development of the theory of plate tectonics, has been further subdivided, as the diagram shows.

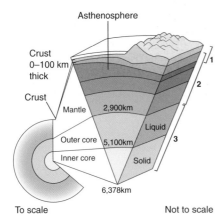

1 Lithosphere (crust and upper-most solid mantle)
2 Mantle
3 Core

This cutaway shows the internal structure of the Earth.
From the USGS website This Dynamic Earth
(http://pubs.usgs.gov/publications/text/inside.html)

The crust is very thin compared to the mantle and core. It is thinnest at the floor of the ocean – only about 5 kilometres thick. The mantle is a dense, hot layer of semi-solid rock, approximately 2900 kilometres thick. The core is actually made up of two distinct parts: a 2200 kilometre-thick liquid outer core and a 1250 kilometre-thick solid inner core. As the Earth rotates, the liquid outer core spins, creating the Earth's magnetic field. How do we know this, since we obviously cannot look into the core? The theory was developed through time, beginning with early observations of sailors and explorers in the late 1800s and more recent measurements of the Earth's magnetic field with sophisticated equipment.

Continents, including the area above sea level and the submerged continental shelf that surrounds each continent, account for about 35 per cent of the surface of Earth. The highest point on Earth is Mount Everest in Nepal, at a height 8850 metres. The lowest place on Earth is the Dead Sea in Israel, which is 394 metres below sea level.

An early controversial theory in geology, continental drift, suggested that the continents were at one point in geologic history all one land mass but migrated to their current positions over millions of years. Later discoveries confirmed the concept, with major changes, and it became generally accepted as the theory of **PLATE TECTONICS** in the mid 1960s. The continents are still on the move and may travel as much as 5 centimetres each year. Once plate boundaries were mapped, it became clear that much of the earthquake activity (see **EARTHQUAKES**) and volcanic activity (see **VOLCANOES**) on the Earth occurs at plate margins.

The **OCEANS** and seas account for about 65 per cent of the Earth's surface, not including the continental shelf. The oceans have a tremendous effect on the environment, not only along coastlines, but also as a moderator of the Earth's climates.

The Earth is a dynamic planet – there is something happening on it all the time. Unlike on Mercury, where a meteorite striking the surface every few years might be the only change, or on Jupiter's moon Io, where a volcano might periodically erupt, nearly everything on the Earth is moving – the air, glaciers, ocean currents, animals and people. Each of these components is important to the environment and has the potential to alter it. Humans, in particular, have a great capacity for altering the environment. Because of the diversity of the planet's environment, and because changes are continually occurring, the study of the Earth is as varied and fascinating as the Earth itself.

WEBSITES

Discussion on the magnetic field's origin:
www.sciencenews.org/sn_arc99/11_13_99/fob5.htm

The University Corporation for Atmospheric Sciences atmosphere information:
www.windows.ucar.edu/tour/link=/earth/Atmosphere/overview.html

The University of Arizona's Nine Planets resource:
http://seds.lpl.arizona.edu/nineplanets/nineplanets/nineplanets.html

EARTHQUAKES

One of the most frightening and destructive events on the Earth is a severe earthquake. An earthquake is a sudden and intense movement of the ground, caused by the abrupt release of strain that has accumulated over a long period of time. This movement is most commonly caused by a slip along a fault, which may or may not coincide with an active plate margin. Earthquakes can also occur in volcanic areas where magma is bursting through the crust on its way to the surface.

Some earthquakes are linked to human activity, such as underground mining, water extraction, or the building of dams and reservoirs. In Denver, Colorado, USA, a series of earthquakes during the 1950s and 1960s were due to the pumping of liquid waste from Rocky Mountain Arsenal, a weapons production plant, into the subsurface.

Approximately 90 per cent of all earthquakes occur at the margins of the world's great plates. This is because the plates are in continuous but very slight motion, creating strain, which eventually causes the rapid movement that creates the seismic waves that cause the earthquakes. An earthquake occurs when the Earth's crust shakes due to seismic waves passing through it. These seismic waves are low frequency sound waves that begin deep within the Earth's crust where a sudden movement or 'snapping' has taken place.

Some people worry about falling into cracks as the Earth splits apart, but ground ruptures are actually fairly uncommon. Most injuries and fatalities from earthquakes are from the collapse of bridges and buildings.

A magnitude 7.4 earthquake in 1964 killed 26 and damaged over 12,000 apartment buildings in Niigata, Japan, after soils liquefied and foundations failed.
Photo: NOAA National Geographical Data Center

This five-story concrete-beam bank building on the main street of Duzce, Turkey, failed on one side. The outer, hollow steel tube system buckled and crumpled. Photo: Roger Bilham, Department of Geological Sciences, University of Colorado

Sometimes the after-effects of the earthquake are worse than the earthquake itself. These include **FLOODS**, fires, **TSUNAMIS**, liquefaction of soils (when waterlogged sediments lose their strength in response to strong shaking) and **LANDSLIDES**.

The most intense earthquake ever measured (9.5 on the Richter Scale) occurred on 22 May 1960 in southern Chile. The series of earthquakes that followed occurred over a 1,000 kilometre-long section of a fault, one of the longest ruptures ever recorded. The number of fatalities associated with the event has been estimated to be between 490 and 5,700. An estimated 2 million people were left homeless and 58,622 houses were completely destroyed. The main shock caused a series of tsunamis along the coast of Chile, and as far away as Hawaii and Japan. Subsidence caused by the earthquake caused local flooding and permanently altered many Chilean shorelines. The Puyehue volcano erupted 47 hours after the main shock.

The Indian Ocean earthquake of 26 December 2004, which measured 9.0 on the Richter Scale and caused devastating tsunamis, had a greater impact than the Chile earthquake. The island of Sumatra, closest to the underwater epicentre, suffered the greatest loss of life and almost total destruction of a broad swathe of coastline. In total, over 200,000 people died in Indonesia, Thailand, throughout South East Asia and as far away as the African countries of Somalia and Tanzania. Millions more were left homeless.

Measuring earthquakes

The Richter Scale

An earthquake can be characterised by magnitude and intensity. The two concepts are often confused. The magnitude of an earthquake is measured using seismographs to gather data that is interpreted by the Richter Scale. The Richter Scale was developed in 1935 by Charles F. Richter of the California Institute of Technology. It compares the size of earthquakes using the logarithm of the amplitude of waves as recorded by seismographs. Because it is not based on a simple mathematical progression, each whole number increase in magnitude represents a tenfold increase in measured amplitude. As an estimate of energy, each whole number step in the magnitude scale corresponds to the release of about 31 times more energy than the amount associated with the preceding whole number value. The scale is as follows:

- Magnitude 1–3: Recorded on local seismographs, but generally not felt.
- Magnitude 3–4: Often felt, no damage.
- Magnitude 5: Felt widely, slight damage near epicentre.
- Magnitude 6: Damage to poorly constructed buildings and other structures within 10 kilometres.
- Magnitude 7: 'Major' earthquake, serious damage up to 100 kilometres (recent earthquakes in Taiwan, Turkey, Japan and California).
- Magnitude 8: 'Great' earthquake, great destruction, loss of life over several hundred kilometres (1906 San Francisco, 1949 Queen Charlotte Islands) .
- Magnitude 9: Rare great earthquake, major damage over a large region over 1000 kilometres (Chile 1960, Alaska 1964, and west coast of Indonesia, 2004).

Source: Geological Survey of Canada: www.pgc.nrcan.gc.ca/seismo/eqinfo/richter.htm

The Modified Mercalli Scale

The intensity, as expressed by the Modified Mercalli Scale, is a subjective measure that describes how strong a shock was felt at a particular location. This scale, developed in 1931, identifies the intensity of an earthquake's effects in a specific area using values ranging from I to XII. The condition of "I – not felt except by a very few under especially favourable conditions', contrasts with 'XII – damage total. Lines of sight and level are distorted. Objects thrown upward into the air.' The intensity of an earthquake's effects does not require any instrumental measurements but requires eyewitness accounts. While the Modified Mercalli Scale is used with recent earthquakes, it also gives seismologists a method to evaluate historical earthquakes using newspaper accounts, diaries, and other historical records to give intensity ratings.

The tsunami reached the Pacific and Atlantic Oceans and was recorded in New Zealand, Antarctica and North America. Entire coastal environments were changed in an instant.

The strongest earthquake ever measured in the USA measured 9.1 on the Richter Scale in Prince William Sound off the coast of Alaska in 1964. The earthquake and tsunami that followed caused 125 deaths, including those as far away as California. The death toll was relatively small because the area in Alaska was sparsely populated.

The most severe earthquake in the world, in terms of lives lost, was a magnitude 8.0 jolt that struck Tianjin (formerly Tangshan), China, on 27 July 1976. The official casualty figure issued by the Chinese Government was 255,000, but unofficial estimates were as high as 655,000. The most destructive US earthquake occurred in San Francisco on 18 April 1906. Though its magnitude was 7.7, and the energy less than one-thirtieth the energy released by the 1964 Alaska event, the San Francisco earthquake caused an estimated 3,000 deaths, largely because of the fire that resulted.

Earthquakes can vary greatly in size – from those that can only be felt by very sensitive

Most destructive known earthquakes on record: more than 50,000 deaths

Date	Location	Deaths	Magnitude	Comments
23 January 1556	China, Shansi	830 000	~8	
27 July 1976	China, Tangshan	255 000 (official)	7.5	Estimated death toll as high as 655 000
26 December 2004	Near Indonesia, Sumatra	c.250 000	9.0	Tsunami
9 August 1138	Syria, Aleppo	230 000		
22 May 1927	China, near Xining	200 000	7.9	Large fractures
22 December 856	Iran, Damghan	200 000		
16 December 1920	China, Gansu	200 000	8.6	Major fractures, landslides
23 March 893	Iran, Ardabil	150 000		
1 September 1923	Japan, Kwanto	143 000	7.9	Great Tokyo fire
5 October 1948	USSR (Turkmenistan, Ashgabat)	110 000	7.3	
28 December 1908	Italy, Messina	70 000–100 000 (estimated)	7.2	Deaths from earthquake and tsunami
September, 1290	China, Chihli	100 000		
November, 1667	Caucasia, Shemakha	80 000		
18 November 1727	Iran, Tabriz	77 000		
1 November 1755	Portugal, Lisbon	70 000	8.7	Great tsunami
25 December 1932	China, Gansu	70 000	7.6	
31 May 1970	Peru	66 000	7.9	$530 000 damage, great rockslide, floods
1268	Asia Minor, Silicia	60 000	~	
11 January 1693	Italy, Sicily	60 000		

instruments to those that do great damage. By plotting earthquakes on the Earth's surface, it is apparent that they do not occur randomly, but they follow the boundaries of the plates that make up the Earth's surface.

While most earthquakes occur near plate boundaries, the most widely felt earthquakes in the recorded history of North America were a series that occurred in 1811–12 near New Madrid, Missouri, far from a plate boundary. A great earthquake, whose magnitude is estimated to be about 8, occurred on the morning of 16 December 1811, with another on 23 January 1812 and the strongest yet on 7 February 1812.

Aftershocks were nearly continuous between these great earthquakes and continued for months afterwards. These earthquakes were felt by people as far away as Boston and Charleston, but the region was sparsely populated at that time. If just one of these enormous earthquakes occurred in the same area today, millions of people and buildings and other structures worth billions of dollars would be affected.

As they are among the Earth's most destructive forces, earthquakes have a great effect on watersheds, slopes, drainage, plants, animals – and hence on the environment.

Further reading:

Bolt, B. A., 1993, *Earthquakes*, W.H. Freeman.

WEBSITES:

British Geological Survey:
www.earthquakes.bgs.ac.uk
USGS National Earthquake Information Center:
wwwneic.cr.usgs.gov
Understanding Earthquakes, from University of California, Santa Barbara:
www.crustal.ucsb.edu/ics/understanding
USGS earthquake publications:
http://pubs.usgs.gov/gip/earthq1
Worst earthquake disasters, from the USGS:
www.usgs.gov/public/press/public_affairs/press_releases/pr1133m.html

SEE ALSO

landslides, plate tectonics, tsunamis

ECOLOGICAL EFFICIENCY

The following paragraphs include three numbers worth remembering: 10 per cent, 1 per cent and 4. Specifically, 10 per cent represents the typical ecological efficiency of transfers down the food chain; 1 per cent represents the typical efficiency of **PHOTOSYNTHESIS**; and 4 represents the amount of energy, in kilocalories (kcal) in each gramme of carbohydrate or protein.

Consideration of trophic structure of communities and the pyramid of energy allows us to visualise **ecological efficiency**. Ecological efficiency is the amount of energy stored at one level as a fraction of the amount of energy stored at the previous level; that is, output per input. Efficiencies in real **FOOD WEBS** vary from a low of 1 or 2 per cent to as high as 30 per cent, but a typical 'ballpark' figure is 10 per cent. Data from some classic studies are shown in the table below.

Typical ecological efficiency of about 10 per cent means that it takes about 100 kilocalories (kcal) of meadow vegetation to make 10 kcal of herbivorous meadow mice or grasshoppers. It takes 10 kcal of grasshoppers to make 1 kcal of fox or hawk. It takes 1 kcal of carnivore to make 0.1 kcal of fox-sucking flea or hawk-sucking louse.

The efficiency of photosynthetic plants or algae at the base of the food web is variable, but averages around 1 per cent. Therefore, 100 kcal of meadow plants represents about 10,000 kcal of sunlight and of that vegetation, maybe only half is edible. So, it takes perhaps 20,000 kcal of solar energy to provide 100 kcal of energy to the food web. Considered this way, it begins to make some sense why fleas and lice are so small and foxes are so few and far between, relative to meadow mice or grasshop-

Three classic studies of productivity (kcal/m²/year) and ecological efficiency: Lake Mendota, Wisconsin (Juday, 1940—data from Lindeman, 1942), Cedar Bog Lake, Minnesota (Lindeman, 1942), and Silver Springs, Florida, USA (H.T. Odum, 1957)

	Lake Mendota		Cedar Bog		Silver Springs	
	Productivity	Efficiency	Productivity	Efficiency	Productivity	Efficiency
Radiation	≤ 1 188 720		~ 1 188 720		1 700 000	
Producers	4 800	0.4%	1 113	0.1%	8 833	0.5%
Primary consumers (herbivores)	416	8.7%	148	13.3%	1 478	16.7%
Secondary consumers	23	5.5%	31	22.3%	67	4.5%
Tertiary consumers	3	13.0%	—	—	6	8.9%

pers. If you prefer to think in terms of mass rather than energy, there are about 4 kcal of food energy in each gramme of carbohydrate or protein; 25 grammes, or about 1 ounce, of dry-weight vegetation corresponds to 100 kcal.

If ecological efficiency is on average 10 per cent, what happens to the other 90 per cent? The 90 per cent is not just a case of 'environmental inefficiency'. The components of inefficiency need to be understood. In part, of course, the 90 per cent is the inefficiency of any system of energy flow. The second law of thermodynamics states that the efficiencies of energy transformation can never equal 100 per cent. Consequently, all processes lose energy, typically as heat, and therefore are not reversible unless this lost energy may be supplied from the environment. But, that 90 per cent also includes the inescapable energy costs of living at any given trophic level. That includes growth, reproduction, movement, and – at the cellular level – costs of active transport, cellular maintenance and molecular synthesis. The table summarises some major pathways of energy flow ('expenses') within given trophic levels that contribute to ecological 'inefficiencies'.

Obviously the term 'inefficiency' is a relative term. From the viewpoint of a meadow mouse, respiration by the plants is an 'inefficiency', but from the plant's 'viewpoint', respiration is about being alive. From the weasel's point of view, mouse respiration and the bits of vegetation growing through cracks in the pavement are 'inefficiency'. Indeed, from the standpoint of the weasel, everything that the mouse does is inefficiency, except when the mouse accumulates body mass (protein, fat) or offspring that the weasel could have for lunch!

WEBSITES

Ecosystem ecology with a discussion of ecological efficiency:

www.sbs.utexas.edu/mbierner/bio406d/lectures/sp03/EcosystemEcology.pdf

Principles of ecology notes from Old Dominion University:

www.lions.odu.edu/~kkilburn/econotes.htm#top

Ecology terms and resources from about.com:

http://biology.about.com/

ECOLOGICAL PYRAMIDS

Concepts such as **FOOD WEBS** or chains are helpful to understand the environment, but they say nothing about the *number* or *amount* of individuals or **BIOMASS** involved at each stage in the process. That's where ecological pyramids come in. Ecological pyramids describe an environmental community in simple terms that allow deep insight into the way nature's economy is organised. Three kinds of ecological pyramids exist – numbers, biomass and energy.

Pyramid of numbers

The plant–insect–bird–hawk food chain can be represented as an ecological pyramid. Plants absorb energy from the sun, the insects eat the plants, the birds eat the insects and the hawks eat the birds. The energy of the sun has therefore been transferred from the sun to the tissues of the hawk. Since the number of individuals in each level usually decreases, the resulting diagram looks like a pyramid.

The number of individuals does not always decrease. Therefore, the diagram for some ecological 'pyramids' is not always in the shape of a true pyramid. If the first level of the pyramid was 'trees' instead of plants, the 'pyramid' would not look like a pyramid because the number of trees is less than the number of insects. Ecological pyramids represent the basic laws of energy. The first law of thermodynamics states that the amount of energy remains constant. The second law states that 'useful energy decreases at each conversion'. Therefore, ecological pyramids represent the transfer of energy from one level to another.

Figure 1 shows two pyramids of numbers – often called 'Eltonian pyramids', after British ecologist Charles Elton who developed the metaphor. Producers are labelled as 'P' and consumers as 'C'. Producers make the food for the consumers. Consumers are organisms that feed on other organisms (the producers) as a source of food. Plant-eaters (herbivores) are primary consumers (C1). Secondary consumers (C2) depend on the biomass of primary consumers for their energy and nutrition requirements; they are carnivores. As Figure 1 shows, in an ecosystem there are much fewer secondary consumers than primary con-

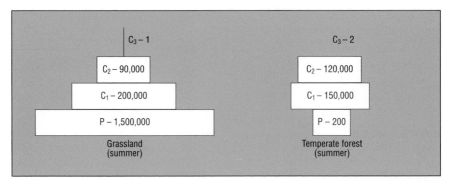

Figure 1 Pyramids of numbers. Figures 1-3 reprinted from EP Odum, *Fundamentals of Ecology*, 1953, with permission from Elsevier

sumers. This is due to the fact that herbivores have, over time, developed defences to prevent them from being eaten by carnivores. These defences include things like the quills on a porcupine and the ability of a snowshoe hare to change colours with the seasons. Finally, tertiary consumers (C3) consume the biomass of primary carnivores and herbivores.

Each pyramid represents 0.1 hectare. On the left is a pyramid for a patch of grassland. Note how each successive layer (called 'trophic levels') of the pyramid consists of fewer individuals of larger size. This is the typical pattern.

However, you may have noticed that something is missing. Feeding on the leftovers at each trophic level (dead grass, dead bugs, dead spiders, dead birds and moles) are a host of minute, soil-dwelling detritus feeders, like fungi, mites and nematodes, and legions of even more minute bacteria. These complicate the picture unduly and they are very hard to count, so many ecologists like to think of such

organisms as constituting a detritus food web of their own. Another problem with Eltonian pyramids is this – you can think of 'pyramids' that stand on their point, or are otherwise quite improper pyramids. The right-hand 'pyramid' in Figure 1 has just such a shape. One-tenth of a hectare of forest might support only a few dozen individual trees and shrubs, which in turn will support tens of thousands of individual insects.

Pyramids of biomass

Following Elton's lead, ecologists tried to develop more instructive descriptions of trophic structure. It was clear that the factor of size was missing from the Eltonian pyramid of numbers, so investigators began to build *pyramids of biomass*.

Figure 2 shows pyramids of biomass for several different communities. Some of the

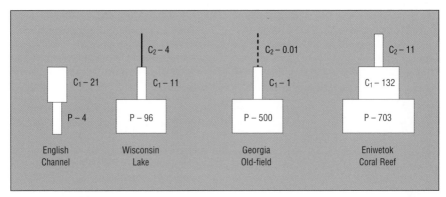

Figure 2 Pyramids of biomass

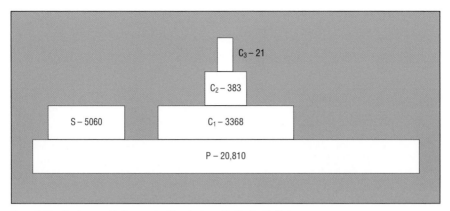

Figure 3 The classic pyramid of energy for Silver Springs, Florida (kcal/m^2/yr)

pyramids are indeed pyramidal and some actually include estimates of the biomass of decomposers. They are limited in that they represent estimates based on a single point in time. Because they do not take account of the rate of accumulation of biomass, they can be inverted. If, as in the case of the ecosystem in the English Channel, you have small short-lived producers being eaten by large, long-lived consumers, you can have an inverted pyramid of biomass. Consumer biomass represents accumulation over time. Some improvement is needed in the pyramid of biomass to gain meaningful insight into the trophic structure of biotic communities. Ecologists overcame the size problem of the pyramid of numbers with the pyramid of biomass. Now they had to overcome the problem of time.

Pyramids of energy

The solution to time lay in something truly fundamental – energy. Pyramids of energy 'tell it like it is', because they account for time. A classic pyramid of energy for Silver Springs, Florida is shown in Figure 3.

Pyramids of energy are necessarily pyramidal *if* the system under study does not import energy across space from adjacent ecosystems or across time (from storage deposits such as coal, oil, peat or natural gas). This follows quite directly from the second law of thermodynamics and resulting considerations of **ECOLOGICAL EFFICIENCY**.

Transfers from one trophic level to the next necessarily involve 'loss' of some energy as heat and some 'loss' to useful work.

Trophic levels

Trophic levels represent successive steps removed from direct dependence on the ultimate energy source in a biotic community. The ultimate energy source is usually the Sun. In typical biotic communities, plants or algae constitute the producer trophic level and a variety of animals, fungi and microbes comprise several levels of consumers. Technically, producers are called autotrophs (from the Greek, 'self-feeders') and consumers in general are called heterotrophs ('other feeders'). Some heterotrophs are adapted to feeding at various trophic levels; these are called omnivores. Skunks, bears and raccoons are omnivores, feeding on both plant and animal matter. Other heterotrophs are specialised herbivores or carnivores, feeding on vegetation or animal flesh, respectively. Vegetarians (primary consumers) often are specialised to eat particular plant life forms. Grazers, like cattle and red deer or elk, eat grasses and other herbs, whereas browsers (such as deer) are specialised to feed on woody plants. Other primary consumers may be specialised to eat seeds, fruit, pollen, nectar, gum or other plant organs and products.   *continued over*

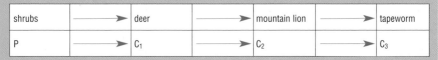

Trophic Levels continued

Most biotic communities have no more than three or four levels of consumers:

shrubs	⟶	deer	⟶	mountain lion	⟶	tapeworm
P	⟶	C_1	⟶	C_2	⟶	C_3

Further, as the classic ecological pyramids suggest, specialised carnivores are destined to rarity by simple energetics. Note also that the third order consumer in this example is of small body size. Again, that is a consequence of energetics and ecological efficiency. At the end of the food chain, too much energy has been dissipated to support large heterotrophs.

Therefore, there is necessarily less energy available on the second step than there was on the first, and so forth. In which case, the pyramid really must be a shaped like a pyramid.

WEBSITES

Chigwell School's resources on ecological pyramids:

www.chigwell-school.org/academic/departments/science/gcsebiology/gcseresources/ecolinternetcourse/PYRAM.htm

Trophic levels, food chains, and ecological pyramids:

www.botany.uwc.ac.za/sci_ed/grade10/ecology/trophics/troph.htm

Arcytech's information on the importance of ecological pyramids:

www.arcytech.org/java/population/facts_foodchain.html

ECOLOGICAL SUCCESSION

Imagine a neglected lawn or an abandoned field. Now 'fast-forward' the scene for a decade or two, and you are imagining ecological succession. Ecological succession is a process of community development in ecological time, from seasons to centuries. Over time, the physical characteristics of a place change, and therefore the resources available to organisms change. Given enough time and change, a particular species may no longer be able to persist – it will eventually be replaced by another species that is better suited to the new conditions.

Ecological succession is often described as plant succession, with a focus on the vegetation. However, as vegetation changes, so do opportunities for change for animals, fungi and microbes. Because plants generally are macroscopic and slow moving, plant succession is simply the most obvious manifestation of a process that is community-wide and community-deep.

At the dawn of the twentieth century Henry Chandler Cowles was one of the prime forces in the emerging study of 'dynamic ecology', through his study of the Indiana sand dunes at the southern end of Lake Michigan, USA. Cowles found that he could attribute the vegetation at any point in the dunes to several variables – the distance from the lakeshore, the estimated age of the dune, and the type of soil that had developed. In 1899, his classic paper 'The ecological relations of the vegetation of the sand dunes of Lake Michigan' appeared in the *Botanical Gazette* and launched the modern study of ecological succession.

Succession in a formerly ploughed field, south-central Texas, USA. Photograph by Joseph Kerski

Today, ecologists often distinguish two types of succession – primary and secondary. When ecological succession starts from scratch on a newly opened site, it is called primary succession. Primary succession is the process of life colonising dead or sterile areas such as volcanic lava flows and new sand dunes, or on rock left behind by retreating glaciers, transforming them into living communities. This is the process where lichens colonise bare rock, or horsetail ('jointgrass') invades the sandbars left with the retreat of a glacier, or organisms colonise a newly cooled lava flow. The Indonesian island of Krakatau exploded in 1883, blowing over 4 million cubic metres of rock almost 30 kilometres into the atmosphere, completely eliminating living things from the island. In 1884 the first macroscopic colonist arrived, a spider. By 1886 there were 34 species of plants on the island. A half-century after the explosion, in 1933, 720 species of insects and 30 species of vertebrates were living on the island.

Seldom has any one ecologist actually seen primary succession happen at a particular place. Primary succession usually takes a long time. It requires thousands of years for lichens and physical weathering to break down granite into particles small enough to produce a soil that will support pine trees. Because we do not live long enough to see most primary succession happen, we come to a view of the past by 'working backwards', or synthesising bits of the present.

Secondary succession begins at least with something to start with – soil. This might be thought of as an ecosystem's response to an injury, similar to the way human skin heals itself after it is cut. Such succession transforms a disturbed or damaged part of a community. For example, a tree that falls creates an opening in the canopy of leaves: succession fills in the opening. A wetland gradually fills in over time and becomes a meadow. After forest fires burn a section of terrain, it can be good and even necessary for the forest community as a whole. The fire returns carbon to the soil. Many plants such as fireweed contain seeds that lie dormant in the soil for years until they are heated by a forest fire. Through the succession that results, the fire recreates diversity, which strengthens the forest by making it more resistant to disease.

Another common example of secondary succession is the establishment of a community of 'wild' organisms after an agricultural field is abandoned. This process has been termed oldfield succession, but it is really just a special case of secondary succession. Oldfield succession is a special case because agriculture often involves such things as soil amendments (known as productive subsidies) and pesticides (known as protective subsidies). These amendments may, to some extent, influence the type of organisms that can successfully invade the field once those subsidies are removed. In the south-eastern USA, an abandoned cotton field is taken over by 'weeds' – invasive plants adapted to the harsh conditions of direct sunlight and desiccation of the bare soil. The weeds shade the soil, cooling it and then slowing drying out. Dead weeds form a layer of organic matter. Eventually, conditions are suitable for invasion by perennial grasses and broad-leaved herbs. This perennial meadow is invaded in turn by woody shrubs, and finally a hardwood forest forms. This process may take a century or more. Therefore, the underlying processes are the same between primary and secondary succession.

As succession proceeds, there tends to be an increase in total biomass, total detritus, stability (resistance to being disturbed), energy flow, community respiration, richness and evenness diversity and food web complexity. At the same time, there is a decrease in net ecosystem production; that is, there is a trend towards a balance between production and respiration.

The first organisms in succession are termed pioneers. In the above cotton field example, invasive weeds are the pioneers. The climax organisms in a succession are the final ones, which, in theory, are the same organisms that dominated the area before the disturbance occurred. In the above example, the climax plants are those of a hardwood forest. The stages in succession are often called seral stages, in the sense of being parts of a series. Collectively, the series of seral stages has been called a sere where it relates to a particular environment such as a sand dune.

The climax community represents a stable end product of the successional sequence. In a sense, the ecological succession has 'stopped'. However, any ecosystem, no matter how inherently stable and persistent, is subject to external disruptive forces such as fires and storms and humans that can retrigger the successional process. Furthermore, over long periods of time climate and other fundamental aspects of an ecosystem change. No ecosystem, then, has existed or will exist unchanged or unchanging over a geological timescale.

WEBSITES

Hamline University Graduate School of Education resources on ecology: http://cgee.hamline.edu/see/questions/dp_transformation/dp_trans_succession.htm Pennsylvania State University's information on ecological succession: www.nk.psu.edu/naturetrail/succession.htm Henry Chandler Cowles and ecological succession: http://en.wikipedia.org/wiki/Ecological_succession

SEE ALSO

wildfires

ECOLOGY

Ecology is the scientific study of interactions between living organisms and their biotic and abiotic (non-living) environments. Ecology often is considered a 'life science', but that sells ecology short. Ecology is interdisciplinary, a 'biogeoscience' – a life and earth science. Not only biology, but also major elements of chemistry, physics, Earth sciences (geology, GEOGRAPHY, meteorology, climatology, soil science, etc) and even astronomy and cosmology inform ecology as well.

The fact is, we humans and our earthly surroundings are – from a cosmological standpoint – remarkable but momentary manifestations of stardust. John Muir, the nineteenth-century Scottish writer, conservationist and founder of the Sierra Club, noted, 'if you really get a hold of anything in the Universe, you find that it is hitched to everything else'. Thanks to Newton's notion of Universal Gravitation, this statement is a literal truth. However, it also is an ecological truth – everything in the environment is connected to everything else.

The term 'ecology' was coined in 1869 by Ernst Haeckel, a German biologist and an early and earnest proponent of Darwin's theory of evolution by natural selection. Haeckel proposed the term 'ecology' for the scientific study of 'the conditions of the struggle for existence' – interactions between individuals and their biological and physical environments, resource acquisition, competition and population growth.

Ecology has been defined and described in various ways and each captures some of the scope and depth of this remarkably complex field. Some have described ecology as 'the study of the ultimate complexity of life'. Ecology studies the most complex levels of biological organisation: individuals, populations, communities, ECOSYSTEMS, the BIOSPHERE, the ecosphere. Ecologists have noted that, 'ecology is not only more complex than we know, it is more complex than we *can* know'. It's simple mathematics: How many species are there? (Probably 10,000,000, tens of millions, or 10^7.) How many interactions can they have? Considering only species-level complexity – just one component of biodiversity – the arithmetic of simple combinations of 10^7 species in the biosphere interacting two or ten or thousands at a time exceeds the number of potential connections of the neurons in the human brain.

At the large scale, ecosystems (and their sum, the ecosphere) are not just levels of biological integration, but also geo-biological entities, phenomena integrating EARTH and life in interconsequential processes. Preeminent ecologist G.E. Hutchinson went further, to describe ecology as 'the study of the Universe'. He described ecologists as 'chartered libertines', suggesting the need to range beyond 'mere' life sciences to draw for understanding and perspective on the principles and insights of the physical and social sciences as well.

A cynic once described an ecologist as 'a naturalist who calls a shovel a "geotome"'. The implication is that ecologists take the simple things of our surroundings and complicate them with jargon. There may be a grain of truth in that exaggerated view. Ecologists do study the familiar, but seldom are their words for the sake of words only. Ecologist R.R. Lechleitner playfully and richly described ecology as '. . . the elucidation of the obvious'. Ecology is about understanding the depth and richness of the structure and function of our familiar surroundings, and of course that is no easy task. As mathematician Alfred North Whitehead observed, 'It requires a very unusual mind to undertake the analysis of the obvious'.

We also understand what ecology is by knowing what it is not. Ecology is not environmentalism and it is not resource management, although clearly ecology must inform both of those endeavours.

Environmentalism is a philosophical position, the value judgement that humans *should* establish a right and sustainable relationship between themselves and the ecological systems that support them. Sciences – like ecology – cannot make '*should* statements' and maintain their integrity as sciences. Environmentalism is also a social and political movement, using political and legal mechanisms to achieve more responsible and sustainable interactions between humans and the ecosystems upon which we depend.

Perhaps an analogy will help to make the point. The well-meaning statement, 'I want to protect the ecology', is almost meaningless. Ecology is a science. Analogous statements such as, 'I want to protect the astronomy' or 'I want to protect the chemistry' are nonsensical.

The distinction between ecology and its application is poorly made in everyday conversation and in the popular media. The terms 'ecology', 'applied ecology' and 'environmentalism' are often used more or less interchangeably. That is inaccurate and maybe even dangerous. Ecology is applied in everyday life, through resource management, environmental remediation and agriculture, to name a few examples. These applications involve the use of ecological insights, data and principles to meet human needs. For one thing, it allows the sloppy or the cynical to divert discussions of *what is* (a scientific study) to what *ought to be* (a value judgement).

WEBSITES

Ecology.com's information bank: www.ecology.com/

The Ecological Society of America: www.esa.org/

Earthwatch – organisation to promote understanding of the environment: www.earthwatch.org/

International Society for Ecology and Culture: www.isec.org.uk/

ECOSYSTEMS

An ecosystem is an arbitrary volume of the environment, including a biotic community and its abiotic (non-living) environment, interacting in exchanges of materials and a flow of energy. An ecosystem is a component of a **BIOME**. A biome can be thought of as many similar ecosystems throughout the world grouped together. Ecosystems represent the dynamic interactions between plants, animals, microorganisms and the environment – all working together as a functional unit.

A biotic community is structured by **SYMBIOSES**, and the abiotic environment is continually shaped by the geophysical forces of weathering. Biotic and abiotic components interact through biogeochemical cycles. Most ecosystems are powered by a flow of solar energy. Ecosystem processes represent the sum of processes at the level of cells and the organelles that compose them – chloroplasts and mitochondria in typical ecosystems.

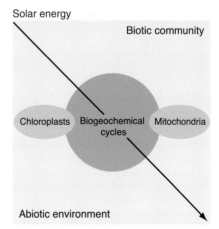

The ecosystem concept: a biotic community (structured by symbioses, mediated by cellular organelles) and the abiotic environment (developing under the influence of geophysical weathering), the components interacting in biogeochemical cycles, the whole powered by a flow of energy.

The ecosphere

The ecosphere is the global sum of ecosystems, the totality of interactions between Earth and life: lithosphere, atmosphere and hydrosphere integrated by biosphere. The ecosphere is a huge concept for a huge area – an invaluable but nearly intractable concept. Still, we can build a mental image of the volume of the ecosphere using some rough numbers and calculations.

The surface of the Earth is about 510 million square kilometres. Of that surface, about 71 per cent is ocean and 29 per cent is land. Nearly all the volume of the oceans is in the ecosphere, although obviously there is more life in some places than others. Tidal pools or mudflats, at the interface between land and sea, are often teeming with life. The average depth of the oceans is about 3600 metres; at great depth in the open ocean there is much less biology happening.

On land, there are some deserts (such as Chile's Atacama) where there is almost no resident life, but such areas are of modest size. Life is also scarce on much of interior Antarctica, although microbes are known from soil and pools of liquid water beneath the ice cap. Elesewhere, how deep is the ecosphere on land? The root zone in many places is several metres deep, with roots most concentrated where moisture is most available – in the surface metre or two. However, many CAVE AND KARST systems teem with life tens to hundreds of metres beneath the surface. Microbes are present in soil and groundwater to tens of metres deep.

On the surface of the land, the average thickness of the vegetation – surface to the top of the canopy – varies. Shortgrass steppe has an average height of perhaps 5 or 10 centimetres. The shrublands of southern California are 1–2 metres tall. The deciduous forest of England and New England is mostly 10–20 metres tall. The only way to deal with this variation is by calculating the proportional contribution to the ecosphere of the various biomes of the world. The average depth of vegetation on Earth is about 4 metres; allowing for a metre of root zone, the average thickness of the ecosphere on land is 5 metres. The volume of the ecosphere is thus $1,300,280,000 = 1.3 \times 10^9$ cubic kilometres, which is a cube about 1100 kilometres on a side, roughly the distance from Chicago to Baltimore or Paris to Rome.

Since the ecosphere is such a huge and complex phenomenon it is too big to visualise whole, let alone to study whole. Ecologists therefore simplify in order to understand. They do this by abstracting from the literally endless web of interactions that typifies the ecosphere – choosing some manageable piece of the whole to study. Such a 'studiable' piece of the ecosphere is called an **ecosystem**. The ecosystem concept is a simplifying notion – a mental tool to make the incomprehensible comprehensible.

The wonderful thing about the Earth's environment is that intriguing exceptions exist to the ecosystem model. These include ecosystems in the deep sea near thermal vents, where chemosynthetic microbes are the base of the food chain. In the absence of sunlight, the source of energy is completely different. Chemosynthesis is the process by which certain microbes create energy by mediating chemical reactions. The animals that live around these vents make their living from the chemicals coming out of the seafloor in the vent fluids.

The ecosystem concept has been invaluable in the development of ecology. The idea was proposed in 1935 by English ecologist Arthur Tansley, who urged his colleagues to look beyond plant ecology and animal ecology to see life and the Earth as a whole:

the more fundamental conception is … the whole *system* (in the sense of physics), including not only the organism-complex, but also the whole complex of physical factors forming … the habitat factors in the widest sense. It is the systems so formed which, from the view of the ecologist, are the basic units of nature on the face of the Earth. These ecosystems, as we may call them, are of the most various kinds and sizes.

However, even before Tansley, the idea of the ecosystem existed. Nineteenth-century ecologists wrote of 'microcosms' – pieces of the larger environment selected for study or even taken into the laboratory for analysis. A greenhouse is an ecosystem and so is a gold-fish bowl or an artificial pond. In Victorian times, every elegant parlour had a terrarium or other contained ecosystem. Some 4500 years ago Sumerians kept aquaria – containers with natural waters stocked with populations of fish and the producers (like algae) to support them. The earliest aquaria probably grew fish as human food, but ornamental aquaria and fishponds are known from China 2000 years ago. Today, numerous companies market sealed, self-contained 'ecosystems' that are claimed to live for years. Therefore the ecosystem is an old idea with profound modern implications: in an ecosystem, everything is literally connected to everything else.

WEBSITES

The Millennium Ecosystem Assessment – an international effort, under the auspices of the United Nations, to evaluate the state of the ecosphere, in particular the state of 'ecosystem services' (contributions of natural ecosystems to human welfare):
www.millenniumassessment.org/en/index.aspx
Ecosystems from Public Broadcasting System.org:
www.pbs.org/earthonedge/ecosystems
Missouri Botanical Garden ecosystem and biomes resources:
http://mbgnet.mobot.org

SEE ALSO

biomes

ECOTONE

We are used to seeing edges on maps that divide such regions as, say, coniferous and

The grassland biome of the Great Plains gives way to the montane forest of the Rocky Mountains, western USA.
Source: Joseph Kerski

deciduous trees, and tend to think these are sharp divisions. However, in the real world, these 'edges' are zones of transitions, ecotones. Ecotones occur in such common places as where a meadow meets a forest, or where a desert meets an oasis. Ecotones are not limited to horizontal changes across the landscape – they also occur with changes in elevation. Temperate plains give way to montane vegetation, which is followed by subalpine plants, and finally, alpine tundra.

An ecotone is a boundary or transitional zone between adjacent biotic communities and refer to transitions between local patches of habitat in the landscape or between adjacent ecoregions. Ecotones may exist on any scale. The term derives from *eco-* plus *-tonus* (Greek for 'tension', as in muscle tone or the tone produced by a taut string). In classical ecology, the belief was that there is a zone of 'tension' between adjacent biotic communities. Today there is more emphasis on the distinctive ecology of edges.

The so-called 'edge effect' has long been known. Ecotones tend to have greater species richness and perhaps higher productivity than do the biotic communities on either side. For example, the shrubby woodland between a meadow and a forest will tend to have some of the species of both communities, plus some species adapted to the 'edge'. Animals of edges have simultaneous access to the productivity of the meadow and the cover of the forest. For centuries, game managers, recognising the productivity of edges for desirable (i.e. harvestable) species such as deer and

Estuaries

Estuaries are ecotones between freshwater and saltwater, a distinctive ECOSYSTEM. Estuaries include many different types of habitats and are vital to many important species of plants, fish and other wildlife. Habitats include shellfish beds, sea grass meadows, salt and fresh marshes, forested wetlands, beaches, river deltas and rocky shores. Estuaries receive their unique characteristics by the interaction between streamflow and tidal action. Water flows downhill, carrying mineral nutrients from the land, so estuaries can be highly productive, with net primary production of 1500 g/m²/yr.

Egrets are one of the species that favour estuary environments. Source: Marine Grafics

This exceeds that of temperate forests. In fact, although they comprise less than 0.5 per cent of the area of saltwater habitats, estuaries account for nearly 7 per cent of marine production.

Rivers also carry organic matter to estuaries, so there is an import of production from upstream. As a consequence, respiration may exceed local production. Producers include both rooted plants (such as salt marshes) and phytoplankton. Consumers include fish and invertebrates, such as zooplankton, larger molluscs, crustaceans and 'worms' of various groups. Estuaries are nurseries for many kinds of marine fish, so the integrity of this shoreline BIOME is critical to the integrity of ecosystems in the open sea.

Despite their importance, estuaries are often looked upon as wastelands – too wet to walk, too shallow to sail, too salty to drink. About 60 per cent of the **Earth's** human population lives within 100 kilometres of a coastline and, because major rivers are important transportation corridors, much of that coastal habitation is along estuaries. These include the cities of London along the Thames, Baltimore along the Chesapeake, Dakar along the Sunderbans at the Bay of Bengal in Bangladesh and Montevideo and Buenos Aires along the Rio de la Plata between Uruguay and Argentina. Human land use near estuaries is often intensely residential, commercial and industrial rather than simply agricultural. This means that estuaries are frequently polluted by industrial wastes and sewage outflow, filled in to build industrial land or dredged to make harbours.

With global climate changes, estuaries will experience species shifts and toxic algal blooms. Warmer, wetter winters, coupled with more moisture year-round may lead to flooding, causing a flushing of sewage and other wastes from urban areas into wetlands and coastal marine waters. With a SEA-LEVEL RISE, salt water intrudes more frequently into estuaries, further altering the delicate balance that these ecosystems experience.

rabbits, have manipulated habitat to increase edges and hence increase numbers of their target species.

More recently, landscape ecologists have helped conservationists to understand the productivity and species richness of edges in a more sophisticated way. Disturbance increases the amount of edge in a landscape. Fragmentation of habitat patches, by building a road, for example, increases the ratio of edge per the amount of interior space. Species adapted to interiors such as vireos, ovenbirds and many species of warblers will be influenced detrimentally by fragmentation. Edge-dwelling species (such as starlings, American robins and nest-parasitic brown-headed cowbirds) will benefit from fragmentation.

WEBSITES

Press article about ecotones:
www.caledonianrecord.com/pages/
hidden_worlds/story/14d9e2f7d

The US long-term ecological research Site at
Cedar Creek natural history preserve,
Minnesota, represents the ecotone between
the tallgrass prairie, the temperate deciduous
forest, and the boreal coniferous forest:
http://cedarcreek.umn.edu/habitats/

US National Estuary Program, coordinated
by the Environmental Protection Agency:
www.epa.gov/owow/estuaries/about1.htm

UK Thames Estuary Partnership,
representing a public/private partnership
dedicated to the health of this major estuary:
www.thamesweb.com/

Chesapeake Bay is the largest estuarine
system in the eastern USA and represents a
monumental challenge in federal, state and
private cooperation in its conservation:
www.chesapeakebay.net/info/

Estuary Live – organisation dedicated to
teaching and learning about estuaries:
www.estuarylive.org

EL NIÑO

Fisherman off the coast of Peru during the late 1800s recognised that during certain years, a warm south-flowing current displaced the north-flowing cold current in which they usually fished. Because this occurred near Christmas, they named this phenomenon El Niño, Spanish for 'Christ child'. Today, it is part of the El Niño-Southern Oscillation (ENSO), a continual but irregular cycle of shifts in ocean and atmospheric conditions that affect the globe. El Niño refers to the more pronounced weather effects associated with abnormally warm sea surface temperatures in the eastern and central Pacific Ocean. Its counterpart – effects associated with colder-than-usual sea surface temperatures in the region – was labeled 'La Niña' (or 'Little Girl') in 1985.

The warm El Niño phase typically lasts for eight to ten months and the entire ENSO cycle between three to seven years. The cycle is not a regular oscillation like the change of seasons, but is highly variable in strength and timing. Understanding this irregular oscillation and its consequences for global climate became possible only after scientists unravelled the intricate relationship between OCEAN and atmosphere. Although meteorologists have long been forecasting daily weather, they had little information about conditions in the 70 per cent of the EARTH covered by oceans until arrays of fixed, unmanned, mid-ocean buoys and orbiting satellites were launched.

Because the sun heats the equatorial regions more strongly than the rest of the globe, air tends to rise from the surface there, replaced by inflow from the subtropics. The Coriolis effect turns these inflows to the right in the northern hemisphere and to the left in the southern, resulting in the great trade-wind belts that blow Equator-ward and westward over the width of the tropical Pacific. Cool water in the east and warm water in the west force winds to blow towards the west, towards warm water. Thus, the winds determine the water temperature but the water temperature also determines the winds. Because of the force of the trades, sea level at Indonesia is actually about half a metre higher than at Peru. As warm water is pushed westward, cooler water underneath in the eastern Pacific is exposed in an 'upwelling', cooling the eastern water, and starting the cycle all over again. At the upwellings, such as in the Galapagos Islands, abundant nutrients and aquatic life exists, because plankton and other aquatic life quickly take up nutrients in the water that are normally near the surface.

All this was good news to the Spanish sailors, who used these trade winds to sail from their South American colonies to the Philippines. Yet El Niño may have affected human history long before the Spanish sailors. Polynesians probably used El Niño to reach Tahiti and other eastern Pacific islands, knowing that the winds would change in a few months to ensure they could return home.

Rising air over the western Pacific cools and can hold less evaporated water, so it falls as rain. In returning to a liquid state, it releases the heat that was used to evaporate it from the ocean surface. This middle-atmosphere heating amplifies the rising motion. This becomes the principal mechanism where heat from the sun warms the atmosphere, as the atmosphere by itself is relatively transparent to solar radiation. As this warm pool pumps great amounts of heat and moisture into the upper atmosphere, this system is one of the major driving forces of world **CLIMATE**. Similar to a large rock in a stream determining the pattern of water flow, this system sets the path of jet streams (storm tracks) that control temperate-zone weather. Therefore, when the warm pool changes shape or position, the effects ripple outward to affect much of the world's weather.

During El Niño events, this entire system relaxes. The trade winds weaken and piled-up water in the west sloshes back east, carrying the warm pool with it. The region of rising air moves east with the warm pool, and so does the pumping of heat and moisture into the upper atmosphere, distorting the usual paths of the jet streams. Weakening trade winds weaken the upwelling in the east and as the warm pool moves east, the upwelled water is also not as cool as usual. The east-to-west temperature contrast is smaller, and so the trade winds weaken even further, leading to a complete collapse with essentially flat conditions across the entire equatorial Pacific.

Many assert that the engine driving climate is the heating and cooling of the tropical part of the largest ocean, the Pacific. Consider the sea breeze. On a sunny afternoon, land heats up faster than the ocean, and air rises. Air over the cooler surface of the ocean flows towards the shore to take its place. Aloft, the warm air returns to the sea, then subsides over the ocean to complete the circuit. The same principles apply to the planet. The Sun's rays strike more vertically in the tropical zones than at midlatitudes or at the poles. The tropical oceans therefore absorb much more heat. As the equatorial waters warm air near the ocean surface, the air expands, rises, carries heat with it and drifts toward the poles. Cooler, denser air from the subtropics and the poles blows toward the equator to replace it. Thus, the atmosphere and ocean together act like a global heat engine, which, modified by the planet's west-to-east rotation, gives rise to the high jet streams and the westward-blowing trade winds. The winds drive large ocean currents. In the tropical ocean, westerly trade winds pick up water vapour over the ocean and deposit it somewhere else. Therefore, the west (the Pacific coast of South America) is generally dry, while the east (Indonesia and New Guinea) contains lush jungles.

By burning **FOSSIL FUELS** at the rate we do, we increase the carbon dioxide (CO_2) concentration in the atmosphere that could cause global climate change by affecting the ENSO. Disrupting the drainage of warm water from the equatorial Pacific would reverberate through the rest of the system in unpredictable ways.

Among El Niño's consequences is increased rainfall in the east, as rainfall follows the warm water eastward. Rain and flooding increase across the southern USA and in Peru, with drought in the West Pacific, sometimes associated with devastating brush fires in Australia. Severe El Niño events have resulted in thousands of deaths worldwide, thousands homeless and have caused billions of dollars in damage. On the other hand, residents on the northeastern seaboard of the USA can credit El Niño with milder-than-normal win-

How much heat?

How much does the temperature of the central and eastern part of the tropical Pacific increase during El Niño? Between 2 and 4°C. This doesn't seem like much, but consider the heat content. Heat content can be estimated by considering the density of water (1030 kg/m^3) and the heat capacity of seawater (4000 Joules/kg/degrees Kelvin). The heat quantity increased by about 3.5e22 J (3.5 times ten to the twenty-second Joules) during the 1996–97 El Niño. This quantity is equivalent to 8e9 kilotonnes, or about 400,000 20-megatonne hydrogen bombs. Each El Niño-related heating is equivalent to the total output of roughly 1,500,000 power plants working continuously for eight months.

This argument may be misleading in that El Niño probably does not actually heat that much water. It *does* redistribute the heat already in the ocean. However, the work required to move that much water is probably similar in magnitude to the work required to heat the water. This calculation is harder to do because we can't measure the currents nearly as well as the temperatures.

ters, lower heating bills and relatively benign hurricane seasons. El Niño adversely affects higher trophic levels of the food chain; fish die off, with consequent hardship for the birds, mammals and people that depend on them. Warmer water near Central America spawns more frequent and stronger hurricanes, which can travel as far west as Hawaii.

Climatologists and oceanographers can now warn of an impending El Niño event several months in advance, providing precious time to take steps to mitigate its worst effects. If they have information about subsurface temperatures in certain parts of the tropical Pacific, they can improve their predictions of the behaviour of trade winds several months hence. Conversely, if they have information about the behaviour of the trade winds, they can predict sea surface temperatures. This may be the first step to providing the climatic counterpart to daily weather predictions that we take for granted.

WEBSITES

The US National Oceanic and Atmospheric Administration's animations and explanations of El Niño and La Niña:
www.pmel.noaa.gov/tao/elnino/
nino-home.html
Climate data from National Data Buoy Center:
http://mob.ndbc.noaa.gov/
National Academy of Sciences El Niño Resources:
www7.nationalacademies.org/opus/
elnino.html
El Niño information from the WW2010 project at the University of Illinois:
ww2010.atmos.uiuc.edu/(Gh)/guides/mtr/
eln/home.rxml
El Niño information from NOVA public broadcasting system:
www.pbs.org/wgbh/nova/elnino/

ENDANGERED SPECIES, TRADE IN

People use wild and domesticated species of plants and animals in many ways. In a typical day there are many ways in which we rely

directly or indirectly on other species. Breakfast might include oatmeal and fruit, or perhaps bacon or sausage. We naturally consume many plants and animals. Also, the material that is used to make many of our clothes comes from animals, and houses almost certainly have wood in them. Imagining a life without relying on other species is challenging.

For many plants and animals, this is not a problem – there are enough of that particular plant or animal species so that the population is not threatened. For some species, however, there are simply not enough of them (or their replacement rate is too slow) to sustain the demand. The current rate of using the species is not sustainable in the long term, or even the short term. This is particularly true for endangered and threatened species. Endangered species are those that are in imminent danger of extinction in all or a large portion of their range. Threatened species are those that are likely to become endangered in the near future. In response to these challenges to the species' survival, international agreements regulate, and in some cases make illegal, trade in endangered species.

The best known of these regulations was crafted in 1975 as the Convention on International Trade in Endangered Species of Wild Fauna and Flora (CITES). Since much of the trade involves more than one country (for example, a wild animal might be imported from an African nation to a collector in Europe), international agreement and cooperation is essential. Currently, 164 countries have signed the treaty. The convention protects to varying degrees more than 5000 species of animals and 25,000 species of plants, not one of which has become extinct since CITES was implemented. Some, however, remain perilously close. For example, there are only 400 to 500 Sumatran tigers left in the wild, and about 50 are poached (captured or killed illegally) each year.

Despite the protections CITES offers, there are still billions of dollars of illegal trade involving the trafficking of endangered species taking place each year. This trade includes both live plants and animals, as well as materials developed using parts processed from the species, such as leather from exotic animals and rhinocerus horns. Some of the best-known species involved in illegal trafficking include tigers (whose bones are used for medicines), parrots (who are captured for the pet trade), apes, rhinos, whales and rainforest plants. In fact, trafficking in endangered species is currently considered to be the third largest illegal trade in the world after drugs and weapons. One of the most famous illegal trades for decades has been in ivory – the material of elephant tusks. Although any international ivory trade must be sanctioned by CITES, a report by the agencies Traffic and World Wide Fund for Nature at the end of 2003 said that investigators found more than 4000 kilograms of 'illegal' ivory on public display in nine cities in Senegal, Ivory Coast and Nigeria. This volume of ivory represents the tusks of more than 760 elephants.

Bushmeat trade

One major category of trade in illegal species involves wild animals being captured and killed as a food source ('bushmeat'), both for local peoples' survival and as a food delicacy. The need for poor local people to eat and the economic incentives others have to capture wild species serve as powerful motivators to hunt beyond what is ecologically sustainable. Whatever the reason, when the demand for a species exceeds its ability to reproduce sufficiently, the population will inevitably decline. This poses a particular problem for threatened or endangered species that are already facing potential extinction. Recent practices expanding forest logging have served to make the problem worse, since logging roads and trucks make it easier to link wild areas with markets where the illegally captured animals can be sold.

To stop this problem, environmentally focused organisations such as Traffic and the Jane Goodall Institute advocate for steps that would reduce the bushmeat trade, including vigorous local and international law enforcement, awareness campaigns designed to reduce demand for bushmeat, and the creation of natural reserves that provide protected space for species that are victims of the practice.

WEBSITES

CITES – Convention on International Trade in Endangered Species of Wild Fauna and Flora:

www.cites.org/index.html

TRAFFIC – Wildlife Trade Monitoring Network:

www.traffic.org

Young People's Trust for the Environment factsheet:

www.yptenc.org.uk/docs/factsheets/env_facts/tradend_species.html

Jane Goodall Institute:

www.janegoodall.ca/chimps/chimps_bushmeat_a.html

BornFree Organization, dedicated to the protection and conservation of animals in their natural habitat and against the keeping of animals in zoos and circuses and as exotic pets:

www.bornfree.org.uk

World Wide Fund for Nature:

www.wwf.org

World Conservation Union:

www.iucn.org

ENERGY CONSERVATION

In 1974 the Organization of Petroleum Exporting Countries (OPEC) flexed its collective muscle by quadrupling the price of oil. This led to a fuel crisis, with long queues at the petrol stations and higher prices. Since then, people have been increasingly concerned about the need to conserve energy.

A staggeringly high amount of energy is wasted each year. For example, in the UK the amount of heat lost annually through the roofs and walls in houses is equivalent to that needed each year to heat 3 million homes. Most of this is not wasted in a deliberate or wilful way, but is simply lost through ignorance or is lost in old buildings, or ones without energy-saving devices. With an energy crisis looming, this is something that has now spurred several governments and other organisations into promoting energy conservation.

Many people believe that the real answer to meeting the rising energy demands of the twenty-first century is for everyone, whether at work or at home, to be more energy conscious. In addition, energy saving measures would help reduce environmental damage and meet many of the recommendations put forward at the Rio Earth Summit, or contained in Agenda 21 documents. (Agenda 21 is a set of ideas or policies for sustainable development for the twenty-first century resulting from the Earth Summit conference.)

However, if energy conservation is to be successful, certain myths and misconceptions need to be removed from the public's mind. Perhaps the largest myth is that businesses, heavy industry and coal-burning power stations are the only contributors to damaging the environment, and are the only ones putting harmful substances into the lower levels of the atmosphere. In fact, the average home emits more carbon dioxide than the average car during a year. Studies have shown that in the UK, one quarter of all carbon dioxide emissions come from the energy used to heat and light people's homes and power the household appliances. The figure is so high largely because of the lifestyle choices and all the trappings of modern consumer society. UK households use £1.2 billion worth of electricity on lighting every year, and a further £1.2 billion worth of electricity a year goes on cooling and freezing food and drinks. A similar pattern occurs in the USA. Therefore, energy conservation, especially in the home, is a key issue facing all more economically developed countries in the twenty-first century.

The good news is that energy conservation measures are starting to reduce the amount of energy wasted. Energy-saving habits began in the 1970s and 1980s with the installation of

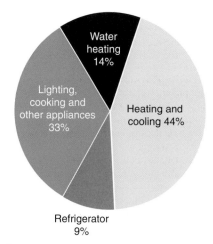

Energy use in homes in the USA. Source: http://www.eere.energy.gov

cavity wall and loft insulation in many homes. The result was a 32 per cent efficiency gain in domestic energy use, based on the amount of energy delivered per household. But even further gains could be made, for example if more people used long-lasting, energy-efficient light bulbs. In fact, if every household in the UK switched to these, it would save the equivalent of the energy produced each year by two power stations.

Some measures are obvious, such as involve turning off lights, televisions, computers and stereos when not in use, or ensuring that water heaters are covered with an insulation jacket. These could reduce energy usage in a typical home by between 5 and 10 per cent. Other simple measures could have a sizeable impact. For example, in the kitchen,

Case study

Save energy, earn money!

The *Honolulu Advertiser*, Hawaii's main newspaper, carried the following report on 9 January 2004:
'The Honolulu city government has earned more than $445,000 in rebates because of energy-saving initiatives over the past seven years', Hawaiian Electric Company said. The savings come from 110 projects that have reduced the city's energy use by 6250 megawatt hours a year, enough to power 780 homes.

Steps include using LED lights instead of traditional incandescent lamps in traffic signals. 'Our goal at the city is to reduce electrical demand in city facilities by 50 percent by the year 2010,' said city Managing Director Ben Lee.

energy can be saved if ovens were turned off a couple of minutes before cooking time ends so that the dish uses the heat already present in the oven. Rather than using tumble dryers, it would be better if clothes were dried outside on a clothes-line. Even on the road, simple energy-saving measures could be used, starting with making sure that the tyres on vehicles are properly inflated. This will improve mileage by a further 5 per cent, and reduce the amount of petrol consumed.

Many of these measures are simply good habits, and measures that were common practices in earlier generations. As with so many environmental issues, it is people that are the real problem. Only when the queues at the petrol station increase or when power cuts lead to blackouts at home do people realise how much we are reliant on energy for so many things.

WEBSITES

www.saveenergy.co.uk/why

www.eere.energy.gov/consumerinfo/
energy_savers

www.est.co.uk

SEE ALSO

combined heat and power, energy crisis

ENERGY CRISIS

Imagine waking up one morning and going into the bathroom to find freezing cold water coming out of the shower. Going downstairs to the kitchen, you find that there is no chilled fruit juice in the fridge, and everything inside has been sitting at room temperature overnight and is starting to spoil. You try to turn on the kettle to make a cup of tea, and nothing happens. There's no newspaper in the letter-box, and neither the TV, radio nor computer can be turned on. The toaster does not work, so you take a bowl of dry cereal before you drive off in the car. It is rather dangerous since none of the traffic lights are working. None of the petrol stations have any petrol so you return home before you run out of fuel.

It may sound a bit like a strange nightmare, but this could be a real scenario if scientists are correct and energy supplies run out. Such a prospect is looming simply because demand for energy, especially those derived from non-renewable sources, has risen quite dramatically since the end of the Second World War. Since this time, more countries have industrialised and modernised, leading to a rise in incomes, an improvement in standards of living, and increased usage of fuels in both the home and at work. Urbanisation and the development of modern infrastructure have also seen motor vehicles replace horse-drawn carts, adding even further to the demands for energy.

People in the developed world rely heavily on **FOSSIL FUELS** – including coal, petroleum and natural gas. Taken together, these meet more than 85 per cent of global energy demands. But the reserves of these fuels are not infinite, and scientists predict that within the next two centuries we will run out of these valuable energy sources. If world leaders are to be believed, we have already entered a global energy crisis. In August 2001 US Energy Secretary Spencer Abraham issued a dire warning 'The failure to meet this challenge (the energy crisis) will threaten our nation's economic prosperity, compromise our national security, and literally alter the way we live our lives'.

This is not the first time that people have faced an energy crisis. Back in the 1970s the Organization of the Petroleum Exporting Countries (OPEC) started to push up the price of crude **OIL**. For example, a barrel of crude oil from Abu Dhabi rose from US $2.54 in 1972 to US $36.56 in 1981. As well as rising prices, the oil crisis of the 1970s was fuelled by political factors, as many of the oil-rich Middle Eastern countries reduced their exports to Western nations as a response to Western involvement in the Arab–Israeli conflicts in the late 1960s. The USA was particularly affected, as around 70 per cent of its oil imports in 1977 came from OPEC countries. Rising prices and a drop in supply led to long queues of motorists at petrol stations, as well as prompting panic amongst investors, and the onset of a trade recession and rampant inflation.

Modern society is so dependent on a regular and uninterrupted supply of energy that anything that threatens the status quo and peoples' standard of living is treated with great concern. However, not everyone agrees with this 'doom and gloom' stance; many point to domestic energy supplies within many more economically developed countries (MEDCs). Indeed, there are vast reserves of oil in areas such as Prudhoe Bay, on the north coast of Alaska. However, the cost of oil drilling in such a fragile and environmentally sensitive area is high, and the Trans-Alaskan pipeline running south to the ice-free port of Valdez was one of the most expensive engineering schemes ever undertaken in the USA.

Consumers in the future may be unwilling to pay such high prices. Many people consider that the key issue is to reduce demand by, for example, embracing measures of **ENERGY CONSERVATION** and trying to reduce the wilful waste of expensive and ultimately finite sources. If demand is not reduced, the only alternative will be higher prices and more expensive schemes such as the Trans-Alaskan pipeline, unless of course **RENEWABLE ENERGY** supplies take off and people turn instead to wind, solar or tidal power.

At some stage in the future, there may be no alternative but to use these renewable sources as the fossil fuels eventually run out, leaving the oil crisis of the 1970s as a distant but painful memory.

WEBSITES

'Energy Matters', with information about the 1970s oil crisis and background details on the Middle East crisis:

http://library.thinkquest.org/20331/history/mideast.html

ABC News, giving a recent American perspective:

http://abcnews.go.com/sections/business/DailyNews/energycrisis_010423.html

SEE ALSO

energy conservation

ENVIRONMENTAL ACTIVISM

Environmental activism is a complex and diverse set of activists, sympathisers, organisations, campaigns and ideas. The environmental movement is political and social action directed towards the preservation, restoration or enhancement of the natural environment. It includes powerful international groups,

numerous national environmental organisations and thousands of local groups. There are full-time activists, occasional participants, financial supporters and passive sympathisers. There are individuals and groups that try to live lifestyles with low environmental impact. While there is an enormous range of viewpoints among environmental campaigners, most seek to protect commonly owned, or unowned, resources for the benefit of future generations.

Much environmental activism is directed towards conservation, as well as the prevention or elimination of **POLLUTION**. The conservation ethic is an ethic of resource use, allocation, exploitation and protection. One concern common to most types of environmentalism is opposing pollution. In this sense, most people in the world are 'environmentalists' since nobody wants to breathe air choked with fumes or drink dirty water.

Environmentalism is activism usually based on the ideology of an environmental movement, often taking the form of public education programmes, advocacy, legislation and treaties. The conservation movement was an invention of American John Audubon and others who invoked Christian reverence for the Creation to protect natural habitats from humans during the 1800s. They lobbied consistently for parks and human exclusion from 'the wild'. The conservation movement gave rise to the global ecology movement in the 1970s, from which came Green political parties in many democratic countries. Its views on people, behaviours, and events centre around the political and lifestyle implications of ecology and the idea that nature has a value in itself. The worldwide Green parties are committed to the following four pillars: ecology, social justice, grassroots democracy and nonviolence.

'Tree hugging', sometimes associated with environmental activism, received its name from the practice of some environmentalists sitting in trees to prevent loggers from chopping them down. This and other nonviolent action used by environmentalists include rallies, street theatre and symbolic actions such as dumping nonrecyclable containers on the steps of the manufacturer, blockading shipments of rainforest timbers, sitting in front of bulldozers and occupying development sites. Conventional techniques are also used, including public education, letter writing campaigns, speeches, lobbying, advertising, drafting legislation, making submissions and sueing polluters. A few environmentalists use sabotage, but with a strong commitment to avoid harm to humans.

The environment is inextricably bound up with politics, society and culture. Therefore, environmentalism is usually linked to broader issues. For example, the global peace movement seeks to end violence, usually through pacifism, nonviolence, diplomacy, boycott and ethical purchasing. Environmentalism is controversial between environmentalist groups and the broader society, but also within the groups themselves when they disagree about methods and agenda.

Direct action is a method and a theory of stopping objectionable practices or creating more favourable conditions using immediately available means, such as strikes, boycotts, workplace occupations, sit-ins or sabotage. Direct action often involves some form of civil disobedience.

Most radical environmentalists take the view that destruction of the environment is fast approaching a crisis and that if serious changes are not made in the relationship between humans and the natural world, the health of the planet and many living species are at grave risk. Deep Ecology promotes a reconnection with ecological values, encouraging actions that can lead to experiences that transform old ways of thinking and develop a more spiritually connected way of living.

Political ecologists are those who take issue with new untested technologies. More extreme members are sometimes called Gaians. Gaians believe that traditional methods of social change are insufficient for achieving necessary changes in the relationship between humans and the environment. A Gaian is a radical Green who claims that since we live as part of one planet's photosynthesis chain and are trapped within its gravity well, we are components of one large body – the global ecology of the Earth itself. Radical environmentalists resort to non-traditional activism that is often illegal and at times may be anarchistic. Anarchism advocates the elimination of the state in favour of a society based on the voluntary co-operation of free individuals.

Green anarchism suggests that ecological unbalance, economism, authority-power, international injustice, war, racism and centralism are a direct result of a cruel, industrialistic rule of civilisation. Green anarchists practise an underground resistance based on autonomous direct actions, uniting different system-critics such as nonviolence, ecoactivism, queeractivism, animal rights activism, anti-capitalism and anti-elitism. Eco-anarchism argues that small villages of no more than a few hundred people are the correct scale of human living.

Some environmental activists assert that capitalism does not work well to handle environmental problems, partly because the costs of environmental impacts are seldom included in the costs of production. For example, there is no simple market mechanism to make automobile manufacturers pay for the costs of ill health due to vehicle emissions, traffic accidents, use of land for roads or greenhouse warming. Profits are privatised (captured by owners and users) while the environmental and health costs are 'socialised' – borne by society as a whole. Others point out that socialist economies such as the former USSR caused enormous environmental problems that were just as harmful as those caused by capitalist societies. They argue that the core problem is modern industry itself and not the economic system in which it grows. Natural capitalism is a set of trends and economic reforms to reward energy and material efficiency.

Another school of thought, whose most prominent exponent is Paul Ehrlich, says that overpopulation is the prime culprit. Still another perspective, championed by Barry Commoner, is that use of new technologies is the driving force behind environmental assaults: even with the same population, new chemicals, for example, cause more far-reaching impacts.

Environmental activism's success stories include preventing the creation of a massive fleet of supersonic transport aircraft, limiting production to a few Concordes, and the shutting down of most of the world's whaling industry. Forestry campaigners have slowed unsustainable and damaging forestry operations across the globe, while anti-motorway protesters have challenged the expansion of road systems. Opponents of nuclear power have slowed the world's nuclear power industry. Campaigners have pushed for controls on the production of carbon dioxide emissions to prevent global warming. Local citizens have stopped innumerable commercial developments, roads, and saved open space.

Rachel Carson's famous book, *Silent Spring*, alerted the world in 1962 to the dangers of pesticides and was a key stimulus for the formation of the environmental movement itself. Critics of pesticide use have used investigation, education, publicity, lobbying, lawsuits, meetings and the promotion of alternatives.

By being concentrated, remote, run by large organisations and backed by state power, nuclear power became an ideal target for nonviolent action in the late 1960s. The movement has been highly participatory and played an important role in increasing the conscious use of nonviolent action.

WEBSITES

Environmentalism examples and definitions: http://encyclopedia.thefreedictionary.com/environmentalism
Environmental activism from About.com: http://environment.about.com/od/activism/
Strategising for environmental activism: www.environmentaladvocacy.org/resources/articles_papers/Bringing_About_Change.pdf
Description of environmental campaigns from *Nonviolence versus Capitalism:*
www.uow.edu.au/arts/sts/bmartin/pubs/01nvc/nvc09.html
Friends of the Earth campaigns: www.foe.co.uk/campaigns/
Planet Ark: www.planetark.com/planetark.cfm
UK version of the Planet Ark site: www.recyclingnearyou.co.uk
Encams: Keep Britain Tidy: www.encams.org/AboutEncams/KeepBritainTidy.asp?Sub=0&Menu=0.26.11.34

ENVIRONMENTAL RESISTANCE

Populations adjust to the size of the resources available to them in the environment. Environmental resistance is the sum of factors that operate to keep observed rates of increase below theoretical maximum rates of increase. These include predators, limitations on food

and water, pollution and natural hazards.

Humans play an increasing part in determining the population of species, either via domestication or as a consequence of their actions. Darwin identified that species tend to produce more offspring than the external environment can support, so effectively, homeostatic control comes into place as the maximum population of a species is the 'norm', taking into account that one of a species' primary reasons for being alive is to reproduce.

However, some species are unable to increase their population levels through breeding. This is most usually because of climatic change, or human-caused environmental damage. As a result of actions such as when humans steal the eggs of rare birds or when rainforests are cut down, species are endangered towards extinction. When species are endangered, the population of a species becomes critically low. The endangering or extinction of a species can unbalance the **FOOD WEB** in an area, directly placing other species in the **ECOSYSTEM** under threat.

For an oak tree in a bottomland forest, environmental resistance would include fox squirrels (which eat the acorns), cottontail rabbits (which strip bark from saplings), wandering cattle (which trample seedlings), leaf-mining insects (which take energy that might have been diverted into reproduction and to make acorns), fire, flooding, drought and the farmer's chainsaw. For the fox squirrel,

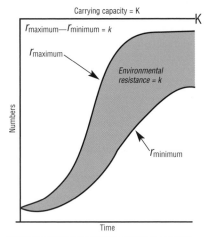

Environmental resistance is the difference between maximal growth rate and observed growth rate

Case study

Venezuela and environmental resistance

An example from Venezuela illustrates the interconnectedness of these environmental concepts. In 1986, a dam in Venezuela was constructed, flooding over 2,590 square kilometres and turning hundreds of hilltops into islands. The islands ranged in size from less than 1 hectare to more than 150 hectares. The larger the island, the fewer species it lost. Within eight years, on the smallest islands, 75 per cent of the species that had previously lived there had died. However, all of the islands, even the largest, lost their top predators, such as pumas, jaguars and eagles. The animal species that did remain, mostly herbivores and small carnivores, greatly increased in population because of a reduction in competition for resources and the fact that predators were no longer eating them. Intense grazing by the increased herbivore populations is now degrading the variety of plant life on the smaller islands.

environmental resistance would include red-tailed hawks and feral cats, ticks, fleas, mites, bacteria, viruses and parasitic worms, late spring snows (which chill and kill the nesting young), the farmer's shotgun and (indirectly) the same factors that hindered the oak. In a way, we may think of reproductive strategy as the population's general evolutionary response to the fact of environmental resistance.

When we think of environmental resistance, we are referring to SYMBIOSIS ('life-together', and especially predation and parasitism), and also the perils and rigours of the physical environment. The sum of these factors is illustrated by the figure on page 105. A population under ideal conditions would be expected to grow as described by the curve labelled $r_{maximum}$. Under real conditions, population growth actually follows a pattern described by the curve labelled $r_{minimum}$. The difference between the curves defines environmental resistance.

As is evident elsewhere in this book, the environment is continually changing. Not surprisingly, environmental resistance changes from time to time and from place to place. It is a property of the population's environment that acts as a selective force, shaping the reproductive strategies of populations by selecting as grandparents those individuals best able to cope.

WEBSITES

Checks on population growth, including discussion on environmental resistance:
http://users.rcn.com/jkimball.ma.ultranet/ BiologyPages/P/Populations2.html

Population Ecology with discussion on environmental resistance:
http://ipmworld.umn.edu/chapters/ecology.htm

ENVIRONMENTAL STUDIES

The increasing concern about changes in the environment and the influence that humans have had on these changes led to the development of environmental studies. Environmental studies grew during the latter part of the twentieth century into a science in its own right in universities. It developed its own research base with its own methodology, although it also has strong linkages to physical sciences, social sciences, humanities and law. Environmental science and environmental studies also began appearing in secondary schools as separate subjects, and environmental studies is even taught to some extent in most primary schools.

Today, the topics studied under the umbrella term of environmental studies are as diverse as the Earth that is being studied. Environmental studies includes all the topics in this book, as well as:

- critical environmentalism
- environmental thought
- ecofeminism
- deep ecology
- social ecology
- environmental education
- environmental communication

- biological conservation
- natural history
- environmental history
- human ecology
- environmental ethics
- risk policy and perception
- hydrology
- environmental policy
- chemistry
- environmental law
- social movements
- political ecology
- environmental sociology
- environmental psychology
- native studies
- animal rights and welfare
- environmental justice
- technology and cultural studies
- gender, labour, race and the environment
- geography
- geographic information systems
- remote sensing
- international development
- sustainability
- development
- public participation
- ecocriticism
- environmental literature.

Environmental studies on university campuses may be housed in departments of geography, sociology, education, natural science, social science or other departments, or exist as its own department. It is, by nature, an interdisciplinary subject that seeks to understand the environment in a holistic way.

Most environmental studies programmes arose from the appreciation of human dependency on the Earth's natural resources, and from the recognition that humans will only preserve and improve what they understand. Proponents of environmental studies believe that if people did not have the opportunity to study the environment they would not understand, appreciate, or feel stewardship toward the environment. Therefore, education is thought by many to be more important than all the pollution regulations and recycling efforts in protecting the environment.

Environmental studies focuses on the relationships between people, culture and nature. These studies recognise that the quality of human interactions with the biosphere is critically important to sustaining a high quality life. Environmental studies is not simply the examination of the Earth as a system. Even more than other disciplines, environmental

Case study

A century ahead of its time: National 4-H

Millions of times each year, students in the USA make the pledge: 'I pledge my Head to clearer thinking, my Heart to greater loyalty, my Hands to larger service, and my Health to better living … for my Club, my community, my country and my world'.

These students are members of National 4-H. In 1902 progressive educators began to emphasise the needs of young people and to introduce nature study as a basis for better agricultural education. National 4-H was established as a part of the USA's Department of Agriculture Cooperative State Research, Education and Extension Service. Boys' and girls' clubs and leagues were established in US schools and churches to meet these needs. To spark the interest of young people, Farmers' Institutes cooperated with school superintendents by promoting production contests, soil tests and plant identification. By 1904 several boys' and girls' clubs had already exhibited projects. Most states organised clubs outside the schools, with rural parents acting as volunteer leaders and US Department of Agriculture County Extension agents providing materials.

Public support and enthusiasm for 4-H grew throughout the country, with its emphasis on education, 'learning by doing' and the development of young people. The form that 4-H takes today addresses the changing issues and diverse backgrounds of today's youth. Nearly 7 million youth participated in 4-H activities during 2002. Besides environmental education, which attracted 1.4 million students, activities included citizenship education, communications, expressive arts, consumer and family services, healthy lifestyle education, personal development and leadership, plants and animals, and science and technology.

studies emphasises values education – encouraging the development of values that honour people, culture and environment. Most people who enter environmental studies as a career are motivated to make the world a better place to live in.

Environmental career opportunities are growing in corporations, non-profit organisations, private industries, and tribal, federal, state and local governments. These careers include environmental law, environmental management, public health and many more.

Much of environmental education develops through grassroots efforts and, therefore, much of it remains outside the walls of formal education. An interesting example of informal environmental education is Earthwatch Radio. This involves a series of two-minute reports that are scientifically accurate and clearly written, given to broadcasters who agree to use the programmes regularly and broadcast individual programmes in their entirety. Currently, Earthwatch reports are heard on over 125 radio stations in Canada, the USA and Costa Rica.

WEBSITES

National Association for Environmental Education UK:

www.naee.org.uk

North America Association for Environmental Education:

http://eelink.net

Society for Conservation Biology:

http://conbio.net

Environmental Studies Association of Canada:

www.thegreenpages.ca/esac

Australian National University's Centre for Resource and Environmental Studies, an example of a well-developed environmental studies programme:

http://cres.anu.edu.au

Earthwatch Radio:

www.ewradio.org

National 4-H:

www.4-h.org

www.national4-hheadquarters.gov

National 4-H Council:

http://n4h.org

EUTROPHICATION

Eutrophication, or nutrient pollution, can be an entirely natural process or it can be caused by human activities. Natural eutrophication, which involves a slow build-up of nutrients, is the process by which lakes develop over time to become more productive and mature ecosystems. In recent times, however, the term eutrophication has been more commonly associated with the impact of chemical fertilisers used in intensive agricultural systems. As excess nutrient-rich chemicals pollute water sources, such as ponds and rivers, water-based vegetation thrives, literally 'clogging up' the system.

The most common result of eutrophication is excessive growth of algae, leading to algal blooms. The water takes on a green or dark, murky appearance. It appears unhealthy, lacking obvious life and movement, as the algae 'suck' oxygen out of the water. Algae are short-lived plants and when they die they sink to the bottom of the water, where their decomposition takes up yet more oxygen. Some water becomes anoxic – completely devoid of oxygen. The decomposition process releases hydrogen sulphide, a gas poisonous to plants and animals. Lakes and ponds suffer more than streams because they are still bodies of water.

All insects and animals require a reliable oxygen source. When this is reduced due to algae and similar plants absorbing it, the wildlife simply dies – slowly. Eutrophication can cause harm to the whole ecosystem. If one element suffers, then so do those that are its dependents.

Today eutrophication is present in the Baltic Sea. Almost totally enclosed by land, water movement is limited. Eighty million people live in the countries on the shores of the Baltic Sea and excessive amounts of nutrients entering it are disturbing its ecological balance. Some of the algal blooms are themselves poisonous and so threaten the sustainability of the Baltic Sea's ecosystems. In the northern Baltic Sea the transparency of the water was reduced by 3–4 metres over the period 1930–80 as a result of algal growth.

Weather conditions influence the speed of the eutrophication process. Mild, wet winters accelerate it as increased rainfall and run-off wash larger amounts of nutrients into water

Case study

Eutrophication in the EU

Denmark and the Netherlands are the countries most affected with eutrophication in the EU. This is because both countries have large river systems and lakes surrounded by intensively farmed land.

In Denmark and the Netherlands intensive livestock systems dominate agriculture. Pig and poultry rearing, and an emphasis on dairying, produce huge volumes of animal manure. Given its high nutrient content, one method of using this material is to spread it on fields. It acts as a fertiliser and improves soil structure, but at the same time autumn and winter rains leach the nutrients out into the local hydrological system.

Simply to replace manure with chemical fertilisers is not a solution – they, too, lead to eutrophication when used in large quantities. Strategies have therefore focused on reducing the amount of leaching. Land left bare allows more water to pass through it, taking nutrients as it goes. Keeping land covered all year round and ploughing in stalks in spring are two workable approaches to reducing leaching.

The ultimate solution must be to move towards alternative farming systems. Organic farming is less intensive and does not involve any chemicals. The inputs of manure are lower per unit area. Crops use up these additional nutrients, so relatively little is washed out of the soil. However, to retain farmers' incomes, such systems would probably need to be subsidised.

bodies. During the following spring algae grow even faster. Floods also increase inflows of nutrients. For example, in 1997 floods washed huge amounts of nutrients into the Baltic Sea. During that summer the surface accumulations of blue-green algae were the highest ever recorded. As a consequence of these floods BERNET (the Baltic Eutrophication Network) was set up in 1999, its aim being to improve the management of eutrophication in the Baltic Sea.

WEBSITES

www.nemp.aus.net

www.bernet.org

www.umanitoba.ca/institutes/fisheries/
eutro.html

www.ucd.ie/~app-phys/stuart/EUTRO.HTM

www.helcom.fi/environment/
eutrophication.html

SEE ALSO

lakes

EXTENSIVE FARMING

The cowboy state in the heart of the Rockies, Wyoming, is the ninth largest state in the area but the smallest in population (498,703) and in population density (1.9 people per square kilometre) of any US state except Alaska. The spirit of the Wild West is alive and well with its open spaces, rugged country and stunning scenery. In Wyoming cattle ranching is an important part of the economy and one of the world's largest rodeos – Cheyenne Frontier Days – is held annually in July. Visitors to Wyoming can stay on working ranches to experience the life of a real cowboy, driving cattle, branding and roping steers, eating around a campfire and sleeping under the stars.

Ranching is an example of extensive agriculture. It involves large areas of land but generally has low inputs of labour, capital and machinery per hectare. In common with other forms of extensive agriculture, such as hill sheep farming in the UK, yields per hectare are low compared with those on intensively farmed land. Often extensive farming systems are an adaptation to hostile environmental conditions, such as poor-quality soils, inhospitable climatic conditions or sparse vegetation.

In Wyoming ranching takes place in the valleys and mountains to the west and also on the lower-lying prairie lands further east. Average rainfall is low, about 350 millimetres per year, and winters can be long and cold, especially in the highlands. The rough higher ground is used for grazing and some hay may be grown for fodder in the lowlands. The cattle tend to be taken to the lowlands during the winter. The carrying capacity of the land is low – each hectare giving one month's grazing for one cow. Most cattle are sold at under one year old to be sent for fattening to other areas. Very few ranchers make large profits and many supplement their incomes with pony trekking and other tourist ventures. The farms are large, often over 4000 hectares, although cattle numbers may only be around 500 per farm, mostly Hereford and Angus breeds.

If farmers try to increase yields on marginal land environmental degradation can occur. Overgrazing or overcultivation can lead to soils becoming friable and soil erosion by wind or water can result. The Dust Bowl in the USA in the 1930s is a vivid example of the consequences of working the land beyond its sustainable limits. In some parts of the less developed world, such as in Kenya, farmers have been forced on to marginal land by commercial agriculture, often supported by governments keen to improve their foreign

Case study

Hill sheep farming in the UK

The uplands of the UK, such as the Pennines, Lake District and Scottish Highlands, are marginal farming environments. These environments are not conducive to crop growing; the growing season is too short for most crops, and temperatures are not high enough in summer for ripening. The high levels of precipitation, often over 1000 millimetres per year, cause most crops to rot, and the steeply sloping land prevents the use of machinery. Soils are thin, acid and stony, and only support poor quality moorland with heather and rough grassland. The land is only suitable for grazing animals and largely restricted to sheep, since the grass is too poor in quality and the slopes are too steep for cattle to negotiate.

The sheep – particularly the hardy breeds such as Swaledales and Herdwicks – are sturdy and surefooted, and able to cope with the steep slopes, the uneven terrain and the extremes of weather. They graze on the open moors and hills and are only brought to the lowlands for lambing, shearing, dipping and to avoid the worst snowfalls of the winter.

The farms commonly comprise three main areas of land:

- The inbye: the best land, usually on the valley floor and close to the farm buildings. This land is more fertile and sheltered than the land on the valley sides. The land is used to graze animals in bad weather and during lambing and shearing. In the summer this land may be used for haymaking or turnip growing to supplement winter fodder.
- The intake: the land on the lower slopes of the fells (higher open land), so called because it is surrounded by dry stone walls and may also have been slightly improved with fertilisers and drainage pipes and hence 'taken in' from the hills. This land is exposed to the weather conditions and is moderately steep, with acid soils leached by the heavy rain.
- The Fell: open land often over 300 metres above sea level. This land has no walls and is beyond the intake land. This is where the sheep graze for most of the year. It is common land, meaning that the grazing rights are shared with other farmers in the parish. As a result, each farm may have access to a land area well in excess of 1000 hectares, but the inputs and the outputs are low.

The farms are usually family run, with the use of hired labour at key times in the year such as when sheep shearing is contracted out to others. There is therefore little investment in labour. Machinery may extend to a quad bike (a four-wheeled motorbike that is capable of negotiating steep and uneven terrain) and the occasional tractor. In general, upland environments can support one animal per hectare of land. On most farms sheep and lambs generate over 90 per cent of the farmers income, by wool, the sale of lambs and meat. In recent years lamb and wool prices have collapsed, leaving hill sheep farmers as some of the poorest workers in the country and requiring considerable subsidies and income support from the EU's COMMON AGRICULTURAL POLICY. Other hill sheep farmers have sold their farms and retired.

earnings by producing cash crops. Extensive farming is a reasonable response to difficult farming conditions, but it is sustainable only as long as farmers do not push the boundaries too far.

WEBSITES

Further information on hill sheep farms and a case study:

www.face-online.org.uk

Information about destroying rainforests for cattle ranching:

www.rainforestweb.org/
Rainforest_Destruction/Cattle_Ranching

Article tracing the history of cattle ranching in the USA:

www.bchm.org/wrr/recov/p3.html

Information about extensive cattle farming in Burkina Faso:

www.burkinafasolink.org.uk/agriculture.htm

SEE ALSO

intensive farming

FAMINE

Periodically the media portray terrible scenes of starvation and misery, often in African countries such as Sudan and Ethiopia. Charity organisations urge us to help people in desperate situations. Famine grabs people's attention when a significant number of deaths result from starvation.

Famine is caused by food shortages, due to conditions such as drought, flood, war, or other human-caused or natural disasters. Regions that are heavily dependent on local agriculture are particularly vulnerable to famine, since anything that upsets the growing process will have a direct effect on the ability of the region to provide the food needed to support it. Without established means of bringing food into a region, many people will suffer. For example, the Sahel region of Africa – a band of dry grasslands at the southern edge of the Sahara desert, encompassing Sudan, southern Ethiopia and Chad – is home

> **Undernourishment and malnutrition**
> While periodic famines generate most media attention, ongoing problems of food adequacy lead to more disease and death than the more publicised instances of famine. Undernourishment (receiving fewer calories than needed) and malnutrition (receiving enough calories but too few specific essential nutrients such as vitamins and proteins) are two separate but related problems deriving from inadequate food sources. Each presents ongoing challenges for aid workers, as each can lead to diseases such as marasmus (emaciation caused by a diet low in calories and proteins) and kwashiorkor (malnutrition caused by protein deficiency). One of kwashiorkor's characteristic symptoms is the swollen belly often associated with starving people. Both marasmus and kwashiorkor can be treated by the restoration of an adequate diet, if the person's health has not declined too much as a result of the disease.

to 50 million people. Twenty years of subnormal rainfall beginning in the 1960s led to drought conditions and ultimately to more than 100,000 deaths by starvation. Subsequent droughts in the mid-1980s and early 1990s led to even more famine, since the already stressed food crops continued to be deprived of much needed water. Ethiopia suffered even more as a result of civil war that disrupted food supplies.

As with many other environmental issues, changes in natural conditions only partially contribute to most cases of famine. Clearly, persistent drought in the Sahel region was a major factor. However, it is also important to consider the role of rapid population growth, which puts more people in need of a limited food source. The issues are complex – rapid population growth is a result of poverty, not just a cause of poverty. Similarly, in the case of the Irish potato famine (described in the case study) food was available within Ireland. Unfortunately, the prevailing economic and political rationale was to export food to those capable of paying a higher price, which played a significant role in contributing to the deaths

Case study

The Irish potato famine

The Irish potato famine is one of the world's most famous famines, which took place between 1845 and 1850. For the preceding 200 years the potato had been a primary food source in Ireland – nutritious, easy to grow (well suited to the cool and wet climate) and easy to store. Unfortunately, blight spread over the potato crops, rotting them in the fields as they grew. This posed a major problem, since many Irish people depended on the potato for their primary food source. Other possible food sources, such as corn, wheat and oats, were exported to Britain for sale at a much higher price than the increasingly poor, starving Irish could pay. Compounding the misery, many Irish citizens were evicted because – as a result of the crop failure – they could not afford to pay the

A family evicted by their landlords. Courtesy of the National Library of Ireland, Lawrence collection ref LROY1767

rent to their British landlords. As conditions worsened, a wave of cholera and typhus spread throughout the land, killing many people. In all, as many as 1 million Irish – one out of every nine – died in the famine, and Ireland lost many other people to emigration, mainly to Britain and the USA.

of many farmers. Balancing individual rights and societal needs is never easy. Environmental issues add yet another degree of complexity to the mix, since natural variations in weather, disease or plant growth can wreak havoc with intended social arrangements.

Further reading

Raven, P., and Berg, L., 2004, *The Environment*, 4th edn, John Wiley and Sons.

WEBSITES

Famine Early Warning Systems Network:
www.fews.net
Food and Agricultural Organization (FAO)
of the United Nations:
www.fao.org
FAO's State of World Food Insecurity Report:
www.fao.org/DOCREP/X8200E/
X8200E00.HTM
World Food Program:
www.wfp.org
United States Environmental Protection
Agency, Irish Potato Famine page:
www.epa.gov/history/topics/perspect/
potato.htm
World Health Organization:
www.who.int/en
World Health Organization Nutrition
Information:
www.who.int/nut

SEE ALSO

population increase

FARM DIVERSIFICATION

Farmers or bed and breakfast providers? Or golf course managers? Or workshop owners? Farmers today are diversifiying – some in a very big way. Diversification describes the ways in which farmers in many developed countries have added new activities to the farm in order to increase their profits and to continue to have a reasonable standard of living. Diversification is needed to strengthen the farming industry and to protect it from variable market prices that make specialist farms concentrating on only one activity very vulnerable. The addition of new activities increases farm earnings and employment opportunities in rural areas.

The new activities carried out by the farmers may or may not be agricultural. Farmers in member states of the EU can access a range of grants that have been made available to support the diversification process and to promote farmers as guardians of the countryside, such as grants from the Rural Enterprise Scheme.

Examples of diversification – not all agricultural – include:

WEBSITES

The Scottish Agricultural College on diversification:
www.sac.ac.uk/management/external/diversification
Diversification in Northern Ireland:
www.ruralni.gov.uk/ruraldev/diversification

growing new and different crops such as oilseed rape and linseed	dry stone walling (field boundary walls characteristic of parts of northern England) and wildflower promotion	sports activities, for example horse riding, quad biking, mountain biking and grass-boarding
tourist activities, for example opening the farm to the public, offering bed and breakfast or farm-based holidays	organic farming	farm shops and small-scale processing industries, for example farmhouse butter and cheese making
		tree planting
rare and different animal breeds, such as deer, goats, rabbits and even snails	fish farms	wind turbines
		clay pigeon shooting, model aircraft flying
golf courses	'pick your own' fruit growing enterprises	arts and crafts workshops, for example candle making, stick dressing and woodturning
		conservation projects such as meadow reseeding and pond building

Case study

Lower Gill Farm in the Forest of Bowland, Lancashire, UK

The owner of Lower Gill Farm, Mr Wilson, describes how he diversified the farm into holiday lets and a joinery workshop:

The farm was beginning to lose money and I was finding the work increasingly difficult. I needed to scale down the agricultural production. I also wanted to improve the landscape and develop the farm for visitors. We've converted derelict farm buildings and a Grade-II-listed farmhouse into five holiday cottages, an indoor swimming pool and a joinery workshop. The farmhouse conversion won an architectural design award, which we were very pleased about.

We have developed areas to attract ground-nesting wading birds and to increase biodiversity. We have also restored 3 kilometres of hedgerow and 60 metres of dry stone wall. Two new ponds have been created and 1.9 hectares of native woodland have been planted. The public can enjoy access to the woodland along a new footpath across the farm. I have managed to reduce the farming side of the business and have increased my income from the workshop and holiday cottages. I am much happier – and the environment has improved too.

FERTILISERS

Yellow leaves? No flowers or fruit? Yellow spots on leaves? Stunted growth? All these symptoms are not necessarily plant diseases but could be the result of a poor growing medium – the soil. If the soil is not in good health then plants do not grow well. One of the main roles of farmers is to keep the soil healthy or plants will not be productive – causing profit margins to fall substantially.

Plants need carbon dioxide and water for **PHOTOSYNTHESIS** but they also need a variety of nutrients. For example, nitrogen is a macronutrient that is needed in relatively large quantities to make amino acids. Others, the micronutrients, are needed in very small amounts. Plants starved of any of these nutrients show deficiency symptoms. For example, plants starved of nitrogen have stunted growth and older leaves turn brown.

When plants are harvested the nutrients are harvested with them. This interrupts the natural system where plants would die and return the nutrients to the soil. Hence over time continuous harvesting depletes the soil of valuable nutrients and the soils progressively become less fertile. As a result farmers need to use **CROP ROTATION** techniques and/or fertilisers to keep the soil healthy and maintain productivity. Fertilisers may be either organic, such as farmyard manure, or inorganic, that is manufactured. In the developed world the drive to intensify food production has meant that most farmers have relied almost exclusively on the use of fertilisers, and predominantly inorganic fertilisers. However, recent trends towards more **ORGANIC FARMING** are leading to a return to the use of crop rotation and organic fertilisers.

While organic fertilisers such as animal manure and sewage sludge may contain important macronutrients, they also add organic matter that improves soil structure and water-holding capacity. This makes the soil less susceptible to erosion by wind and water. The use of manure is also a good way of recycling the manure produced on mixed farms.

Inorganic fertilisers, such as ammonium nitrate, are manufactured products and contain concentrated amounts of macronutrients. They can be applied very easily by machine and, because they are more concentrated, they can be applied in small amounts. However, because these fertilisers do not contain organic matter, soils become starved of humus, making them drier and more prone to erosion. (Humus is decomposed vegetation, often black in colour and very rich in nutrients.) The fertilisers may also contain impurities such as lead, zinc and arsenic. These may build up to harmful levels in the soil.

Inorganic fertilisers are much more easily leached or washed out of the soils and this leads to the danger of high concentrations of fertilisers in watercourses, which causes eutrophication and its consequent adverse effects on freshwater ecosystems. **EUTROPHICATION** is the excessive growth of certain aquatic plants caused by an increased supply of nutrients. Fertilisers are a primary source of the nutrients nitrogen and phosphorus in run-off from agricultural land. The resulting increase in algae and aquatic plants that multiply with this extra nitrogen and phosphorus deprives other plants and animals of oxygen, light and space. Some invertebrates are better adapted to survive low oxygen levels than others. Stonefly nymphs need well-oxygenated water, while chironomid larvae can tolerate low oxygen levels. These and other indicator species can be used to test water quality in streams and rivers. In warm, shallow seawater, excessive nitrate levels can lead to algal blooms that may threaten both the ecosystem and human health.

Despite all our knowledge about the soil, plant nutrient requirements and fertilisers, mistakes still happen. Sometimes, as in the African Sahel, people have few resources readily available to them to improve soil fertility. Soils tend to become impoverished, the vegetation disappears and **DESERTIFICATION** occurs. In more developed countries the drive to increase production and profits blinds some farmers to the threat of reduced fertility, leading to problems of soil erosion such as in the Dust Bowl in the USA during the 1930s and more recently in East Anglia and Yorkshire in England.

WEBSITES

Information on the impact of fertilisers on the environment from the UK's Environment Agency:

www.environment-agency.gov.uk/yourenv/eff/business_industry/agri/fertlisers/?lang=_e

Greenpeace:

www.greenpeace.org

Friends of the Earth:

www.foe.org

SEE ALSO

crop rotation, organic farming

FERTILITY

"Kids will change your life!" is a familiar refrain in virtually all societies. Photograph by Joseph Kerski

The decision to have children is a personal choice, but one that has tremendous impact on entire societies and also on the environment. Fertility refers to the reproductive behaviour of women – how many children they have. The causes and consequences of fertility are of major importance to the world today. In the past, much of the world had high reproductive rates because high fertility was necessary to counterbalance the high death rates. However, as death rates began to fall, fertility rates changed. Today, fertility rates vary considerably from place to place.

The most common measure of fertility is the crude birth rate. Fertility is the number of live births in one year per 1000 population:

$$\text{Crude birth rate} = \frac{\text{Number of births in one year}}{\text{Population}} \times 1000$$

The crude birth rate can be calculated fairly easily with a minimum amount of data. For example, a birth rate of 20/1000 (or 2 per cent) would mean that 20 babies were being born per every 1000 persons. This measure takes in the whole population, including men and women who are too old to bear children. Between 1995 and 2000 the crude birth rate for the world was 22.7 per 1000, for the UK 12.3 per 1000, and for the USA 14.7 per 1000.

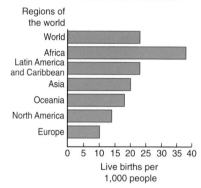

World crude birth rates vary widely around the world. Africa has the highest birth rate. The country with the highest rate has a rate over six times higher than the country with the lowest rate. Chad, a country in Central Africa, has a birth rate of 49, the highest in the world in 2003, for many reasons. High infant mortality means that couples need to bear more children in the hopes that some will survive to adulthood. More children may be needed to gather food and fuel wood for the family and communities. The lowest birth rates are found in Europe, especially Eastern Europe. Bulgaria and Ukraine both have birth rates of eight, the lowest in the world because of a reduced desire for large families and past political upheavals. The rest of world falls between these two extremes. The two most populous countries in the world, China and India, have rates of 13 and 25, respectively. China's one-child family policy has obviously had a great effect on its birth rate.

The total fertility rate is the average number of babies born to women during their reproductive years. A total fertility rate of 2.1 is considered the replacement rate; once a rate of a population reaches 2.1, the population will remain stable, assuming no immigration or emigration takes place. When the total fertility rate is greater than 2.1, a population will increase. When the rate is less than 2.1, a population will eventually decrease, although because of the age structure of a population, it will take years before the decrease is realised. In 1998 the world total fertility rate was 2.9. However, for more developed countries it was 1.6, and for less developed countries it was 3.2.

Fertility rates have a major impact on the environment all over the world, because population growth depends on fertility, and it is population growth that places demands upon the environment – its land, energy, mineral and water resources. Less developed countries have been growing at a much faster rate than more developed countries, which puts more strain on their environment. In developing countries environmental problems such as water quality, erosion from **OVERGRAZING**, air pollution and **DEFORESTATION** are exacerbated because of continued pressure on the landscape from an increasing population. The effect is prolonged, because as children in countries with a high fertility rate age, there are eventually a greater number of women in their childbearing years. Even if the fertility rate declines, more young women means more births, and sustained pressure on the natural environment.

Within most countries, fertility rates change over time. American women averaged more than seven children until the early 1800s. The total fertility rate began declining to 1.74 children per woman by 1976, interrupted only by the 'baby boom' following the Second World War. Delayed marriage, contraceptive use, abortion, high divorce rates and other social trends were the main reasons for the decline. Between 1990 and 2001 the rate varied between 1.97 and 2.08.

The fertility of a population is the result of a number of complex interacting factors. They can be biological, social or economic, and operate differently from time to time or place to place. Biological factors associated with fertility include age, health and nutritional status. The age of women is obviously an important factor. The primary childbearing years for women are between 15 and 49, although modern science is helping older women extend their childbearing years. Health and nutritional status can play an important role. A variety of diseases can affect reproduction, while nutritional problems,

especially severe malnutrition, can lead to temporary infertility.

A variety of social factors affect fertility. Marriage is an important social institution throughout the world and can have a major impact on fertility. Of particular importance is the age at which people marry. Typically, countries with high fertility have an average marriage age that is fairly young, whereas nations with low fertility usually have a much older average age at marriage. An important step China has taken to lower fertility in the past two decades has been to encourage people to postpone marriage.

Contraception also plays a very important role in fertility. Although modern contraceptive practices are more common in the wealthier industrialised countries, there has been an increased influence of family planning in much of the poorer nations of the developing world. Contraception is being practised in areas where it had never been practised before. Estimates are that more than one-third of married couples in the world today practise some form of contraception.

Family planning has played a critical role in reducing fertility worldwide. During the 1960s and 1970s, the US Congress specifically allocated funds to support family planning programmes throughout the world. The United Nations also established programmes through the United Nations Fund for Population Activities (UNFPA), an organisation dedicated to family planning in order to improve the health and well-being of individuals and families worldwide. During the 1980s a rising tide of political conservatism in the USA under President Ronald Reagan led to the withdrawal of the USA's support for UNFPA and other international family planning programmes. The Clinton Administration, however, reversed those policies during the 1990s and once again the USA supported UNFPA. In 2000 the Bush Administration once again reversed those policies, no longer supporting international family planning programmes.

Fertility is also related to a variety of economic determinants. A couple's decision to have a child is heavily influenced by the costs and benefits the couple perceives from having that child. These costs and benefits have both economic and non-economic components.

WEBSITES

Studies in Family Planning journal from the
Population Council, giving recent
information on fertility and contraceptive
practices:
www.popcouncil.org/sfp
United Nations Fund for Population
Activities:
www.unfpa.org
Birth rates around the world from the
EarthTrends Database:
http://earthtrends.wri.org/searchable_db/
index.cfm?step=countries&ccID=0&theme=4
&variable_id=368&action=select_years
Population Reference Bureau's Report on
Fertility:
www.prb.org/Template.cfm?Section=PRB&te
mplate=/ContentManagement/ContentDispl
ay.cfm&ContentID=9789

SEE ALSO

mortality, population increase

FLOODING

In February 2000 a cyclone swept across
Mozambique, which left some 950,000 people
homeless as floods devastated huge areas of
low-lying land. Roads, homes, bridges and
crops were destroyed. Electricity supplies were
disrupted and towns left without clean water
supplies after their pumping stations were
swept away.

Journalist Greg Barrow flew over the strick-
en area and filed this report for the BBC:

> From above it looks as if a huge tidal wave
> of brown water has swept through the
> Save river valley. Trees have been
> uprooted, houses lie in ruins and debris is
> floating in the floodwater. Those who
> survive the flooding have been stranded
> on rooftops and in trees. Beneath them
> the bloated corpses of livestock float in
> the waters.
>
>   Two South African helicopters
> chartered by the Mozambican
> Government are doing what they can to
> rescue the survivors. Pilots say some

> people have been trapped in trees for
> days without food or water. As the
> helicopters fly above them the people
> wave frantically, motioning to their
> mouths and stomachs with their hands to
> show that they are hungry. When the
> rescue helicopters come to winch them to
> safety there's a desperate scramble for
> whatever belongings the people can grab.
>
>   The helicopters have lifted around 300
> survivors to higher ground but their work
> is difficult, fuel supplies are low in the
> area and pilots say there's a lack of
> coordination in caring for the flood
> victims. The South African national
> defence force says it has more helicopters
> and planes on standby to fly up to the
> Save area, where up to 40,000 people are
> believed to be trapped by the floods.

Flooding on a similar scale to that which
occurred in Mozambique in 2000 is rare.
However flooding itself is very common
throughout the world and it is a perfectly
natural response of a river that has too
much water to cope with. Heavy rainfall,
often combining with snowmelt, causes
channels to be overtopped, and floodwaters
surge over the neighbouring floodplain.
Apart from injury and loss of life, flooding
may damage or destroy buildings and inundate agricultural land, ruining crops and
drowning livestock. Transport systems such as
roads and railways may be disrupted, and
services such as gas, water and electricity
may be cut off. Sewerage systems are often
severed, posing serious health risks. One
of the most ironic consequences of flooding
is that fresh safe water is often in short
supply.

For all the negative aspects of flooding, it
is important to recognise the role that the
annual flood has had in promoting settlements based on thriving agricultural communities in river valleys. For example, the
annual flood of the River Nile in Egypt gave
rise to whole civilisations who depended
absolutely on the fertile silt and the water
for irrigation. Flooding is essentially natural
and, by settling on a river floodplain,
people are inherently putting themselves at
risk.

Case study

Flooding on the River Yangtze, China

The River Yangtze has frequently hit the headlines for its appalling flooding, resulting in the deaths of thousands of people. In the 1990s alone nearly 2000 people lost their lives and over 130 million people were made homeless.

Yangtze disasters of the twentieth century		
Year	Deaths	Homeless
1931	140 000	40 million
1995	1 300	100 million
1998	300	30 million
1999	240	2 million

In the 1998 flood the Yangtze reached its highest level ever – and it stayed there for over two months. Several flood peaks occurred over this period. In one day the city of Wuhan received as much rain as London, UK, has in six months. Several smaller cities were totally submerged for at least a month. It is hardly surprising, therefore, that 5.6 million homes were washed away, accounting for the large numbers of homeless. Huge areas of crops were destroyed, affecting people far beyond the valley itself. It is these regular disasters that have led the Chinese authorities to a decision to build the highly controversial Three Gorges Dam, currently under construction.

WEBSITES

Full coverage of the Mozambique floods on the BBC's website:

http://news.bbc.co.uk/1/hi/world/africa/655510.stm

UK Environment Agency:

www.environment-agency.gov.uk

Public Broadcasting Service, with an archive collection:

www.pbs.org/amex/flood

USGS Flood Hazards:

www.usgs.gov/themes/flood.html

SEE ALSO

dams, levees and embankments, water supply

FLUE GAS DESULPHURISATION

Old films set in London often set the scene with the sound of Big Ben chiming away in the background and a strange low-level mist hanging over the historic buildings lining the banks of the Thames. While the historic clock, the old buildings and the famous river are still there today, the strange mists, or more correctly 'pea-soup' fogs, are definitely a thing of the past. The Great Smog of London in the winter of 1952–3 and the CLEAN AIR ACTS hastened their departure. During the past 30 years, a series of anti-pollution devices have

been installed that remove nitrous and sulphur oxides from the emissions from factories and coal-burning power stations. One example is flue gas desulphurisation.

The most widely used technology for the desulphurisation of flue gases is called wet scrubbing. This involves an alkaline slurry, usually lime-based, which is placed downstream of the boiler and flue gas cleaning plant. This can remove 90 per cent of sulphur dioxide from the flue gases. The resulting solid is predominantly calcium sulphite, which can be further oxidised and sold as a by-product to the building industry. Wet scrubbers exist in over 300 coal-fired plants worldwide, and there are plans for another 100 units. However, fitting the equipment is expensive, and the materials must be of a high quality to handle the slurry and by-product.

The other method currently employed is dry scrubbing, which involves an alkaline-based slurry that is sprayed into the scrubbing vessel in very fine droplets and dries as it contacts the flue gas. At this point, it reacts with the sulphur dioxide and the dry deposit is picked up by the gas stream and carried into the fabric filters of the dust-collection plant. Dry scrubbing technology has been installed in over 80 power stations, especially those burning low to medium-sulphur coals. The equipment costs less to install than the wet scrubbing systems, but operating costs are

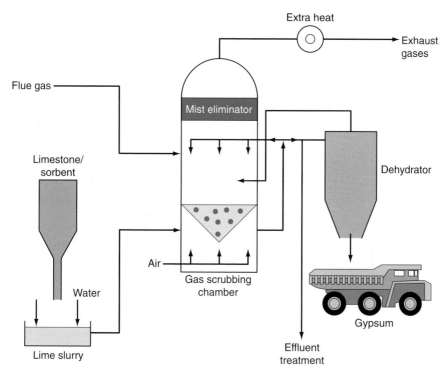

Flue gas desulphurisation wet scrubbing. © The State of Queensland (Department of Energy) 2004

noticeably higher. Another disadvantage is that there is no commercial use for the dry by-product.

In recent years, experiments have taken place with sorbent injection methods. This could be a cheaper alternative to the scrubbers, and involves adding a dry calcium sorbent (material to absorb the gases) with the fuel, or directly into the boiler. Trials have shown that because of a relatively poor contact between the sorbent and the sulphur dioxide, the removal efficiency is much lower, usually in the range 30–60 per cent. Modifications are taking place with humidification of the flue gas or adding other substances into the gas stream, and if these prove successful efficiencies may rise up to 80 per cent.

The UK has been particularly active in introducing wet and dry scrubbers, and in 1988 it complied with the EU Large Combustion Plant Directive by fitting flue gas desulphurisation (FGD) equipment in some of the largest plants. From 2005 National Air Quality Standards targets came into force, and in the past few years various companies have been updating or replacing their FGD equipment in order to meet these new targets.

An example came in July 2001 when British Energy announced that it would be spending over £50 million on its Eggborough plant in the Aire Valley. This coal-fired power station, which began supplying the National Grid, has a maximum output of 2000 megawatts, and produces enough electricity for 2 million people. The improvements resulted in FGD equipment and a new lining being placed on the station's 180-metre chimney stack.

WEBSITES

Encyclopedia of the Atmospheric Environment:

www.doc.mmu.ac.uk/aric/eae/Acid_Rain/Older/Industrial_Emission_Controls.html

TNO Environment, Energy and Process Innovation:

www.mep.tno.nl/Informatiebladen_eng/212e.pdf

SEE ALSO

nitrogen oxides, pollution

FOOD WEB

Food market in Los Angeles, California
Source: Joseph Kerski

A food chain illustrates how each living thing gets its food. Some animals eat plants (herbivores) and some animals eat other animals (carnivores). For example, a simple food chain links the trees and shrubs, the giraffes that eat trees and shrubs, and the lions that eat the giraffes. Each link in this chain is food for the next link. A food chain always starts with plant life and ends with an animal.

'Food chain' is a familiar ecological concept, but the metaphor of a 'chain' – a linear series of links – is too simple for most environments. Even in the Arctic, a classic simple system like lichen → caribou → grey wolf is complicated by the fact that the caribou's diet shifts with the season, and an occasional wolverine or polar bear sometimes enters the equation. Almost always, feeding relationships form not a simple chain but a complex food web.

Why are there more herbivores than carnivores? In a food web, energy is passed from one part to another. When a herbivore eats, only a fraction of the energy that it gets from the plant food becomes new body mass. The rest of the energy is lost as waste or used up by the herbivore as it moves. Likewise, when a

carnivore eats another animal, only a portion of the energy from the animal food is stored in its tissues. In other words, organisms along a food web pass on much less energy in the form of body mass than they receive. Because of the large amount of energy that is lost at each part of the web, the further along the food chain you go, the less food (and hence energy) remains available.

Most food webs are illustrated as a pyramid, with fewer and fewer numbers toward the top of the pyramid. In the above example, many trees and shrubs provide food and energy to giraffes. However, on the next level up, fewer giraffes exist than the number of trees and shrubs, and even fewer lions exist than giraffes. In other words, a large mass of living things at the base is required to support a few at the top. There cannot be too many links in a single food web because the animals at the 'end' of the web would not get enough food (and hence energy) to stay alive. However, once again, we are not served well by this traditional food *chain* type of thinking. Most animals are part of more than one food chain and eat more than one kind of food in order to meet their food and energy requirements. These interconnected food chains form a food web.

A change in the size of one **POPULATION** in a food web will affect other populations. This interdependence of the populations within a food web helps to maintain the balance of plant and animal populations within a community. For example, when there are too many giraffes, there will be insufficient trees and shrubs for all of them to eat. Some giraffes will starve and die. Fewer giraffes means more time for the trees and shrubs to grow to maturity and multiply. Fewer giraffes also means less food is available for the lions to eat and some lions will starve to death. When there are fewer lions, the giraffe population will increase.

Scavangers and decomposers ultimately eat/consume animals at the apex of the food web. Scavangers are any of the creatures that eat dead plants and animals. They are important in getting rid of plant and animal remains. What they leave behind is then used by the decomposers. Decomposers, predominantly bacteria and fungi, are the organisms that break down the final remains of living things. They are important in freeing the last

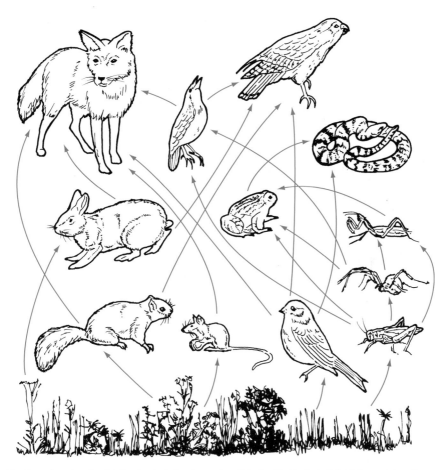

The figure above is a simplified food web of a brushy meadow. (It is simplified by the fact that it does not show the fleas on the fox or the lice on the hawk or the microbes in the soil.) From Ehrlich et al, *Ecoscience*, 1977

of the minerals and nutrients from the remains and recycling them back into the food web.

Consumers that cannot make their own food eat the producers (plants). Organisms that feed on primary consumers are called secondary consumers. Those who feed on secondary consumers are tertiary consumers. When a squirrel eats acorns or fruits, it acts as a primary consumer; when it eats insects or nesting birds, it is a tertiary consumer.

Most species are connected to more than one other species. Note that the arrows all point in the same direction. This is to indicate the direction of *energy flow*. According to the second law of thermodynamics, energy can move only one way – 'downhill' – because at every step in energy transformation from sun to parasite or decomposer, some energy is dissipated as heat. Energy flow through a community can be evaluated in terms of ECOLOGICAL EFFICIENCY.

A food web, therefore, is the network of feeding relationships in a biotic community. Food webs are structured by SYMBIOSIS, particularly the trophic ('nutritive') symbioses of predation and parasitism, where individuals of one population live at the expense of individuals of another.

An important way to simplify food webs is by consideration of trophic levels, which are successive steps removed from direct dependence on the Sun (or other ultimate energy source): producers, or autotrophs, and several levels of heterotrophs – primary consumers (herbivores), secondary consumers (carnivores), tertiary consumers ('top carnivores'). Obviously, some species (so-called omnivores, like most humans) operate at more than one trophic level. Even some plants do that – insectivorous plants like Venus flytrap are both producers and secondary consumers. Most decomposers are 'omnivores', cleaning up the leftovers from all other trophic levels. Trophic levels are often depicted as **ECOLOGICAL PYRAMIDS**.

Pollution in the environment can disturb food webs. For example, acid rain can cause phytoplankton in lakes to die. Insects, which rely on phytoplankton for food, now have less food to eat and they begin to die. These insects are a source of food for many other animals, such as fish, birds, frogs and salamanders. As the insects die, there is less food for these animals. This process continues up the entire food web. Environmental disturbances, many of which are caused by humans, can affect the entire food web of which we are also a part.

WEBSITES

Detailed food chain and food web discussion: http://users.rcn.com/jkimball.ma.ultranet/ BiologyPages/F/FoodChains.html
Ocean life food web guide: www.oceanoasis.org/teachersguide/ activity10.html
Cycling through the food web: www.bigelow.org/bacteria/
Marine ecosystems information with a thorough food web discussion: www.oceansonline.com/oceanicfoodwebs.htm

FORESTS AND FORESTRY

This book is made of paper. You may be sitting on a wooden chair and your stove may burn wood. Trees are a major part of our life; the organisation American Forests describes their importance thus:

Trees are the oldest and largest living things on the earth, and they are a good measure of the health and quality of our environment. Trees are the original multi-taskers; they provide social, ecological, and economic benefits. Their beauty inspires writers and artists, while their leaves and roots clean the air we breathe and the water we drink.

Forests are extensive areas of land covered mainly with trees. Forestry includes the science of planting and caring for trees as well as the management of forests. Large areas of the world surface would naturally support forest vegetation, although the current area of forested land is much reduced because of human activities (see **CLEAR FELLING** and **DEFORESTATION**).

The pattern of vegetation, especially the distribution of forest types, is largely controlled by climate. On the global scale different climates give rise to large-scale global ecosystems or biomes. The major forest biomes are tropical rainforest, northern coniferous forest (taiga), temperate deciduous forest, Mediterranean forests and tropical deciduous forest (see **WOODLAND ECOSYSTEMS**).

Urban forests

As urban areas have expanded to make way for housing and infrastructure, so vast numbers of trees have been felled. Many US cities have seen a decline in natural tree cover by as much as 30 per cent over the last few decades. Some people now believe that urban authorities should reverse this trend by planting urban forests. There are several benefits associated with the provision of more green areas in cities. Aesthetically, trees bring a sense of calm to people as they go about their busy daily lives. They provide areas for rest and relaxation, as well as colour and texture in often drab urban environments. Trees are also important habitats for birds and other creatures. Hydrologically, trees help to reduce storm flow, as water can more readily infiltrate into the soil. Trees also help to reduce the impact of global warming by taking carbon dioxide out of the air.

Case study

Forests and forestry in the West Coast of South Island, New Zealand

The West Coast of South Island, New Zealand, is home to two main forest types: mixed hardwoods and southern beech. The mixed hardwoods are warmth-loving trees with links to sub-tropical species and grow at low levels, mostly below 500 metres above sea level. The beech represents the cold-tolerant species that grow above 500 metres above sea level and on mountain slopes.

Until recently the most valuable trees were the hardwoods, such as the rimu, and so forestry activities naturally concentrated in lowlands. These areas were also under threat from clearance for farmland, although high rainfall, over 2000 millimetres per year, and relative isolation saved the area from much exploitation. The logging companies also preferred the more accessible forests on the North Island.

Before 1860 the forests in South Island were largely untouched. Only a few stumps could be seen, relics of trees chopped down by Captain Cook at Astronomer's Point in Dusky Sound. The 1864 Gold Rush changed this and trees were felled for use in the gold-prospecting era to allow towns to be built and tracks to be cut. Forests were then cleared to expand farmland and for timber to export.

The earliest conservation activities began in 1905. The beech forests became a 'protection forest' and the Fiordland area was designated a National Park. However, the lowland forests remained under threat, except for small areas designated as scientific reserves. The creation of the Westland National Park in 1960 still concentrated almost exclusively on the high altitudes.

Conservation activities began in earnest in 1965 when a hydroelectric power scheme was proposed at Lake Manapouri in Fiordland. The proposal mobilised the formation of a mass conservation movement that fought against the Government and other interested parties for more extensive and better protection of the lowland forests. The power of the opposition led the Government and the New Zealand Forest Service to improve standards of timber extraction and prevent logging in and around ecological and scientific reserves. A much greater emphasis was placed on conservation.

By 1986 the Government was finally convinced of the need to protect the remaining forests. The Paparoa National Park was designated, giving permanent legal protection to 58 reserves. A North Westland Wildlife corridor was set aside. This has since been strengthened by the creation of a 2.2 million hectare site proposed as a World Heritage Site.

Source: L. Wright, *Update*, 'Environmental systems and human impact,' Cambridge University Press

WEBSITES

American Forests, an organisation concerned with improving the health and coverage of trees:

www.americanforests.org

Information on urban forests in the USA, including case study on Washington DC:

www.americanforests.org/graytogreen

SEE ALSO

afforestation, deforestation, woodland ecosystems

FORESTS, TROPICAL, TEMPERATE AND TAIGA

Forests are some of the most loved places on Earth. There are three distinct BIOMES of forests: tropical forests, temperate forests and taiga.

Tropical rainforests

The tropical rainforest biome is a response to high temperatures and precipitation (up to 4.5 m or more per year), with little seasonal variation. The biotic response is a complex forest with high ecological diversity (both richness and evenness), at least where the forest is undisturbed. Trees are often buttressed for stability in wet soils and epiphytes (such as orchids and bromeliads) and lianas (vines) may be common. Unbroken tropical rainforest tends to have little vegetation on the deeply shaded forest floor and little downed timber or other litter. Warm, moist conditions mean that microbial recycling is very rapid and litter mass is low (only about 3 per cent of the global total). Nutrient cycles are tight and shallow feeding roots of the trees absorb decomposed matter very quickly. As a consequence, when tropical rainforest is removed, the exposed soils that supported such luxuriant life tend to be nutrient-poor. Disturbed rainforest forms 'jungle' or *yungas* – with exuberant growth of understorey plants. The disturbance can be due to human activity or to natural tree fall.

Tropical rainforest covers about 11 per cent of the Earth's land surface but is responsible for nearly one-third of global net primary production, one-third of the animal biomass, and some 42 per cent of the Earth's plant biomass. It has been estimated that tropical rainforests include between one-half and three-quarters of all living species. Tropical rainforest is an extremely important biome.

Animal life includes many arboreal forms and numerous specialised herbivores, especially beetles. However, because of the prevalence of trees (hence wood), consumption of primary production is only about 7 per cent, less than in grasslands and only half as high as in the savanna biome. Hence, tropical rainforests often are 'biotic hotspots', targeted for conservation. Despite current efforts, however, the area of the biome is diminishing by 1–2 per cent each year, as forests are cleared for subsistence agriculture, timber, grazing land and plantation agriculture.

Not all tropical forests are rainforests or evergreen. Rainforests grade into seasonal forests (including deciduous forests) in their edges, or **ECOTONES.** Tropical seasonal forests have about half the extent of rainforests and only about one-third of the production. On the slopes of tropical mountains, rainforests may grade into cloud forests, where rainfall is low, but 'cool greenhouse' conditions are maintained by daily fog.

Temperate forests

The temperate forest biome occupies about 2.4 per cent of the Earth's land surface, provides about 7 per cent of terrestrial productivity and 11.4 per cent of global productivity. Dominant producers are mostly deciduous trees and an understorey of shrubs and annual and perennial herbs contributes to high species richness for the temperate zone. Browsers (such as deer

and rabbits) make use of the woody vegetation. Fruit-eating (frugivory) is common among birds and mammals, and many insects are leaf-eaters (folivores). Saprobic (decomposer) fungi also are present and diverse.

The climate is moist but highly seasonal, with cold winters and hot summers. Properly managed, forest soils are productive and have reliable rainfall. As a consequence, much of the temperate deciduous forest of Europe, North America and Asia has been cleared for agriculture. Some of those areas have been allowed to revert to forest, as is the case in large parts of New England and the Mid-Atlantic States of the USA.

In addition to temperate deciduous forests, locally moist and moderate conditions along western coasts of continents sometimes allow temperate evergreen forests. Well-known examples include the temperate rainforests of the Olympic Peninsula in Washington, USA, coastal British Columbia, Canada and comparable climates in Chile and Tasmania.

Taiga

The taiga biome, or northern (boreal) coniferous forest, occupies about 8.2 per cent of the Earth's land surface and contributes about 8.3 per cent of terrestrial production. Some map projections exaggerate areas toward the poles, which make this biome look even more widespread than it is. These are the forests of sub-Arctic Canada, Alaska, Scandinavia and Siberia. The climate is strongly seasonal, cold in winter, moderately warm in summer.

Dominant plants are conifers, whose pyramidal shape sheds snow readily. The taiga is the only biome dominated by gymnosperms, which were the dominant seed plants during the Mesozoic Age (the Age of Reptiles), but now represent less than 1 per cent of plant species.

Taiga is home to a disproportionate share of terrestrial plant biomass, some 13 per cent, but a modest amount of the animal biomass (<6 per cent). Trees are the principal resource in the forest, but being mostly cellulose, they are not a particularly usable diet for most animals. Shrubs occur in openings and the herbaceous understorey is of modest extent. Some conspicuous herbivores (such as moose and hares) exploit trees, especially deciduous alders and birches, as food sources. Other consumers rely on bark (porcupines) or seeds (mice and squirrels).

Although it contributes only about 8 per cent of terrestrial production, taiga contains 43 per cent of terrestrial litter, in the form of dead materials such as branches, needles and fallen trees; the largest litter stock of any biome. This build-up occurs because only 4 per cent of net primary production is harvested by animals, and the cold climate slows microbial and biological action. Things cannot break down as quickly as in tropical rainforests or in temperate forests.

Forest versus woodland

Strictly speaking, a woodland is a biotic community dominated by trees, but – in contrast to a forest – the trees are discontinuous. A savanna might be considered a low-density woodland, because the trees are widely spaced. Woodlands may have an understorey of shrubs or herbaceous plants or both. Often, woodland is maintained by physical factors in the environment, such as drought or fire. When fires are suppressed, woodlands may become a forest. When fires in grasslands are suppressed, woody plants may increase, producing a woodland or a shrubland. Woodlands can sometimes be achieved and maintained by artificial thinning.

WEBSITES

A long-term ecological research (LTER) site studies the rainforest of the Luquillo Experimental Forest near San Juan, Puerto Rico:
www.lternet.edu/sites/luq
The Up-Island Regional School District, Martha's Vineyard, Massachusetts hosts a biomes site, including resources on tropical rainforests:
www.blueplanetbiomes.org/rainforest.htm
La Selva, Costa Rica, biological station in the lowland tropical rainforest:
www.ots.ac.cr/en/laselva
The Hubbard Brook Experimental Forest temperate forest landscape, the longest-running such study in the USA:
www.hubbardbrook.org

Susan Woodward's information on taiga:
www.radford.edu/~swoodwar/CLASSES/GEOG235/biomes/taiga/taiga.html

FOSSIL FUELS

It is remarkable to think that the swamps, mudflats and shallow, warm seas that existed millions of years ago have provided us with one of our most valuable resources. The three fossil fuels – **COAL**, **OIL** and **NATURAL GAS** – are all products of decayed organic matter that built up in swamps, bogs and marshes, as well as on the sea floor around 360 to 280 million years ago – an era now known as the carboniferous period. The word carboniferous is derived from carbon, the basic element in coal and other fossil fuels. Together, these three fuels satisfy 75 per cent of our current energy demands.

The legacy of these ancient environments has also influenced the location of heavy manufacturing industry, which has usually been located in places rich in these fossil fuels. Fossil fuels have also helped vast economies to grow and have allowed many individuals and corporations to accumulate enormous wealth – proving the old saying that 'where there's muck, there's brass'.

As far as scientists are concerned, the value of fossil fuels can be measured in more than just economic terms, because the fuels also give valuable clues about the environments that existed hundreds of millions years ago, as well as the natural process of climate change. Since much of the original plant material remains in coal, scientists have been able to isolate chemically identifiable plant parts such as spores and cuticles, which provide information about the type of land and climatic conditions in which the plants were growing. As well as giving important clues about these ancient environments, these fossil plants help scientists to study the changes in the amounts of oxygen and carbon dioxide in the atmosphere.

While many fossil fuels have literally gone up in smoke, studies of these fossil plants have greatly helped Earth scientists understand more about the workings of the biosphere-atmosphere system, as well as giving proof that the Earth's climate has fluctuated quite dramatically in the past, between warm and cold spells – long before the arrival of humans and harmful pollutants.

Evidence of palaeo (fossil)-environments

The levels of oxygen and carbon dioxide in the atmosphere are controlled by the activities of plants. Therefore, the evolution of photosynthesising plants caused major changes in the concentration of these gases. Green plants utilise sunlight to extract carbon from the atmosphere, and via the photosynthesis process incorporate the carbon into their organic structures, releasing oxygen as a by-product. Geologists believe that oxygen concentrations in the atmosphere rose significantly in the Precambrian era (over 550 million years ago), and after the rapid geographical spread of plants in the Devonian period (around 400 million years ago), the levels of oxygen reached its current level.

The amount of carbon dioxide in the atmosphere can also be gauged from the number of stomata, or pores, on the plant leaves. These vary depending on the amount of carbon dioxide, so changes in the density of pores can be used as a surrogate measure of carbon dioxide levels. As these levels affect global climate, they can help identify periods when carbon dioxide levels were low and global cooling took place.

WEBSITES

http://library.thinkquest.org/20331/types/fossil

www.energyquest.ca.gov/story/chapter08.html

SEE ALSO

coal, natural gas, oil

FUELWOOD

Fuelwood is wood from shrubs or trees that is burned for heat or for cooking. Despite the widespread use of fossil fuels and various forms of renewable energy, fuelwood is still one of the most important energy sources in the modern world. This is especially the case in many less economically developed countries, with Honduras being a typical example. In this Central American nation, the burning of fuelwood accounts for 65 per cent of the country's energy. Every day, people may have to walk around for up to three hours, gathering fuelwood.

Fuelwood is not only found in the less wealthy parts of the world. In more economically developed countries many rural areas have

Case study

Fuelwood in India

In many countries the selling of fuelwood is a major element in the nation's economy. In India the business of fuelwood, primarily conducted by villagers in and around forests, has an annual turnover of US $16.54 billion, according to estimates from the Ministry of Environment and Forests National Forestry Action Programme. This is equivalent to nearly 25 per cent of India's foreign exchange reserves.

But fuelwood in India is not just a billion-dollar commercial enterprise. For many people in rural parts of India the collection and use of fuelwood is the only way of surviving. Basumati Tirkey, a 35-year-old resident of Bangamunda village, some 40 kilometres from Rourkela in Orissa, is typical of the many millions of Indians who will probably spend a lifetime collecting fuelwood. Each day she walks 9 kilometres, carrying a load of 35 kilograms of fuelwood, to earn just 15 rupees – the equivalent of just 18 pence.

It is estimated that some 11 million people in India, many of whom are in hand-to-mouth subsistence communities, rely on fuelwood. Researchers at the National Botanical Research Institute in Lucknow estimate that, in an average village, a woman walks more than 1000 kilometres a year to collect fuelwood alone. They are prepared to do so because without the energy it provides it is questionable whether they would survive.

Case study

Fuelwood in Cambodia – good or bad?

The arguments for and against the use of fuelwood, and the role of demographic factors, are exemplified by Cambodia. Fuelwood and other biomass energy are the main sources of energy in the nation that at long last is starting to move forward after years of warfare, depravity and abject poverty. In 1999 fuelwood represented 96.3 per cent of national fuel consumption, with around 84 per cent of households and small enterprises (such as brick kilns, bakeries and food processing) using fuelwood and charcoal as their main fuel for firing and cooking.

Using fuelwood is therefore an essential part of Cambodian life, but it is one that requires careful control. Between 1973 and 1997 the country's forest resources declined from 13 million hectares to 10.6 million hectares, and, once again, demographic factors accelerated the rates of deforestation.

Cambodia's recent population growth has put increasing pressure on wood supply with annual demand for fuelwood increasing from 1.8 million cubic metres for the period 1961–71 to 6 million cubic metres for 1991–4. In the light of such statistics, the Cambodian Government has made a concerted effort to outlaw illegal logging, and also encouraged reforestation schemes and programmes that improve the energy efficiency of cooking stoves in rural communities.

charcoal burners that provide energy to remote and isolated communities. However, in these peasant communities, fuelwood is still the primary source of energy for cooking and heating. It is estimated that around 3 billion people currently rely on fuelwood. Since 1960 fuelwood consumption has increased by over 250 per cent, and this figure shows no sign of dropping.

Despite being a cheap and plentiful source, fuelwood is not entirely an eco-friendly supply of energy, especially as the cutting down of trees can cause significant amounts of environmental damage and endanger many threatened habitats. A recent example occurred in the Democratic Republic of Congo (formerly Eastern Zaire) where the influx of 750,000 refugees put pressure on the country's infrastructure, and led to the collection of fuelwood from Virunga National Park, the mountain gorilla reserve. In just a few months, over 8100 hectares of park were cleared for fuelwood and building material, significantly reducing the size of the gorilla's habitat.

Elsewhere the removal of tree cover has led to accelerated rates of soil erosion. Studies in the Sahel nations of West Africa have shown that desertification has been one of the negative spin-offs caused by the increased use of fuelwood, because of the removal of vegetation that would otherwise nourish the soil and help to bind it together. As in the earlier example, the delicate balance between people and the environment on the fringe of the Sahara

Desert has been affected by an increase in population, which has led to an increased demand for fuel for heating and cooking.

WEBSITES

www.wenetcam.net/fuelwood.php

www.cifor.cgiar.org/docs/_ref/publications/newsonline/33/fuelwood.htm

SEE ALSO

desertification, soil erosion and conservation

GEOGRAPHIC INFORMATION SYSTEMS

The earliest preserved map is a Babylonian clay tablet, estimated to be from 900 BC, showing Babylon at the centre of the Earth, which is a circle. Since ancient times, maps have always helped us to understand the world in which we live. Maps show where things are, and therefore help us understand how phenomena are related to each other. Patterns that would be difficult to detect in a table of data or paragraphs of text may be readily apparent on a map. Maps are powerful sources of information.

Given that environmental issues occur somewhere on the planet, it is useful to construct one or more maps to describe each of these issues. However, most environmental issues, problems and phenomena are complex – they occur at a variety of scales and they may change over time and space. Therefore, using traditional paper maps to understand the environment will always have limitations.

Geographic information systems (GIS) provide a technology and method to analyse map data using a computer. By analysing phenomena about the Earth's hydrosphere, lithosphere, atmosphere and biosphere, GIS help people understand patterns, linkages and trends about our planet.

The term geographic information systems has three elements. Geographic refers to a representation of the Earth – a topographic map, aerial photograph, satellite image, routes in a watershed, slopes, land use, or a variety of other themes. Information includes the features on the map, stored in one or more tables in a relational database. This could be the per cent of forest cover of a particular tract of land or the depth of a landfill. The power of GIS is in the systems, which allow the data user to analyse the map and the information about the Earth simultaneously. With GIS, it is possible to do so at a variety of scales, using different symbols and even in three dimensions.

GIS are more than just computer software. They include spatial data, methods and people that make daily decisions using these tools. GIS are more than computer mapping. They are about helping people analyse the Earth with a problem-solving set of tools. Since the 1960s, GIS have quietly transformed environmental decision-making in universities, government and industry by bringing digital spatial data sets and geographic analysis to desktop computers. Geographic information sciences include GIS and the disciplines of geography (examining the patterns of the Earth's people and physical environment), cartography (mapmaking), geodesy (the science of measuring and surveying the Earth), and remote sensing (studying the Earth from space). GIS are used in conjunction with global positioning systems (GPS) for mapping the environment. GIS are also applied to a host of environmental issues, from applying an application of phosphorus to an agricultural field,

to understanding an archaeological site, managing natural resources, restoring rivers to their natural state, managing coastal zones, caves and forests, and planning for future urban growth. One of the ways GIS technology helps farmers is to project crop output by analysing soil classifications and their resulting fertility. GIS can produce maps that show farmers how to fertilise a given field allowing for differing levels of fertility within that same field. GIS provide fertility data on a field that drives another tool called variable rate technology, which controls the amount of fertilisers applied to a field where it is needed.

GIS are all around us. An alarm clock may have awakened you this morning. The household electricity powering the alarm clock is from the local electric utility. Electric utility companies serving millions of customers use GIS to manage their complex infrastructure consisting of tens of thousands of kilometres of transmission and distribution lines and hundreds of thousands of utility poles, as well as thousands of employees maintaining optimal service at hundreds of sites. In the kitchen you might pour some fresh fruit juice. The fruit trees were grown with water provided by an irrigation district serving the agricultural community. The district serves thousands of farmers and maintains hundreds of kilometres of waterways. It uses GIS for engineering and operations and for powerful digital mapping.

Next, you put on a pot of coffee. The water for the coffee is provided by a water utility operating a water distribution system that consists of thousands of kilometres of water mains. The utility uses GIS for customer service, emergency response, water distribution, infrastructure maintenance, automated mapping, network tracing, flow analysis and other aspects of engineering, operations, administration and finance. You go outside and pick up the morning newspaper. The wood that was the source for the paper and for the timber in your house was provided by wood product companies that use GIS for sound forest management practices. GIS enable analysis of property boundaries, vegetation, soil, roads, streams, public land surveys, contours, watersheds and sensitive areas, allowing forest managers to make the best-informed decisions. In addition, the newspaper circulation department uses GIS to

Case study

Using GIS to save African elephants

Joyce could use some help. A 62-year-old grandmother and matriarch to an extended family, she struggles each day to keep her clan intact. Joyce is an elephant. She is one of approximately 500,000 threatened African elephants, a number down from over 1 million just 20 years ago.

With the help of GIS, Joyce and others like her are now receiving some help. The Amboseli Elephant Research Project (AERP) in Kenya uses GIS to visualise how elephant distribution changes over time, how elephants respond to changing human behaviours, and patterns of intergroup dominance. Dr Cynthia Moss, a natural scientist and author of several popular books about elephants, co-founded the AERP with Dr Harvey Croze. Moss, Croze and their assistants use GIS to analyse spatially a dataset of over 1800 elephants that extends back to 1972.

Using GIS, the researchers found, for example, how a few male elephants, which usually live apart from the females, stay in the safe areas during the day and venture outside their boundaries at night to raid farmers' crops. Spatial analysis with GIS also helps researchers to negotiate precise land-use planning decisions. Using high-resolution (61-centimetre) satellite imagery from DigitalGlobe Corporation, the researchers have been able to identify current positions of elephants from space and can even determine the size of each elephant. This allows them to study the elephants in a non-invasive way; tracking them with radio collars was discontinued because the technique was found to traumatise the elephant.

understand the dynamics and demographics of carrier routes.

When you leave the house in your car, GIS are there too. GIS technology integrates all kinds of petroleum information and applications into a common system and lets the oil companies view that information in context on a map for exploration, operation and maintenance, production, environment, land-lease management and data management. Before the oil becomes petrol it needs to move from the oilfields to the processing plant via pipelines. The pipeline industry uses GIS for assisting route planning and construction, operations, supply market analysis and reporting functions.

WEBSITES

GIS.com:

www.gis.com

Environmental Systems Research Institute, manufacturer of the most widely used GIS software:

www.esri.com

The annual global GIS Day:

www.gisday.com

USGS GIS poster and information:

http://erg.usgs.gov/isb/pubs/gis_poster

An example of GIS on the web:

http://nationalatlas.gov

GIS journals:

www.library.umaine.edu/science/gisjournals.htm

The Amboseli Elephant Research Project:

www.esri.com/news/arcnews/fall02articles/african-elephant.html

GEOGRAPHY

US National Park Service

Geography is the only subject that gives you carte blanche to look at the whole world and try to make sense of it. The field never stops being exciting, and that's what geography is all about – trying to understand the world. (Peirce F. Lewis, Professor Emeritus of Geography, The Pennsylvania State University)

The study of geography involves the study of major problems facing humankind, such as environmental degradation, unequal distribution of resources and international conflicts. It prepares one to be a good citizen and an educated human being. (Risa Palm, Dean, College of Arts and Sciences, University of North Carolina-Chapel Hill)

Many of us think of geography as the names and capitals of countries, the distribution of their people and resources and their imports and exports. This may be because of an emphasis on memorising facts and figures in a geography class in primary school.

However, geography is one of the world's oldest disciplines, held in high esteem by the ancient Greeks. Geography comes from the Greek words *geo* (Earth) and *graph* (describe) – the science of describing the Earth.

Unlike geologists, who study the Earth itself, geographers consider the impact of people in most of their work. 'How does the environment affect people, and how do people affect the environment?' is a typical question a geographer asks. Geography can be divided into three subfields: physical-environmental (for example climatology, biogeography and geomorphology), human (for example urban, population, political and cultural geography) and technical (for example cartography (map-making), geographic information systems and remote sensing).

Geography is chorology, the study of place, whereas history is chronology, the study of time. Geography does, however, consider time in all its analyses. For example, how long has a certain

The Royal Geographical Society

On many historical maps there are expanses of white space simply labelled 'unexplored territory'. *Terra incognita* has captured the imagination of centuries of geographers who sought new discoveries to help understand the world. In 1830 the Royal Geographical Society (RGS) was established in London, UK, as the first professional organisation devoted to the advancement of geographical science.

The RGS has supported some of the most famous explorers of all time. New Zealand explorer Edmund Hillary led expeditions to the South Pole and other remote corners of the Earth, but it was his and Norgay's first ascent of Mount Everest in 1953 for which he is most known. David Livingstone was an explorer, scientist, missionary, doctor and anti-slavery activist, who during the mid-1800s spent 30 years exploring almost one-third of the African continent. Charles Darwin served as naturalist aboard HMS *Beagle* on a British science expedition around the world. Irish explorer Ernest Henry Shackleton commanded an attempt to reach the South Pole in 1907, coming within 179 kilometres of the goal in 1909, and from 1914 to 1916 attempted to reach the Ross Sea from the Weddell Sea by crossing the Antarctic continent.

The RGS remains active today regionally and internationally, with a membership of over 13,300. It supports research, education and training, together with the wider public understanding and enjoyment of geography. The RGS hosts over 150 lectures and conferences annually, maintains one of the most prized library and map collections in the world, and publishes three scholarly journals, a research bulletin, a newsletter and a popular magazine (*Geographical*). It runs three overseas field research programmes and operates an Expedition Advisory Centre for training young field scientists.

Even today, although there are no continents that have never been explored, *terra incognita* still exists. Unknowns are ever present, whether it is the biodiversity of the Front Range of Colorado, the potential for volcanic activity under Lake Yellowstone, or the effect of groundwater withdrawal under London. With each discovery, thousands of new questions arise each day. Geographers, and the RGS, are still exploring.

Geographers testing water in a stream in southern Alabama. Photo: Dr M Fearn, University of South Alabama

glacier been retreating? Does its ice record show any advance in the past? Historical geography is even a recognised specialisation in the field.

All environmental characteristics and phenomena take place somewhere on the Earth's surface. Therefore, geography is an important discipline in understanding the environment. Geographers play a crucial role in addressing local, national and global environmental concerns such as acid rain, hazardous waste, housing and world population growth. Geographers might conduct climate research at the National Oceanic and Atmospheric Administration or the UK's Meteorological Office, delineate enumeration districts at the UK National Statistics Agency, plan for new housing and parks for Kent County Council in England, or plan flight routes for a major airline. Although geographers can be found in a wide variety of government, private industry and non-profit jobs, they all share a holistic approach to the Earth – examining the Earth as a system.

Scale is central to the study of geography – how much land area does a certain phenomenon affect? Does a certain local issue share characteristics with a global issue? Another important theme to geography is spatial relationships. More than knowing the absolute location of people and places, spatial relationships are the patterns, linkages and trends of phenomena. For example, what is the spatial pattern of the decline in the water level of the Ogallala Aquifer on the Great Plains of the USA? What is the relationship between the past century of precipitation records and certain types of farms, feed lots and rangeland? What water laws exist that make this sit-

uation better or worse? What conservation or legal action can be done to reverse the decline?

WEBSITES

Royal Geographical Society, with the Institute of British Geographers:
www.rgs.org
Association of American Geographers:
www.aag.org
Geographical Association, the world's largest geography education organisation, based in Sheffield, England:
www.geography.org.uk
National Council for Geographic Education, a North America-based geography education organisation:
www.ncge.org
National Geographic Society:
www.nationalgeographic.com
Geography resources at About.com:
http://geography.about.com
'Geography matters!' Paper from Environmental Systems Research Institute:
www.gis.com/whatisgis/geographymatters.pdf

GEOTHERMAL ENERGY

The temperature at the centre of the Earth is around 6000°C – hot enough to melt rock. Temperature rises 1°C for every 36 metres down from the surface, so that even a few kilometres under the Earth's surface the temperature can be over 250°C. Geothermal energy – from the Greek *geo* (Earth) and *therine* (heat) – has been used as an energy source for thousands of years in some countries, for cooking and heating. During the twentieth century, as concern for the environmental effects of using fossil fuels grew, large-scale geothermal resources began to be tapped.

Why is the Earth, originating from a completely molten state millions of years ago, still warm? Why has it not cooled and become solid many thousands of years ago? The ultimate source of geothermal energy is decay of radioactive materials deep within the Earth. In most areas, this heat reaches the surface in a very diffuse state. However, due to certain geological processes, some areas are underlain by

Wairakei Geothermal energy plant in New Zealand, photograph by Joseph Kerski. Because of its location along the boundary of the Pacific Plate, New Zealand contains many hot-spring areas and several active volcanoes. In the early 1950s, with no oil and little hydropower on the North Island, the world's second large hot-water field was developed at the Wairakei field. By 1960, the power plant was generating 69 megawatts electrical (MWe) of electricity and now produces 157 MWe

relatively shallow geothermal resources. These areas are where geothermal energy is feasible.

Geothermal energy generation involves pumping water down via 'injection wells' to the hot rocks below and using recovery wells that draw up hot water and steam. The steam is then purified, and may be used to drive turbines that power electric generators, or it may be passed through a heat exchanger to heat water to warm houses.

Three main types of geothermal plants exist:

- Dry steam plants: use the steam from the geothermal reservoir as it comes from wells, and route it directly through turbines and generators to produce electricity. Dry steam plants were the first type of geothermal power generator built. An example of a dry steam generation operation is at the Geysers in northern California.
- Flash steam plants: use water above 182°C that is pumped under high pressure to the surface generators. Upon reaching the generation equipment the pressure is suddenly reduced, allowing some of the hot water to convert or 'flash' into steam. This steam powers turbines and generators to produce electricity. The remaining hot water not flashed into steam and the water condensed from

the steam are generally pumped back into the reservoir. Flash steam plants are the most common type of plants in operation today.
- Binary cycle plants: here the water or steam from the geothermal reservoir never comes in contact with the turbines and generators. Rather, water is used to heat another 'working fluid' that is vapourised and used to turn the turbines and generators. The advantage of binary cycle plants is that they can operate with lower temperature waters by using working fluids that have a lower boiling point than water, and produce no air emissions.

The first geothermal power station was built at Landrello, Italy, followed by the Wairakei Station at New Zealand and others in Iceland, Japan, Philippines and the USA. Geothermal energy is most useful when the rocks are very hot and close to the surface. It is most economically viable on a large scale in volcanically active places such as Iceland and New Zealand. Researchers have also experimented with hot dry rock projects in England, France and Germany to see if water pumped into hot dry rocks can be heated economically.

Advantages of geothermal energy are that geothermal fields produce only about one-sixth of the carbon dioxide that a relatively clean natural-gas-fuelled power plant produces, and very little, if any, of the nitrous oxide or sulphur-bearing gases. Only excess steam is emitted by geothermal flash plants. No air emissions or liquids are discharged by binary geothermal plants, which are projected to become the dominant technology in the near future. Geothermal energy is available from these fields every hour of every day throughout the year. Geothermal power stations do not take up much room, because they do not need to store or transport fuel. Once the stations are constructed, the energy is almost free. It may need a little energy to run a pump, but this can be taken from the energy being generated. The energy is renewable as long as not too much cold water is pumped down to the underlying rocks.

Salts and dissolved minerals contained in geothermal fluids are usually reinjected with excess water back into the reservoir at a depth well below groundwater aquifers. This recycles the geothermal water and replenishes the reservoir. For example, the City of Santa Rosa,

California, pipes the city's treated wastewater up to the Geysers power plants to be used for reinjection fluid. This system will prolong the life of the reservoir as it recycles the treated wastewater.

The chief disadvantage to geothermal energy is that while it is confined to reservoirs of steam or hot water – hydrothermal resources – there are not many places in the world where such a power station can be built. A geothermal site may run out of steam, and hazardous gases and minerals may come up from underground, which can be difficult to dispose of safely. Some of these solids are now being extracted for sale (such as zinc, silica and sulphur), making the resource even more valuable and environmentally friendly.

Hot rocks must also be fairly close to the surface and of a type that can be drilled through. However, Earth energy can be tapped almost anywhere with geothermal heat pumps and direct-use applications. Other enormous worldwide geothermal resources, such as hot dry rock and magma, are awaiting further technology development.

Geothermal resources range from less than 90°C to greater than 150°C, with the highest temperature resources usually reserved for electric power generation. Uses for low and moderate temperature resources can be divided into two categories: direct use and ground-source heat pumps:

- Direct use: the heat in the water is used directly (without a heat pump or power plant) for heating buildings, industrial processes, greenhouses, growing fish (aquaculture) and resorts. Geothermal heat is used to heat buildings as diverse as those in Iceland or the Town Hall in Southampton, England.
- Ground-source heat: pumps use soil or groundwater as a heat source in winter and a heat sink in summer. The heat pump transfers heat from the soil to the house in winter and from the house to the soil in summer.

In the USA current production of geothermal energy is third among renewables, following hydroelectricity and biomass, and ahead of solar and wind. At the Geysers in Santa Rosa, California, USA, power is sold at US $0.03–0.035 per kilowatt. A power plant built today would probably charge about US $0.05 per kilowatt. Some plants can charge more during peak demand periods.

Despite the environmental advantages of geothermal, the current level of geothermal use pales in comparison to its potential. Current US installed capacity of direct use systems totals 470 megawatts, or enough to heat only 40,000 average-sized houses. Wider geothermal use depends on greater public awareness and financial and technical support.

Case study

Southampton geothermal scheme

Launched in 1986, serving an initial core of consumers from a geothermal well, the district of Southampton, England, has developed a £4-million multi-source heating and cooling system for residents, commerce and industry. The original well, which currently provides 18 per cent of the heat input, operates alongside combined fuel oil and natural gas heat and power generators. These use conventional fuels to make electricity. The geothermal source is 1.6 kilometres underground, with a temperature of 76°C at its source and 74°C by the time it reaches the surface. The water rises naturally in the well to within 100 metres of the surface, and is then pumped to the heat station.

Through an 11-kilometre-long network of tunnels, water is pumped around the city within a 2-kilometre radius of the heat station with just 0.5°C/kilometre in temperature loss. The scheme now delivers more than 30,000 megawatts of heat each year, and serves 20 major consumers in Southampton, including the civic centre, four hotels, the Royal South Hants Hospital, the Southampton Institute of Higher Education and an Asda superstore. The scheme offers substantial capital and operating cost savings to all consumers in the area, and is an excellent example of the combination of geothermal energy with traditional sources.

WEBSITES

Geothermal Education Office, including a map and description of geothermal energy worldwide:
http://geothermal.marin.org/aboutgeo_1.html
International Geothermal Association:
http://iga.igg.cnr.it/index.php
US Department of Energy Geothermal Resources:
www.eere.energy.gov/geothermal
Geothermal Resources Council:
www.geothermal.org
Southampton geothermal scheme:
www.southampton.gov.uk/environment/energy/default.asp

Above, a GPS receiver recording a position at exactly 52 degrees north latitude and 0 degrees longitude, on the Prime Meridian in Hertfordshire, England. Photograph by Joseph Kerski

GLOBAL POSITIONING SYSTEMS

Working out where you are has perplexed humankind for centuries. The Greeks grouped the stars into constellations that represented gods, animals and other shapes, which are still recognised today. Global positioning systems (GPS) use human-made 'stars' – an artificial constellation of 24 satellites, controlled by five ground stations, which orbit the Earth at approximately 20,000 kilometres high. The satellites were developed by the US Department of Defense to provide all-weather, round-the-clock navigation capabilities for military ground, sea and air forces. These satellites broadcast a radio signal that is read by tens of thousands of small receivers that people use to pinpoint their position on the Earth's surface – sometimes down to less than a centimetre of accuracy.

A GPS receiver determines the position on the Earth's surface by measuring the distance from it to the satellites in the sky, using the formula that many people memorised in school mathematics classes: distance = rate × time. The rate that the signals travel is the speed of light. The GPS receiver measures the travel time that the radio signals took to reach the receiver from the GPS satellites. Because it takes only a fraction of a second for the signals to travel 20,000 kilometres, highly accurate atomic clocks are used. A GPS receiver needs at least three satellites overhead to triangulate the signal travel times and therefore determine the position of the receiver. It also has to know exactly where the satellites are in space, and correct for any delays the signal experiences as it travels through the atmosphere.

GPS receivers are now used in mobile telephones, watches, boats, laptop computers, construction equipment, aeroplanes, farm machinery, automobiles and an increasing number of new devices every year. Now that every square metre on the planet has a unique address, the planet can be monitored as never before. Every environmental issue has a geographic component – it occurs somewhere – and therefore, GPS have become a key tool to understanding and protecting the environment.

Combined with geographic information systems (GIS), GPS also have the potential to help people identify and manage natural resources. Intelligent vehicle location and navigation systems will help to avoid congested motorways and find more efficient routes to our destinations, saving millions of dollars in fuel and tonnes of air pollution. GPS, together with satellite imagery of farmland, have ushered in the science of precision agriculture. Farm equipment can apply the correct amount of fertiliser or pesticides on the precise spots where needed, reducing needless use of these substances and thereby helping the environment. Farmers can now manage every aspect of an agricultural operation to improve overall productivity and efficiency – from planting to harvesting and land levelling – and literally work their land by the square metre instead of the square kilometre.

Case study

Using GPS to monitor Mexico's oceans

Tourists flock to Mexico for its beautiful beaches and blue waters. A large portion of Mexico's economy is derived from tourism and from fish obtained from these waters. What would happen if the fish supply suddenly disappeared? Dr Stacey Lyle, a professor in geographic information science at Texas A&M University in Corpus Christi, Texas, collected data in Mexico's Bay of Campeche, west of the Yucatan Peninsula, where decreasing fish catches had been noted. Dr Lyle studied biotic marine life around natural and artificial reefs, such as drilling platforms or old bridges, and created a model of biotic mass around these reefs.

Because the project needed to collect over 100 variables, such as pH, toxins, biotic mass and minerals at different depths, and return to the exact same locations every month for two years, GPS were essential to the study's success. Latitude and longitude measurements were recorded on boats, while in the water, images from digital cameras were tied to a GPS position via a 'time stamp'. The variables are quite different a few metres away from a reef, so the precise locations were critical to the accuracy of the final models. The models built from the GPS-generated locations are now providing insights into how artificial reefs are affecting the environment and the fish in the Bay.

WEBSITES

GPS World Magazine:
www.gpsworld.com/gpsworld/
GPS resource library:
www.gpsy.com/gpsinfo/
The Geographer's Craft Project GPS
information:
www.colorado.edu/geography/gcraft/notes/
gps/gps_f.html
GPS Information Net:
http://gpsinformation.net/
Precision agriculture from the University of
Georgia's National Environmentally Sound
Production Agriculture Laboratory:
http://nespal.cpes.peachnet.edu/PrecAg
Precision Agriculture links from Iowa State
University:
www.soybeans.ae.iastate.edu/
precision-ag-links.html

GLOBALISATION

Globalisation means the growing interdependence and interconnectedness of the modern world. (UK Government White Paper on International Development, 2000)

Globalisation is a process of rapid economic integration driven by the liberalisation of trade, investment and capital flows and by rapid technological change. It affects the clothes we wear, the music we listen to, the food we eat, the jobs we do and the environments we live in. (Oxfam, Oxford, UK)

With the internet spearheading a global communications revolution, the world appears to be getting ever smaller. In a matter of seconds we can communicate by e-mail with almost every corner of the world. The massive growth in international tourism resulting from cheaper flights and higher disposable incomes has opened up parts of the world previously unknown to the West.

Part of this trend of internationalism has been the spread of business and commercial companies as well as cultural values. In some of the world's most remote regions there will be a McDonald's restaurant (McDonald's have 25,000 outlets in 120 countries), a Hilton hotel, CNN on the television and local people wearing Nike shoes and Levi jeans, drinking Coca-Cola. Young people, unable to speak English, will have heard of the footballer David Beckham and will sing the words of English-speaking pop songs. This is globalisation.

The term globalisation was first coined in the 1980s to describe the growing internationalisation of economic development. The concept, however, stretches back over many centuries to the trading empires established by European countries such as Spain, Britain and the Netherlands. This era of exploitation brought huge benefits to the 'mother' countries, but left in its wake many economic and social scars, particularly in Africa and Latin America. During the nineteenth century, international trade expanded greatly as shipping improved. Nation states were increasingly looking outwards to link up with other countries. International migration began to take place on a large scale, as people sought opportunities elsewhere.

After the Second World War Western states resolved to build and strengthen international ties. Over the decades this has led to diminishing national borders and the fusing of individual markets. The fall of protectionist barriers has stimulated free movement of capital and made it possible for companies to establish several bases around the world. Most recently, this trend has been boosted by the development of the internet and massive improvements in telecommunications.

On the face of it, globalisation would appear to be a good thing. Knowing more about different people, their cultures and traditions is enormously enriching. The ability to reach and communicate with people across continents in a matter of seconds promotes friendship, understanding and the sharing of information. Economically, the potential to make use of a global market is clearly hugely beneficial, as countries such as China have demonstrated in recent years. International trade has made us wealthier and has allowed us to lead more diverse lifestyles; it also has the potential to enable the rich to assist the development of the poor.

However, there are many who argue that increasing globalisation is a bad thing and that, in particular, it is increasing the development divide between rich and poor. Legions of demonstrators – a coalition of environmentalists, anti-poverty campaigners, trade unionists and anti-capitalist groups – congregate to protest about globalisation at the annual meetings of the International Monetary Fund and the World Bank. They suggest that multinational corporations are exploiting people in developing countries by, for example, setting up factories to make use of cheap labour. This is particularly true for the manufacturing of clothing in the Far East. Manual workers in the West have also suffered as companies have shifted their production lines overseas to low-income economies. There is real concern that these huge transnational companies are becoming more powerful and influential than democratically elected governments, putting

Case study

Coca-Cola accused of exploitation in India

An article in *India West* in 2003 accused the Coca-Cola bottling plant in Palakkad in the Indian state of Kerala of damaging the environment and exploiting its workforce. Farmers claim that the plant has tapped into the local groundwater supplies, reducing the water in irrigation wells. Fields that used to produce bountiful harvests of rice and coconuts now produce pitiful amounts, threatening farmers' livelihoods. The company claims that lower rainfall totals are to blame, but this fact is disputed by local pressure groups. Ironically, the drinks produced at the bottling plant cannot be afforded by most of the poor local people.

An additional concern has involved the waste produced by the plant. The foul-smelling slurry was originally marketed as a fertiliser; however, farmers developed sores on their skin and found that their coconut palms were dying. Now the waste cannot even be given away.

In the factory itself, which employs only 134 people, pay and working conditions are poor and several workers have been forced to leave due to ill health. One of the attractions of allowing the plant to locate in Palakkad in the first place was the creation of many jobs for local people.

The local authority recently threatened not to renew the plant's licence, but pressure from the USA, which involved bringing to the Indian Government's attention how significant Coca-Cola's investment was in India, led to an appeal. It is thought most likely that the big corporation will eventually win the day and that Coca-Cola will continue to be bottled in Palakkad.

shareholder interests above those of local communities.

While some people believe that globalisation provides opportunities for lifting people out of poverty, others say that it will increase the gap between rich and poor. They say that the West's gain has been at the expense of developing countries. The already meagre share of the global income of the poorest people in the world dropped from 2.3 to 1.4 per cent in the last decade of the twentieth century. National cultures and identities are also under threat, due to the spread of satellite TV and international media networks.

Environmentalists suggest that as large companies strive to be ever more profitable they tend to disregard the environment. Natural resources, such as hardwood timber, are being exploited in countries where regulations are less easily enforced. Huge areas of rainforest are cleared to make way for commercial ranching, and fertile land is being used for the production of cash crops rather than food crops. Intensive farming methods and the use of chemical fertilisers is leading to soil erosion and the pollution of watercourses. Increased use of fossil fuels to power industry is increasing atmospheric pollution and contributing towards global warming.

It is perhaps too easy to view globalisation as an uncontrollable trend. Instead, we should seek to manage it to the benefit of the whole world. As Kofi Annan, UN Secretary General, said in April 2000: 'The central challenge we face today is to ensure that globalisation becomes a positive force for all the world's people, instead of leaving billions of them behind in squalor'.

WEBSITES

World News Network:
www.worldexploitation.com
Guardian Unlimited:
www.guardian.co.uk/globalisation/
0,7368,408592,00.html
Globalisation Guide:
www.globalisationguide.org
Centre for Research on Globalisation:
www.globalresearch.ca
Coca-Cola:
www.cocacola.com

India West:
www.indiawest.com
The Business and Human Rights Resource Centre:
www.business-humanrights.org/
Categories/Sectors/
Agriculturefoodbeveragetobaccofishing/
Foodbeverage

SEE ALSO

migration

GRASSLAND

Grassland along sandhills formed from outwash debris from long-gone continental ice sheets, Nebraska, USA.
Source: Joseph Kerski

The 'typical' vegetated surface on the EARTH might very well be grasslands, not rainforest or coniferous forest. Grasslands are found on every continent except Antarctica. The temperate grassland BIOME occupies some 6.2 per cent of the Earth's land surface. It accounts for about 4.7 per cent of terrestrial productivity and less than 1 per cent of terrestrial plant biomass, but about 6 per cent of terrestrial animal biomass. Grasslands are the largest single component of the Earth's 117 million km^2 of vegetated lands. Grassland climate is extreme, with hot summers and cold winters. Precipitation ranges widely across the grasslands, from < 30 cm/year in semi-desert, shortgrass steppe, to 130 cm/year in some tallgrass prairie systems.

Dominant plants of the grasslands are – as the name implies – grasses, but broad-leaved herbs may be diverse as well. Grasses are adapted to success in highly seasonal environments. These areas are shaped by fire, drought and wind. This has to do with the way grasses grow. Grasses and some of their relatives – generally called 'graminoids' (grass-like) – differ from most other plants in the position of the growing point, the embryonic tissue called meristem. In a typical broad-leaved plant, the meristem is 'apical' (that is, it occurs at the apex – the end – of the shoot). By contrast, graminoids have a basal meristem. When a zebra plucks the top of a broad-leaved herb, it eats the meristem, the point from which new growth is arising. When that same grazer plucks the top of a graminoid, the meristem is left behind, at or beneath the soil surface. When a fast-moving (hence 'cool') fire moves across the grassland, it kills the meristems of broad-leaved plants but generally leaves the meristems of graminoids unharmed. When drought shrivels the leaves of broad-leaved plants, it shrivels the growing point as well, but the basal meristem of graminoids provides protection against desiccation.

Consumers of grasslands include grazers of all sizes, from grasshoppers, through meadow mice, to hoofed animals like bison and pronghorn in North America, and various wild cattle and true antelope in Eurasia and Africa. The grazers of Australian grasslands are kangaroos – not hoofed, but mobile and quite capable of chewing grasses (which are protected from chewing by silica – basically minute sand grains – sequestered in stems and leaves). Many grasses are prodigious producers of seeds, which provide excellent nourishment for mammals (such as kangaroo rats and pocket mice), a variety of birds, and also insects and fungi. The majority of the biomass of a grassland is underground, in roots and various other storage tissues. Hence, there are subterranean resources for consumers who can get to them. The Earth's grasslands are populated by a variety of burrowing consumers, such as pocket gophers in North America, tuco-tucos in South America and molerats in Africa and the Middle East.

Grassland soil tends to be deep and fertile. Therefore, grasslands are the Earth's agricultural 'breadbaskets'. The grass family, including wheat, rice, maize, barley, sorghum, millet, oats and rye, provides some 80 per cent of calories consumed by humans. Because of extensive conversion for agriculture, the most productive grasslands represent vanishing

Savanna

Photo: Reuben Heydenrych

Savanna is a grassland with scattered individual trees or clumps of woodland, a biome with a distinctly patchy structure. This is a fire-prone system with seasonal drought. Savannas are mostly found in the tropics and subtropics, grading into woodland towards the wet end of the ECOTONE and into scrub desert toward the dry extremes. Savannas cover about 10 per cent of the Earth's land surface and contribute about 12 per cent of net primary production. The patchwork of grasses and trees provides great opportunity for specialised herbivores, and savannas support more animal biomass (some 22 per cent of the global total) than any other terrestrial biome except tropical rainforests. That high load of consumers eats about 15 per cent of net annual production, the highest rate of any major biome.

Savanna supports many well-known animals, including the giraffe, zebra, antelope, water buffalo, kangaroo, cheetah, rhinoceros, hippopotamus, monkey, lion, warthog, dwarf antelope, bearded lizard, baboon, kookabura, ostrich and hyena. Burrowing animals found in the savanna include mice, gophers and ground squirrels. Less well-known large herbivores of the savanna include kudi, pryx, margay, capibara, tapir and brindled gnu. In addition, small but abundant termites also are important to the function of the savanna biome.

ecosystems and are perhaps the most endangered of any natural biome. Only about 1 per cent of the original tallgrass prairie of the American Midwest (illustrated in the photograph), remains intact and unploughed.

The University of Minnesota's Bell Museum's build-a-prairie:

www.bellmuseum.org/distancelearning/prair ie/build/

The Illinois Natural History Survey resource on the tallgrass prairie:

www.inhs.uiuc.edu/~kenr/tallgrass.html

The Konza Prairie in the Flint Hills of eastern Kansas is part of the US long term ecological research (LTER) programme:

http://climate.konza.ksu.edu/

Another grassland LTER site studies the shortgrass steppe, a semi-desert grassland near Nunn, Colorado:

www.lternet.edu/sites/sgs/

GRAZING

Some 1.2 billion cattle, together with a billion sheep and half that many pigs, nibble away at the Earth. The number of cattle equals or exceeds the number of people in several countries, including Australia, Argentina, Brazil, Colombia, Nicaragua and Costa Rica. About 60 per cent of the world's agricultural land is grazing land.

The original domestication of wild cattle in the Middle East was probably for ritual or ceremonial purposes, and their use for milk,

Central Montana rangelands, USA. Photograph by Joseph Kerski

horns, hides, bones and dung (for fertiliser) and meat came later. Cattle raising became a mobile form of wealth. The very word 'capital', which comes from Latin, may be derived from cattle. The Spanish *ganndo* originally meant 'that which has been gained' – property (*ganaderia*). Keeping cattle is still considered an especially prestigious activity in some cultures. With the exception of people who kept llamas and alpacas in the Andes, Native American populations had little interest in grass. However, deforestation began in the late 1400s when cattle were introduced to the New World on Columbus's second voyage. During the twentieth century the Pan-American Highway opened up the Central American rainforest to pioneer settlement and grazing, to be repeated later in the Amazon Basin.

As the proportion of animal products in the diet of Western nations has increased, modern governments have seen beef exports as an irresistible source of much needed foreign exchange. Most of the world's beef is consumed by Western cultures; the USA alone produces one-fourth of the world's red meat, but still imports it. Two-thirds of the Central American forest is believed to have been converted to pasture. The coastal plains of Columbia have been converted into a largely manmade grassland in one generation, while the great Amazon rainforest (*selva*) appears more and more like a green carpet full of holes and frayed at its edges, driven by nationalism and political pressure.

Proponents of grazing cite the fact that livestock ownership currently supports and sustains 675 million rural poor in the world. Through the microbiological flora in their second stomach cattle can convert otherwise unusable cellulose (grass) to food. Thus lands that are too steep, too rocky, too dry, too wet, too salty and too infertile to permit cultivation for crops can be used to produce protein by grazing animals. Grazing animals can improve the diversity of grasses by dispersing seeds with their hooves and in their manure. They can also break up crusty ground and stimulate the growth of grass. Arid rangelands are a dynamic and highly resilient ecosystem, provided that the number of people and animals that the land supports remains in balance with their environment. Indeed, the ability to recover after drought is one of the main indicators of long-term environmental and social sustainability of arid grazing systems.

Despite the benefits, 'overdoing' grazing has led to soil erosion, increases in invasive weeds, floods, deterioration of the plant cover, desertification and deforestation, especially the destruction of tropical forests. Largely because of the clearing of tropical forests for pasture (and wood products), the world has become sensitised to large-scale environmental degradation. Deforestation for pasture also influences regional and global climate, and threatens the native peoples who depend on the tropical ecosystem for their survival. Thus far, land has been cheap and labour abundant, providing little incentive for intensification or efficiency, but the situation is changing.

To make matters worse, much non-native vegetation is often planted as feed for livestock, such as the planting of African grasses in Central and South America, disturbing the ecosystems in entire regions. Although palatable to cattle, plants such as pangola and kikuyu became aggressive colonisers and now cannot be eradicated. To compound matters, since the Second World War cattle have been spending the last few months of their lives in feedlots eating grain and agricultural by-products such as cottonseed and soybean cake, sugar-beet pulp, and molasses. Such 'finishing' draws a better price but involves transforming edible protein in the form of grain or soybeans into fat simply to make the meat taste better. Waste from enormous feedlots often goes into large ponds that often rupture and pollute surface and groundwater.

The World Bank estimates that half the world's production of cereal grains – 600 million tonnes – is fed annually to animals. This is an amount that could feed the world's hungry three times over. Two out of every three hectares of cropland in the USA is farmed for animals, with corn, sorghum, alfalfa, other hay and soybeans leading the way. Significant amounts of these crops are exported to developing countries to feed livestock and poultry.

A National Academy of Sciences study found that one to two-thirds of all rainfall in the Sahel comes from evaporation from soil and plants, rather than from the oceans. Hence loss of vegetation from overgrazing and loss of water-holding capacity from soil erosion produces less evaporation and less rainfall. Less rainfall causes a lower carrying capacity, resulting in more overgrazing and erosion – a downward spiral. In Africa, centuries of overgrazing have converted most rangelands to annual grasses that are largely unpalatable, or bare ground. In the USA, where large-scale overgrazing is only about 150 years old, the US Natural Resources Conservation Service has estimated that less than 50 per cent of the West's original rangeland topsoil remains. Most arid rangelands have been overgrazed for over 50 years. The UNEP Global Assessment of Soil Degradation Survey found that 11 per cent of the Earth's vegetated land has been seriously degraded since 1945. Overgrazing by livestock accounts for 35 per cent of this degraded land. Unfortunately, the answer is not for us all to become vegetarians – most grazing lands cannot function as productive cropland.

Where's the beef?

Beef is seen not only as a source of protein, but a symbol of the mass middle-class good life in many countries. But it is an expensive symbol. Its preeminence in our diet is of recent origin. Before 1875 pork was Americans' preferred daily meat, and ham was the choice for honoured guests. For the English, in earlier times it had been mutton and even rabbit. As late as 1950 more pork than beef was consumed in the USA. Although more beef than pork is now eaten, we still talk of living 'high off the hog' and 'bringing home the bacon' and we accuse politicians of 'pork barrel legislation'. It takes 6 to 9 kilograms of grain or the by-products of food and fibre processing to produce 1 kilogram of weight gain in cattle, for a net gain of only 0.1 kilogram of protein. Only half of an average cow becomes beef. An average cow in the USA consumes over 5000 kilograms of range plant material during its lifetime. Sheep eat 20 per cent as much; goats eat 15 per cent as much. Per-capita beef consumption has begun to drop with medical warnings of cholesterol, saturated fats and mad cow disease scares. Worldwide, however, the beef consumption continues to increase as population climbs and Western middle-class values spread.

The same land that cattle and sheep damage today often supported millions of wild grazers less than 200 years ago. Under natural conditions, grazers' hooves help dormant seeds to germinate and establish; they break soil crusts that keep seeds from growing; they trample standing vegetation into mulch that protects the soil and keeps it moist; they turn vegetation into fertiliser; in short, they are 'nature's gardeners'. An estimated 60 million bison helped build prairie soils up to 3 metres deep across North America's plains. The success of this system, however, depends on mobile herds and predators. Mobile herds stay together for protection against predators, graze and trample an area intensively, and then move off to escape their own dung. This gives plants time to recover to regrow their tops as well as their roots. Without predators, even wild grazers bite the same plants again and again, and cause desertification just like domestic livestock.

Holistic range management may offer a solution, seeking to emulate the benefits of wild grazing, with portable fencing, high stock density, frequent moves and limited grazing periods so that plants can fully recover. It also treats the ecosystem as a whole, supports local land ownership and seeks greater involvement by the community. However, several studies have shown that some of these methods still have negative effects on plant succession, range condition and erosion.

WEBSITES

Research on the environmental impact of livestock farming in Europe:
http://agriculture.de/acms1/conf6/ws4sum.htm?&template=/acms1/conf6/tpl/print.tpl
Photographs and essays about the effects of ranching on the western USA:
www.mikehudak.com/PhotoEssays/PhotoEssayDirectory.html
Negative impacts of grazing:
www.sierraclub.org/grazing/livestock/ecological.asp
Benefits of grazing:
http://managingwholes.com/land-grazing.htm
Quivira Coalition's progressive range management:
www.quiviracoalition.org
Savory Center's holistic land management:
www.holisticmanagement.org/index.cfm

SEE ALSO

overgrazing, soil erosion and conservation

GREEN HEAT

Each year, millions of tonnes of household waste is produced by developed nations. In the UK alone, around 20 million tonnes of household waste is generated annually – enough to

Case study

Hot water in Sheffield

One of the most successful burning rubbish schemes has been in Sheffield, UK, where since 1996 Onyx Sheffield (previously Sheffield Heat & Power Ltd) has been incinerating domestic waste to generate steam that is then converted into hot water. This is pumped through a network of underground pipes to supply heat and hot water to over 120 city-centre buildings, including many offices, shops, parts of the university, the Ponds Forge International Sports Centre, the Weston Park Hospital, as well as 2800 houses.

The concept of pumping hot water around a city is not a new one – the Romans first developed the concept of 'district heating' using this method, and in 1851 part of the Great Exhibition at the Crystal Palace in London used district heating. Successful schemes have subsequently been developed in European cities such as Berlin, Copenhagen, Stockholm and Hamburg.

Work began in Sheffield in 1988 with an experimental system that burned household refuse. Since 1996, four incinerators have been burning 115,000 tonnes of refuse each year, producing 36 megawatts of thermal energy and 6.8 megawatts of electrical energy. By switching from traditional fossil fuel energy sources, Onyx Sheffield's customers have helped significantly to reduce harmful polluting emissions – estimated at 31,000 tonnes of carbon dioxide each year. The scheme has cost over £13.5 million to set up, but since 1998 Onyx has had a turnover of £2.7 million, so the scheme has paid for itself.

Case study

Energy from animal waste in India

The concept of producing heat from waste material in less economically developed countries was first developed in Persia in the sixteenth century, in order to heat bath water. Since the late ninetenth century, containers called dung digesters have been used in India, and their usage was given a boost in 1981 when the country's government launched the National Project on Biogas Development.

Waste produce from farms, domestic waste and animal manure is placed in digesters, and with temperatures above 25°C, this mixture starts to ferment, producing a flammable gas called biogas.

Since 1981 over two million digesters have been installed in India villages, with most being designed to produce enough power to serve the daily cooking needs of a household of between four and seven people. As a result, animal dung accounts for around 21 per cent of rural energy use in India, although in some states this figure is as high as 40 per cent.

fill the stadium built in Manchester for the 2002 Commonwealth Games over 350 times. Traditionally, this rubbish has been dumped or buried at landfill sites. However, in the past 20 years experiments have taken place looking at the possibility of burning this rubbish and household waste to produce 'green heat', largely because of the high calorific value of the plastics incorporated within the rubbish.

Environmental concerns had to be taken into account, especially the release of toxic substances into the atmosphere. But by the mid-1990s scientists had designed incinerators that minimised the amount of harmful pollution, and schemes began in several cities to harness the power of rubbish.

WEBSITES

Planet Energy, with an overview about energy from burning rubbish:

www.dti.gov.uk/energy/renewables/ ed_pack/waste.html

Details about Onyx's work in Sheffield:

www.greenenergy.co.uk

SEE ALSO

incineration

GREEN REVOLUTION

Since the beginnings of human civilisation famines have been a constant and deadly threat, when food supplies have failed to meet the needs of people. The world's worst recorded food disaster of recent times happened in 1943 in British-ruled India. Known as the

Bengal Famine, an estimated 4 million people died of hunger that year alone in eastern India. It was a human disaster that scientists vowed should never be repeated.

In the 1790s Thomas Malthus predicted that human populations could grow geometrically (1, 2, 4, 8, 16 and so on), but food production could only grow arithmetically (1, 2, 3, 4 and so on). Famines, therefore, would hold future population growth in check. For nearly 160 years following Malthus, his predictions were partly true. During the 1950s and 1960s, as death rates fell in countries like India, populations began to grow rapidly. There were real concerns about the ability of these countries to feed themselves in the future, and famines seemed more and more likely to occur. Research in Mexico, funded by the Rockefeller Foundation, the World Bank and the United Nations Food and Agriculture Organisation (FAO), led to the discovery of new types of crops that, if introduced into developing countries, could increase food production significantly. The name given to the programme to introduce new technologies into agriculture was the Green Revolution.

In India the Green Revolution was introduced in the late 1960s. It involved several technological changes, which together turned India into one of world's leading agricultural producers. High yielding varieties (HYVs) of seeds were introduced. One such variety was a drought-resistant Mexican rice called IR8, which had an average yield of 5 tonnes per hectare, compared with the traditional yield of 1.5 tonnes per hectare. The IR8 crops also ripened in four months instead of five, allowing a second crop to be grown during a single year, a practice called double cropping. In addition, HYVs had shorter and stiffer stems, making them more resistant to wind and rain and allowing more crops to be planted per unit area.

The advent of new seeds was accompanied by the introduction of tractors, mechanised ploughs and harvesters, which replaced traditional ploughs drawn by water buffalo. Grants and loans helped farmers introduce the new crops.

The high yielding varieties and the advent of double cropping meant that much greater demand was made on soil nutrients. As a result, soil fertility had to be maintained by the application of fertilisers, and new irrigation systems had to be built to ensure adequate water supplies.

The introduction of the Green Revolution has had significant economic, social and environmental consequences, and several advantages and disadvantages can be identified. Its advantages were:

- Farmers who could afford the HYVs and fertilsers increased yields by at least three times. Higher incomes allowed the successful farmers to invest more money in their farms, especially on fertilisers and machinery.
- Faster growing crops allowed double or triple cropping.
- Farmers could increase the variety of crops they grew, for example wheat, rice, vegetables and maize.
- The increased output created a surplus that could be sold, raising farmers' incomes and standard of living. It also alleviated food shortages in the cities and reduced the costs of imports of food, thereby reducing the cost of food for those living in the cities.

Case study

The Green Revolution in India – the experience of three farmers

Mr Shah

I own 1.5 hectares of irrigated land. I had to borrow some money to buy a pump to distribute the water to my rice crops. I am able to grow two crops of rice a year now because of the extra water and the new HYV seeds I use. I also buy fertilisers and pesticides to use on the crops. I learnt all about these new methods from a demonstration farm nearby, paid for by money from the Overseas Development Association in Britain. There were expert farmers to teach us the new methods. I keep enough food to feed my family – my wife, seven children and me. I sell the rest in the nearby town and use the money to buy other foods. I also save some to buy more land so that I can grow more rice. I do most of the work on the farm by hand and the family helps at busy times. I own two water buffalo and a plough to help work the land. The Green Revolution has really helped – I grow more and make more money. My family is better fed and my farm is growing in size.

Mr Patel

I own 0.1 hectares of land. I borrowed the money to buy it from a money-lender because I could not get credit from the bank. I can only produce one crop of rice a year because I have to wait for the monsoon rains – I can't afford irrigation. All the work is done by hand. The family helps out and we have a simple plough. Most of the rice is used to feed the family, so I also work for a neighbouring farmer to earn some money needed to repay the money-lender and buy seeds. The seeds aren't the best type but they would need large amounts of fertilisers that I can't afford anyway. I can't afford the HYVs, irrigation and fertilisers, so the Green Revolution is not helping me.

Mr Bagchi

I don't own any land so I work for another farmer who owns 5 hectares. I live on the farm with my family and we work the land as if it was our own. I don't get paid any wages but I get a share of the crop – enough to feed the family, but there is rarely any left over to sell. The Green Revolution hasn't helped me because I don't own any land and I don't get any benefit from the bigger yields.

Disadvantages were:

- Many poor farmers could not afford to buy the HYVs and other new technologies, so their yields remained the same. Social inequalities increased as rich farmers became richer and poor farmers remained poor.
- Poor farmers who borrowed money found it difficult to pay back their debts. Some of them moved into the cities to look for work, thereby fuelling urbanisation, with its associated environmental problems. Overcultivation became a problem in some areas, leading to land degradation.
- The HYVs needed the application of more fertilisers and pesticides, which are expensive. Overuse of chemicals can lead to pollution of groundwater, which is a major issue in areas dependent on aquifers for their drinking water supplies.
- Irrigation schemes led to some farmers losing their land and homes. In India irrigated land rose from 32 per cent in 1970 to 43 per cent in 1990. In some areas, such as the Punjab, this led to a lowering of the groundwater table. In more arid regions, excessive irrigation can increase the concentration of salts at the surface. The water dissolves salts in the ground, which then become deposited on the surface as water evaporates. A hard 'salt pan' is extremely infertile.

Overall, the Green Revolution raised total output of grains. Rice production tripled in South Asia in 20 years. In India grain imports fell from 10 million tonnes in 1967 to 0.5 million tonnes in 1977. Today India has a food surplus and imports very little wheat and no rice. This is all the more impressive considering the population has grown from 450 million in 1967 to 1.2 billion today. Therefore, the Malthus' prediction that food production could not keep pace with population expansion has not held up during recent decades. However, there has been a steep environmental price that we may have to pay for decades to come.

The disadvantages of the Green Revolution and other intensive forms of agriculture mean that these agricultural systems are now being challenged. The term Green Revolution is being increasingly used to refer to organic farming and the growing movement away from artificial inputs of fertilisers, pesticides, energy and machinery to more sustainable forms of agriculture.

WEBSITES

Examples of how some developing world countries are increasing agricultural production using sustainable methods: www.essex.ac.uk/ces/Researchprogrammes/susag
A report about India's Green Revolution: www.indiaonestop.com/Greenrevolution.htm
The impact of the Green Revolution in India: http://edugreen.teri.res.in/explore/bio/green.htm

SEE ALSO

agriculture, famine

GREENHOUSE EFFECT

Greenhouses have long been used to keep plants warm enough to grow, whatever the is weather outside. The greenhouse glass lets in the short-wave energy from the sun but traps the long-wave energy emitted from within the greenhouse. In this way the greenhouse retains a warm temperature, thereby encouraging plant growth. The same principal is used in passive solar energy systems.

The greenhouse effect refers to the part of the Earth's radiational balance involving the absorption and reradiating of the long-wave radiation emitted from the Earth's surface. The Earth's atmosphere is transparent to the shorter wavelengths we associate with visible light, but not to the longer wavelengths of infrared radiation. We can see the sunlight streaming in, but cannot detect the infrared emission from the Earth's surface. Our only way of detecting the radiation is to look at images taken from a satellite sensor that detects these invisible infrared wavelengths. Only about 20 per cent of the heating of the atmosphere occurs through the absorption of sunlight on its way down to Earth. Much of the rest results from heat moved from the ground surface to the atmosphere by ground radiation, and the processes of conduction, convection and the latent heat released when water vapour condenses.

Temperatures on the Moon are very different to those on the Earth. They can range from 123°C to –153°C, because there is no atmosphere and therefore no greenhouse effect on the Moon. On the Earth the heating of the atmosphere helps temporarily 'store' heat in the Earth-atmosphere system, allowing temperatures to remain moderate when the Sun goes down. Without this feature, temperature extremes on the Earth would be like that of the Moon, rendering the planet inhospitable to most forms of life. Therefore, although much of the recent publicity surrounding the greenhouse effect is negative, without the greenhouse effect there would be no life on the Earth.

Certain gases in the atmosphere are particularly good at absorbing long-wave terrestrial radiation, and where they are more prevalent, more heat will be cycled around. These are the so-called greenhouse gases. The most important greenhouse gas is water vapour. This varies in quantity from place to place and time to time. When the water vapour content of the atmosphere is low, such as in a desert, more long-wave radiation is lost and temperatures plummet rapidly after sunset. In very humid environments, such as a tropical rainforest, much of the day's heat is retained by the large amount of water vapour in the air. This partly accounts for the fact that cloudy nights are usually warmer than clear nights.

Why is each day's coldest temperature right before dawn? The back-and-forth energy transfer cycle of the Earth's atmosphere is not completely efficient – some long-wave terrestrial radiation is lost to space. The lack of absorption at certain wavelengths is called a window. With each daily cycle of long-wave energy absorbed and emitted by the Earth, some goes 'out of the window' and is not reradiated to the surface to heat it again. The steady drop in temperature during the night is caused by the fact that no new energy is received by the Sun. Less and less of the long-wave energy from yesterday's sunlight is available to cycle around to heat the surface.

An important greenhouse gas is carbon dioxide. Although the amount of carbon dioxide in the atmosphere is very small compared to the other gases, this small amount is very efficient at cycling long-wave radiation around in the atmosphere. Carbon dioxide has always been part of the atmosphere and is very

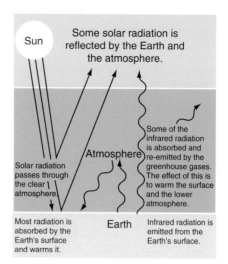

Greenhouse diagram, from the University of Maryland College Park

important because it is used in photosynthesis, providing the base for all the Earth's surface food webs. Amounts of carbon dioxide in the atmosphere ebb and flow with the seasons. When increased photosynthesis in the northern hemisphere occurs during the summer, carbon amounts in the atmosphere decline. When winter comes to the northern hemisphere, levels increase again because the plant life is not storing as much carbon. The northern hemisphere is more influential than the southern hemisphere, because it has more land surface.

Much carbon dioxide dissolves in water. Therefore, oceans that hold huge amounts of water act as a sink for the carbon dioxide. Some of the carbon dioxide in the oceans ends up in carbonate rock, which removes it from the cycle for the long term.

Human activities are increasing the atmospheric content of carbon dioxide in the atmosphere, causing alarm that the additional heat held will warm the planet. Raising the Earth's average temperature, it is believed, will result in multiple adverse consequences. Most of the additional carbon dioxide in the atmosphere has been released through the burning of fossil fuels and from land use changes. One land use change is deforestation, where the burning of the trees releases their stored carbon, and their elimination removes plants that would have provided long-term storage of carbon. Carbon in coal, oil and natural gas was stored there during periods in geologic time when there was more carbon dioxide in the atmosphere and when ancient environments allowed much of it to be sequestered in rock layers.

Another important natural greenhouse gas is methane, which is the natural gas we use to heat homes and generate electricity. Land use practices such as draining swamps, cultivating paddy rice, land-filling of organic waste and increased animal husbandry all produce increasing amounts of this gas. This gas acts like carbon dioxide, absorbing and reradiating more and more terrestrial radiation, thereby increasing temperatures.

WEBSITES

EPA information on the greenhouse effect: http://yosemite.epa.gov/oar/globalwarming. nsf/content/Climate.html

Public Broadcasting Service greenhouse information: www.pbs.org/wgbh/nova/ice/ greenhouse.html

University Corporation for Atmospheric Research: www.ucar.edu/learn/1_3_1.htm

Physical Geography.net's greenhouse effect resources: www.physicalgeography.net/fundamentals/ 7h.html

UK Met Office: www.met-office.gov.uk

SEE ALSO

climate change

GREENWAYS

Cairo on the Nile. Paris on the Seine. London on the Thames. New York on the Hudson. Many towns grew to become major world cities because of the commerce that their rivers provided. The chances are, though, that if you stand on the banks of a river running through a city today you would be sharing the embankment with litter, an abandoned building, a busy road or an industrial site. As cities evolved during the nineteenth and twentieth centuries, the very rivers that nurtured their growth often became symbols of the worst products of urbanisation – derelict landscapes ridden by crime and pollution.

People have begun to take action, turning river embankments into urban greenways – linear open spaces containing parks, paths, beauty and recreational opportunities. Rivers are the perfect places to start, but it does not end there. Soon, local people reflecting a consensus of community needs and concerns began converting canals, abandoned railways, ridges, hedgerows and other linear features. Urban greenways grew from individuals and cities to regional, national and international movements. Some greenways span hundreds of kilometres through every major **BIOME** on the Earth. All are unique, created through local initiative and reflecting a consensus of community needs and concerns.

Most greenways are either ecological, recreational or cultural/historical. Ecological greenways provide habitats for wildlife, help maintain biodiversity, enhance water quality, or serve other functions such as carbon storage. Recreational greenways comprise networks of trails on land or water, while cultural/historical greenways include corridors that serve an array of functions, from preservation of historic or scenic corridors to flood storage, economic development, housing and transportation.

The move towards greenways, according to the urban planner Searns, can be traced to city planners who promoted boulevards and parkways, leading the 'city beautiful' movement beginning in the late 1800s. The Bronx River Parkway of the 1910s awakened people to the possibility of creating beauty and recreation in cities. Unfortunately, this type of thinking also caused major roads to be built during the twentieth century for the purpose of clearing slums, which fragmented dozens of cities and became a key factor in urban decay. However, trail-oriented, recreational greenways followed the parkway movement, and planners even now are trying to link corridors in regional, national and continental networks.

Many railroad lines in the USA were abandoned after the 1960s, following the rise of interstate trucking and the demise of passenger trains. In 1983 the US Congress established 'railbanking', the preserving of rail lines for trails, leading to a US Supreme Court

Sunlight filtering through the trees along the Sparta-Elroy Trail, Wisconsin, USA. Photo by Robin and Ted Kline

decision that affirmed the statute's legality. Since then, two federal authorisations have shifted a small portion of federal transportation spending to bicycle and pedestrian projects. Although the concept is still disputed by some landowners with adjacent properties, since 1990 US $3.8 billion in federal funding has created more than 1000 trails, covering 17,700 kilometres.

One of conservation's principles is that utilising natural resources does not, in itself, pose a threat to the environment. It is the manner in which people use resources that dictates whether their activities are detrimental or benign. New patterns from urbanisation can threaten ecological functioning, causing loss and fragmentation of natural spaces, degradation of water resources, and a decreased ability for nature to respond to change. Advocates of greenways state that if we are to understand our impact on the landscape, we must learn to see it not as independent pieces, but as intricately connected parts of a larger whole. Greenways must be planned to consider their impacts on natural processes. On the positive side, they can provide recreation, beauty, plant and animal

habitats, and conduits for plants, animals, water, sediments, chemicals and seeds. They can also be a barrier preventing movement, a filter allowing some things to pass while inhibiting others, and a sink for trapping sediment, toxins or nutrients.

The greenways movement was aided by increasing recognition that wildlife needs to move across large areas searching for food, nesting sites and mates. Urbanisation not only reduces open space, it fragments open space. Encouraging young animals to move out to seek new territories avoids overcrowding of existing habitats and allows recolonising areas from which animals have disappeared. People and governments have begun setting aside linear wildlife corridors to increase the effective amount of habitat that is available for species, to aid biodiversity and allow populations to interbreed, improving long-term genetic viability. This is especially important for migratory animals and those with large home ranges. Corridors and greenways are particularly beneficial along riverbank corridors, where they provide connectivity for land and water-based mammals, reptiles, birds, amphibians and fish. Ideally, urban and rural parks and open spaces

Case study

Corridors and the red squirrel

Creating wildlife corridors to allow animals to migrate between otherwise isolated patches of protected areas has been a popular although somewhat controversial idea ever since the eminent Harvard biologist E. O. Wilson first proposed it during the 1960s. However, new studies indicate that such corridors may indeed help improve some species' chances of survival. Red squirrels are a good example of how corridors can work. These animals, the only native squirrel to the UK, used to inhabit much of the islands, but widespread deforestation in the nineteenth century pushed them into small patches of Scotland forest. There the red squirrels faced pressures from their larger cousins, introduced American grey squirrels, and their numbers dwindled.

For more than 100 years red squirrels remained isolated, breeding largely within their own immediate groups and creating genetically distinct populations. Then in the 1950s, some of the land separating the squirrel groups was replanted with trees. By the 1980s the trees had grown to a squirrel-friendly size, creating a corridor – albeit fragmented – of habitat.

However, the habitat corridor was enough for the squirrels. During the next 20 years they used the new patches of forest cover as stepping stones to spread from one patch to the next. Now, according to a recent study published in the journal *Science,* comparisons of DNA samples taken from red squirrels preserved in the 1920s and modern red squirrels show that populations have bridged the gap and are once again genetically mixed.

Other studies also corroborate the beneficial effects of wildlife corridors. A 1998 review of 32 studies in the journal *Conservation Biology,* for example, found that all sorts of animals make use of corridors, even birds and butterflies.

should be linked to form wildlife corridors that can then be joined to outlying larger reserves.

While corridors and greenways may make our communities more pleasant places to live, many caution that we must remain diligent in protecting open space and the environment. The total amount of land in wildlife corridors and greenways is still dwarfed by the amount of open space paved over each year for commercial, industrial and residential development. Corridors and greenways also cannot substitute for large areas of protected habitat, like those in major reserve systems such as **NATIONAL PARKS**. Because of the complexity of land use and ownership, greenways are politically and economically difficult to develop, but the benefits to people and the environment are real.

WEBSITES

Case studies, plans, and history of greenways from American Trails:

www.americantrails.org/resources/greenways

Reports and resources from Halifax Urban Greenway:

www.halifaxurbangreenway.org/reports/links.html

Rails to Trails Conservancy:

www.trailsandgreenways.org

National Trails UK:

www.nationaltrail.co.uk

Information on red squirrels in the UK:

www.redsquirrel.org.uk/RED_ALERT/html/aboutreds.htm

www.swt.org.uk/documents/Information/Information_red_squirrel_conservation.pdf

GROUNDWATER AND AQUIFERS

All water stored underground or moving within the bedrock is called groundwater. Some water moves quickly through cracks in rocks to find its way back to the oceans or to re-emerge at the surface once again. Some, on the other hand, can remain underground in store for centuries, or move very slowly through tiny pores in the rocks. The average 'stay' of water underground is 40 years.

All groundwater began as precipitation – rainfall, snowfall, sleet, hail, or even dew. As such it is one segment of the global **HYDROLOGICAL CYCLE**. The amount of underground water is a function of the rock type. All rocks are classified as either impermeable or permeable. The former simply do not allow water to pass into them, but the latter can both retain water or allow it to pass through. There are two types of permeability. Porous rocks have a structure akin to that of a sponge, with interconnecting pores or small holes. Chalk is one such rock, quite common in south-east England. Pervious rocks, in contrast, allow water to pass through their cracks only, and not through the body of the rock itself. Limestone is a good example and is found in many parts of the USA, such as south-east New Mexico and from Indiana to Tennessee.

Layers of rock that are porous and that hold a reliable store of underground water are known as aquifers. Aquifers are of great use to humans as water resources. The word aquifer comes from two Latin words: *aqua*, or water, and *ferre*, to bear or carry. Aquifers have many uses. Not only are they a water supply waiting to be tapped by people, but also they act as a water store during times of high rainfall, and so help prevent flooding. The upper level of the underground water store is called the water table. Its level varies according to the amount of rainfall, time of year and the shape of the land surface.

Where the aquifer meets ground level, water will flow out on to the surface as a spring. This can be the start of streams and rivers or simply an area prone to flooding. An easily accessible water supply is created, which is an important factor in the location of settlements. Humans must obtain water either from surface water (rivers, lakes and desalinated from oceans) or from groundwater.

Aquifers have proved especially important in the distribution of settlements in arid and semi-arid zones. Indeed, in such areas settlement is only possible by tapping into aquifers. In southern Algeria there is a limestone plateau 600 metres above sea level, inhabited by the Chekba people. They dig wells into the pervious rock which supply both domestic and irrigation water. In the same region of north-west Africa, the Algeria–Tunisia border is traditionally populated by the Soufas. They are

Overuse of groundwater from aquifers

Overuse of aquifers has a great influence on groundwater levels, and this is currently an important component in the UK's water supply and demand. The more water that is abstracted for domestic, industrial and agricultural uses the lower the water table is likely to fall, potentially reducing supplies for subsequent years. South-east England is relatively low-lying with many permeable rocks, so reservoir construction is not easy. Groundwater extraction from chalk aquifers is therefore the preferred water supply system in this region.

Groundwater levels underneath London have dropped dramatically in the last 150 years, but the situation there is not nearly as dramatic as in parts of the USA and Thailand. Land in states such as California, Texas, Arizona and Florida have all suffered damage due to land subsidence resulting from overuse of aquifers. Land subsidence occurs when large amounts of groundwater have been withdrawn. Aquifers made of softer rocks or from unconsolidated materials compact because the water is partly responsible for holding the ground up. In other words, the rock collapses in on itself. Often such subsidence is not really recognised, because it happens over a large area and a long period of time. In fact the damage can cost vast sums of money. In California's San Joaquin Valley land levels dropped by around 10 metres between 1925 and 1977. This is one of the USA's most agriculturally productive areas, so pumping groundwater for irrigation went on throughout the twentieth century, and is still continuing today – at an increasing rate and with corresponding consequences.

Looming large as a global issue is the unsustainable exploitation of aquifers. For example, in parts of the Middle East water is extracted from 'fossil' aquifers that were formed thousands of years ago when the climate was wetter than it is today. With minimal natural recharge, over-exploitation of these reserves mainly for agriculture will leave future generations with limited water supplies. Demand (and waste) has to be reduced, or means of artificial recharge must be employed, in order to prevent disaster in the future.

fortunate in that their aquifer lies close to the surface so that it is easy to exploit.

All over North Africa people have settled where aquifers provide their water supply. One of the best known is the Qattara Depression in North West Egypt. It actually lies below sea level and so is easy to find in an atlas because its shading is always distinctive. Approximately 96 per cent of the Egyptian population lives along the River Nile but a significant proportion of the rest is in the Qattara region.

WEBSITES

UK Environment Agency:

www.environment.agency.gov.uk

Lovearth Network:

www.aquifers.net

National Atlas of the United States:

www.nationalatlas.gov/aquifersm.html

The US Geological Survey:

http://water.usgs.gov

SEE ALSO

precipitation, water supply

HURRICANES

Hurricanes are among the world's most powerful and destructive storms. They are capable of bringing tremendous destruction to coastal communities – an average of 15,000 people

Hurricane Bonnie, which made landfall in North Carolina USA during August 1998, as taken from the Space Shuttle. NASA

are killed every year through hurricanes. The main hazards associated with hurricanes include extremely high winds (often gusting in excess of 200 km/h), torrential rain (often exceeding several hundred millimetres), storm surges and even tornadoes. In addition to causing death and injury, hurricanes destroy crops and infrastructure, resulting in loss of livelihoods and homelessness. However, despite their destructive nature, hurricanes are an important part of the environment, bringing much moisture to subtropical and mid-latitude lands and cleansing some low-lying marshes of accumulated biomass and sedimentation.

Hurricanes are tropical low-pressure systems spawned over warm ocean surfaces. These destructive storms differ from low pressure systems outside the tropics in that the air that flows into them is uniformly warm and moist. Thus they have no fronts associated with them and are very symmetrical. Satellite images of clouds associated with these storms show that they are about four times smaller than mid-latitude lows, and have a pinwheel rather than comma-shaped cloud shield.

Hurricanes drift westwards with the trade winds, picking up more and more moisture and energy from the warm oceans. Each year there are 50–60 hurricanes in the northern hemisphere and 20–30 in the southern hemisphere. When hurricanes hit land, their full force is felt by the coastal communities. However, as the hurricane moves over the land, it becomes cut off from its source of energy – the water – and it gradually fades and dies. This explains why hurricanes are not land phenomena like tornadoes.

Disturbances in the tropics or a trough in the westerly winds may trigger a hurricane. Latent heat provides lift, and if an area of thunderstorms becomes intense, long-lived and organised, a tropical depression results. Disturbances begin in summer and last into autumn, occurring only when the sea temperatures in the tropics reach 26.5°C. The outflowing air, rather than flowing down and suppressing updrafts, exits the storm high aloft. These tropical disturbances create conditions of lift, triggering rapid convection and the formation of thunderstorms. Downdrafts, which kill thunderstorms, do not exist, so the storm is fed warm, moist air

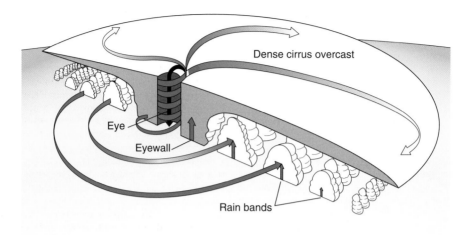

Cross section of a hurricane. From US National Weather Service

by air flowing in symmetrically all the way around. A deep storehouse of heat exists from the 26.5°C water, penetrating 60 metres down, to power the storm. By now, the winds may have risen to 63 km/h, and the storm has achieved Tropical Storm status and is given a name. As the air starts to rise, it cools and condenses releasing latent heat that powers the storm. As more and more air is sucked into the base of the storm, the clouds grow ever taller. The spinning of the earth causes the feature to spiral and form the characteristic Catherine-wheel cloud shape. Lifting air maintains low pressure, and the Coriolis effect speeds up air as it flows in a curved motion.

When winds top 119 km/h, the storm becomes a hurricane. The strongest hurricanes exhibit extremely rapid conversion of water vapour to liquid water. This releases huge amounts of latent heat in 'hot towers' within the wall of cumulonimbus clouds surrounding the calm 'eye'. Because the descending air is warmed and compressed, the eye remains calm and free of clouds.

The direction of storm travel and storm forward speed is determined by overall surrounding pressure conditions and winds; hurricanes cannot steer themselves. Such surrounding conditions in the trade wind area of the subtropics usually favour slow westerly travel to about 30° north or south and then a northward curving and speeding up of the storm as it reaches the area of the prevailing westerlies. If the hurricane moves far enough northward, it will eventually change its direction to the east. Usually these changes in path occur slowly over many days and over broad areas of sea, but sometimes hurricanes will shift quickly and they have even been known to loop.

Strength of hurricanes is measured on the Saffir-Simpson Scale of Hurricane Intensity, which uses wind speed as the factor that determines the category. Weak hurricanes are considered category 1, with the strongest being category 5:

- Category 1 (64–82 knots, 119–53 km/h): No real damage to building structures. Damage primarily to unanchored mobile homes, shrubbery and trees. Some damage to poorly constructed signs.
- Category 2 (83–95 knots, 154–77 km/h): Some roofing material, door and window damage of buildings. Considerable damage to shrubbery and trees, with some trees blown down. Considerable damage to mobile homes, poorly constructed signs and piers.
- Category 3 (96–113 knots, 178–209 km/h): Some structural damage to small residences and utility buildings, with a minor amount of curtainwall failures. Damage to shrubbery and trees, with foliage blown off trees and large trees blown down. Mobile homes and poorly constructed signs destroyed.

Case study

The Galveston Hurricane of 1900

At the dawn of the twentieth century, Galveston in Texas, USA, was booming. It was the country's third busiest port, and the second most heavily traversed entry for immigrants arriving from Europe. The city had more millionaires, street for street, than any other in the USA. Downtown was packed with ornate office buildings, many on the Strand, known then as the 'Wall Street of the South West'. Galveston was the hub of a booming cotton export trade because it had the only deep-water port in Texas at the time. City streets led to imposing Greek revival, Romanesque and Italianate mansions. Streetcars ran along the beach. Bathhouses jutted out like sentinels in the gulf. The nation, too, was bursting with optimism and confidence, with victory in the Spanish–American War and American engineers preparing to take over construction of the Panama Canal. However, Galveston was perched precariously on a barrier island in the Gulf of Mexico, just offshore from the Texas mainland.

The local climatologist had dismissed as absurd the notion that a hurricane could devastate the island city. During the morning of 8 September 1900, people played in a surf that they never had seen so high, under a sky tinged with wonderful colours. By noon, the climatologist, Isaac Cline, was warning people to evacuate. By 1.00 p.m., people were dying in the rising floodwaters, driven by high winds that broke up most of the homes like matchsticks. Cline later described the storm's aftermath as 'one of the most horrible sights that ever a civilised people looked upon' and lost his own wife to the disaster.

It was to be the worst natural disaster ever to strike the USA. Historians contend that between 10,000 and 12,000 people died during the storm, at least 6000 of them on Galveston Island. The saddest part of the tale is that all the citizens could have been evacuated, had they listened to hurricane warnings issued by the Cubans. More than 3600 homes were destroyed on Galveston Island and the added toll on commercial structures created a monetary loss of US $30 million, about US $700 million in today's dollars. Although it began rebuilding right away, Galveston never did recover its status among US cities. Despite the massive engineering work done to raise much of its elevation, dredge the bay and construct a sea wall, it remains vulnerable to another storm.

■ Category 4 (114–35 knots, 210–49 km/h): More extensive curtainwall failures with some complete roof structure failures on small residences. Shrubs, trees and all signs blown down. Complete destruction of mobile homes. Extensive damage to doors and windows.

■ Category 5 (greater than 135 knots, 249 km/h): Complete roof failure on many residences and industrial buildings. Some complete building failures, with small utility buildings blown over or away. All shrubs, trees and signs blown down. Complete destruction of mobile homes. Severe and extensive window and door damage.

Hurricanes die when they move away from the conditions that can maintain them. Eventually most hurricanes will hit land or move over areas of cold water. In the case of land fall, wind speed is lowered and wind direction is changed, destroying the symmetry of air inflow. When parts of the storm receive more air than others, the storm begins to wobble like a kicked child's top. The air coming off land is not as moist as that off the warm ocean surface, so the fuel provided by latent heating is diminished, and cold water deprives the storm of energy because not enough evaporation is going on to maintain the upward flow of air. Most storms that make landfall or drift over cold water retrograde back to tropical storm and then tropical depression status; ultimately they end up, if they are far enough poleward, as extratropical lows.

WEBSITES

Hurricane Bonnie impact studies and oblique photography:
http://coastal.er.usgs.gov/hurricanes/bonnie
USGS Hurricane Mitch reports, maps and satellite images:
http://mitchnts1.cr.usgs.gov

Hurricane information from the University of Illinois:
http://ww2010.atmos.uiuc.edu/(Gh)/guides/mtr/hurr/home.rxml
Images of hurricanes and typhoons from the UK Met Office:
www.metoffice.com/sec2/sec2cyclone/sec2cyclone.html
Hurricane information from the Federal Emergency Management Agency:
www.fema.gov/hazards/hurricanes
1900 Galveston Hurricane:
www.1900storm.com

SEE ALSO

thunderstorms, tornadoes

HYDROELECTRIC POWER

The Ice Ages of recent geological time have left their mark on the landscape in many ways. For some upland areas one of the greatest benefits from recent phases of glaciation is the creation of steep-sided glacial troughs, hanging valleys (where tributaries now plunge down the sides of the overdeepened glacial valley) and large basin-like hollows called corries that can be used as reservoirs. As far as energy is concerned, the greatest benefit left by these Ice Ages has been the steepening of valley gradients, so that the fast-flowing rivers and the plunging waterfalls create a head of

water. This head can be harnessed to generate hydroelectric power (HEP) by allowing the water to pass through a turbine via a natural or manmade channel, often in between two lakes.

The first recorded use of HEP was a clock, powered by a column of falling water, built around 250 BC. By the Middle Ages, watermills had been constructed on fast-flowing rivers in order to provide power for the grain mills. The first use of moving water to produce electricity was a waterwheel on the Fox River in Wisconsin in 1882, two years after Thomas Edison had unveiled the incandescent light bulb.

Since that time, HEP plants have been completed that generate anything from a few kilowatts (enough for a single residence) to thousands of megawatts, power enough to supply a large city. Hydroelectric power is currently the world's largest renewable source of electricity, accounting for around 6 per cent of worldwide energy supply, or about 15 per cent of the world's electricity.

The majority of these HEP plants involved the creation of large dams, both in upland glaciated areas as well as in more lowland areas, where water needs to be stored and there is a high demand for a constant supply of electricity. Some are quite ingenious, such as the one in Snowdonia, North Wales, where so much HEP is produced at the Dinorwic HEP station, opened in 1984 near Ffestiniog, that water is pumped back up through 16

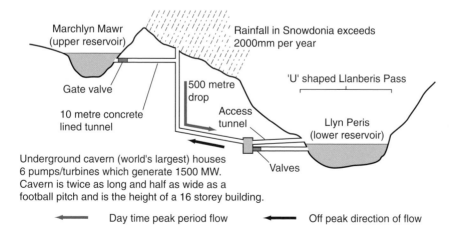

Marchlyn Mawr (upper reservoir)

Rainfall in Snowdonia exceeds 2000mm per year

Gate valve

500 metre drop

'U' shaped Llanberis Pass

10 metre concrete lined tunnel

Access tunnel

Llyn Peris (lower reservoir)

Underground cavern (world's largest) houses 6 pumps/turbines which generate 1500 MW. Cavern is twice as long and half as wide as a football pitch and is the height of a 16 storey building.

Valves

⟵ Day time peak period flow   ⟵ Off peak direction of flow

From David Waugh, *The British Isles*, Nelson Thornes, 1990

kilometres of tunnels to an upper lake, from which it can flow through the turbine again the following day. In recent years, however, there has been some opposition to the construction of large dams, such as the Great Whale (James Bay II) Dam in Quebec, the Gabickovo-Nagymaros Project on the Danube River in Czechoslovakia, and the Three Gorges Dam in China.

Until quite recently it was widely believed that hydro power was a clean and environmentally safe method of producing electricity. Certainly, HEP plants do not emit any of the atmospheric pollutants such as carbon dioxide or sulphur dioxide given off by power plants using **FOSSIL FUELS**. In this respect, hydro power is better than burning coal, oil or natural gas to produce electricity, because it does not contribute to global warming or acid rain.

But recent studies of large reservoirs created behind hydro dams have suggested that decaying vegetation, submerged by flooding, give off quantities of greenhouse gases, equivalent to those from other sources of electricity. Another negative impact is the flooding of vast areas of land, much of it previously forested or used for agriculture. For example, the La Grande Project in the James Bay region of Quebec submerged over 10,000 square kilometres of land. The Three Gorges Dam will result in the creation of a 600-kilometre long reservoir, which will lead to 13 cities, 140 towns and 1300 villages being submerged. In all, 1.2 million people will be forced to find new homes.

Large dams and reservoirs can also have a number of harmful impacts on a drainage basin. Damming a river will alter the amount and quality of water in the river downstream of the dam, as well as preventing fish from migrating upstream to spawn. The silt that is normally carried downstream to the lower reaches of a river will be trapped by a dam, as well as deposited on the bed of the reservoir. This silt can slowly fill up a reservoir, decreasing the amount of water that can be stored, as well as damaging the equipment used for electrical generation.

The tide seems to be turning against large-scale dams, and the Three Gorges Dam may be the last megadam to be constructed. When completed in 2009, its 26 generators will produce about 18,200 megawatts of power, equiv-alent to 10 per cent of China's current needs – the same amount of power that could be produced by 18 nuclear plants or burning 40 million tonnes of coal. But the other environmental impacts could be equally as large, forcing the Chinese to turn to alternative renewable supplies of energy.

WEBSITES:

Further information about water power:
www.ga.usgs.gov/edu/hyhowworks.html
www.energyquest.ca.gov/story/chapter12.html
Further information about the Three Gorges
Dam on the China Online site:
www.chinaonline.com/refer/ministry_profile
s/threegorgesdam.asp
International Rivers Network, with a critical
appraisal of large-scale dam projects:
www.irn.org

SEE ALSO

dams, reservoirs

THE HYDROLOGICAL CYCLE

Water exists on the Earth as a solid (ice), liquid or gas (water vapour). Oceans, rivers, clouds and rain, all of which contain water, are in a frequent state of change. Water evaporates from the surface, water in the air falls as precipitation, water percolates into the ground, and other forms and locations of water are in flux. However, the total amount of the Earth's water does not change. The circulation and conservation of Earth's water is called the hydrological cycle.

Water is crucial to the functioning of all living systems on the Earth, yet only 1 per cent of it circulates in land and weather systems. The rest is in storage, often for thousands of years at a time – in the oceans (97 per cent); ice sheets, glaciers and permanent snow (1 per cent); as well as underground (1 per cent). So much happens with so little – though, of course, it may not seem like a little when it is pouring down upon us.

The Sun is the powerhouse of this huge system, providing the energy for its flows and processes. Through the energy provided by the Sun, water evaporates, primarily from the oceans but also from any wet surface on the land – rivers, ponds, lakes, open ground and buildings. The vapour is transported in the air by the wind, which is also the product of the Sun's differential heating of the Earth's surface. When moist air from over the oceans is blown onshore it is forced to rise up over the surface of the land. As the air rises, it cools. Once the air reaches its dew point (the temperature at which it becomes saturated), it can no longer hold any more moisture at that temperature. Saturation point is reached and condensation occurs to form clouds.

Rainfall, snowfall, sleet, hail and dew are collectively known as precipitation, a word derived from Latin and meaning 'to fall headlong'. With the help of gravity, the water then eventually makes its way back towards the oceans.

Vegetation plays an important role in the hydrological cycle. It intercepts precipitation, 'catching' some of it before it reaches the ground. Water then moves as stemflow down the surfaces of the plant in the direction of the ground. However, some never gets there – it evaporates straight back into the atmosphere. Plants absorb water through their root systems, and at the same time take in soluble nutrients. Moving up through the stems or trunk as sap, the water then emerges through thousands of stomata, tiny holes on the undersides of leaves. This process is called transpiration. It is often combined with evaporation, known as evapotranspiration – and it is remarkably difficult to measure.

Plants transpire much more moisture than we might think. A hectare of maize can give off 35,000 to 50,000 litres of water every day. A mature oak tree transpires approximately 200,000 litres of water per year. Transpiration from plants is therefore one of the most important sources of moisture in the air, although by far the largest source is evaporation from the oceans in the warmest parts of the world.

Water infiltrates into the ground quickly or slowly depending on the permeability – the ability to let water through – of the rock and soil, the relief of the land and the amount of moisture already there. Any remaining water becomes surface run-off (or overland flow), possibly contributing to flood events.

Rivers obtain their water by collecting precipitation that falls within the watershed (or boundaries) of their drainage basin. Water moves towards the channel either on the sur-face or, more usually, underground by percolation and throughflow. Throughflow can be a slow process; it takes water several days, weeks, or even months to make this journey. In more extreme drought conditions rivers can still be fed from the groundwater store and keep flowing. This water movement is known as baseflow.

The hydrological cycle both affects and is affected by the environment. Clearly, the workings of the hydrological cycle are not the same in every environment in the world. Some regions are much wetter or drier than others. The amount of water cycling in the system is the most obvious difference. In deserts, for instance, evaporation and the formation of dew are much more important than in humid areas. All forms of precipitation, river flow, percolation and groundwater flows are much less important in desert zones, simply because there is so much less water in the system.

The fact that the hydrological system is a closed cyclic system has important implications for the environment. Any pollutants that enter the system are not disposed of ('out of sight, out of mind') but simply transferred along the system. So, for example, chemical spills and sewage in rivers ends up in the sea, and agricultural fertilisers washed into the soil eventually find their way into groundwater supplies and drinking water. These interconnections need to be fully appreciated by those involved in activities that have the potential to harm the environment.

WEBSITES

USGS water science:
http://ga.water.usgs.gov/edu
Global change in the hydrological cycle:
www.glowa.org
University of Illinois:
http://ww2010.atmos.uiuc.edu/(Gh)/
guides/mtr/hyd/home.rxml
Biospheric aspects of the hydrological cycle:
www.pik-potsdam.de/~bahc
Mediterranean hydrological cycle:
http://medhycos.mpl.ird.fr

SEE ALSO

precipitation

INCINERATION

One of the most traditional ways of getting rid of rubbish is to burn it. Over time, two basic types of incinerators have been developed that convert waste into smoke. 'Mass-burn' plants are those where all the waste is shredded into tiny pieces before as much as possible is burnt. The other type of incinerator is known as a 'refuse-derived fuel source' plant. This is where the waste is sorted beforehand in order to remove most of the incombustible or recyclable waste before it enters the incinerator. It is more expensive than the mass-burn type, and it is also a rather unpleasant task to sift through the waste. But the advantages are that it can reduce the problems of air pollution or corrosion to the burner.

Despite these differences, both incinerators have several common features. First, the waste is burnt with a flame of between 1500 and 2000°C. Second, the combustion gas then passes through a series of filters so that any toxic and particulate matter is removed. In the case of 'waste-to-energy' plants, this heat can also be harnessed to produce steam that heats buildings, or drives a turbine to generate electricity.

However, there are conflicting opinions over whether or not incineration is a practical and cost-effective means of large-scale waste disposal. For example, landfills are consider-

Waste incinerators worldwide

Country	Number of plants	% of municipal waste incinerated
Sweden	23	55
Denmark	38	65
Germany	47	30
Netherlands	12	40
France	170	42
Spain	22	6
Italy	94	18
UK	34	8
USA	168	16
Japan	1893	72
Canada	17	19

Source: K. Byrne, 1997, Environmental Science, Stanley Thornes

ably cheaper because incineration can require a large capital outlay of up to £100 million. This high initial capital outlay means that incineration is best applied to high volumes of continuous waste. In addition, incinerators do not encourage communities or businesses to reduce their waste production.

Burning waste produces smoke, and if care is not taken in urban areas this can lead to air pollution. Concerns also exist about burning the chlorine-based compounds in plastics, as

Case study

Opposition to incineration in South-East Asia

On 15 March 2004 citizens' groups from no less than 41 countries, including representatives from the Global Anti-Incinerator Alliance (GAIA), lobbied the Japanese Government to encourage it to stop promoting and funding incineration projects in South-East Asia. The focus of their concern was in Malaysia, where a $395 million waste burner is planned in an environmentally sensitive location at Broga. The project will cost around $53 million annually to run and maintain, and could threaten the nearby communities with environmental and health risks. 'We object to the use of Japanese loans, grants, export subsidies and guarantees to promote and/or fund Japanese incinerator projects in Broga and elsewhere', stated Von Hernandez, coordinator of the Manila-based GAIA.

In future, the Stockholm Convention on Persistent Organic Pollutants (POPs), which came into force in May 2004, should compel governments, funding institutions and companies to shift to cleaner and safer technologies, which do not generate hazardous releases into the environment.

well as increasing the volume of nitrous oxides in the air from burning organic waste. Another issue is the creation of residual ash, sometimes containing dioxins and heavy metals.

Despite these concerns, incineration reduces in volume by between 60 and 90 per cent the amount of waste for final disposal. The waste-to-energy methods can also create capital that can, over time, pay for the costs of building the plant in the first place. The ash residue can also be sold and recycled as aggregate material or foundation hardcore in road construction.

These arguments, both in favour of and against incineration, were considered in 1993 by the UK Royal Commission on Environmental Pollution. It concluded that incineration was the best option for waste disposal, and subsequently further improvements have been made to pollution control technology.

WEBSITES

Pyrotechnix Incineration Systems:

www.incineration.com

Global Anti-Incinerator Alliance:

www.no-burn.org

SEE ALSO

waste disposal

INTENSIVE FARMING

BATTERY FARMING, FERTILISERS, PESTICIDES, high-yielding crops, the **GREEN REVOLUTION**, hedgerow removal – all these are a response to the need to grow more food to feed more people, but at what cost to the environment? Are these practices sustainable?

In Western Europe and the USA following the Second World War farm systems were encouraged to intensify, to increase food production to alleviate the shortages and rationing experienced during and after the war. Increased pressure on land to produce crops led to intensive farming, through the use of chemicals (fertilisers, pesticides and other agrochemicals), improved breeding of crops and animals and the increase in the land area under cultivation by removing hedgerows, clearing woodland and draining marshes. This intensification encouraged the development of 'agribusiness' where large companies took over the operation, of farms. Farms became increasingly industrial in their operation, making use of the best scientific research in order to optimise production. Large companies such as Birdseye and Sunkist bought up more small farms.

Intensive farms are characterised by high inputs of labour and/or capital and high outputs per hectare. Farms are usually small and profits high. Traditional farming systems, such as market gardening, dairying, horticulture, battery farming, arable farming in East Anglia and **RICE** cultivation in South and East Asia are all considered to be intensive forms of agriculture.

WEBSITES

Agribusiness Association of Australia:

www.agribusiness.asn.au

Case study

Intensive agriculture in the Central Valley of California, USA

The Central Valley of California is naturally an unproductive semi-desert but since the 1940s it has been transformed into one of the world's most efficient and intensive farming areas. The key to its success has been IRRIGATION – the artificial watering of the land.

A complex system of dams, canals and aqueducts carries water to the Central Valley from the Sacramento River – even to areas that were previously barren. Today there are over 2 million hectares of irrigated land – over 60 per cent of California's total farmland. The irrigation schemes are expensive to construct and maintain and the farmers pay high prices for the water. As a result the crops are expensive to produce and farmers need to recoup their costs by either the sale of expensive crops or through selling large quantities of produce. Over 200 kinds of crops are grown, using the most modern machinery and scientific techniques.

continued over

California is perhaps best known for its citrus fruits, for which cooperatives such as Sunkist have developed. These cooperatives help fund massive electric propellers used to prevent frost developing at blossom time, or large-scale 'smudging' – heating the air by burning oil in pots. Grapes are the state's second-ranking crop by value and about 90 per cent of US wine is produced in California. Raisin grapes are also grown and about 20 per cent of production is exported: a large proportion of this comes to the UK. The Central Valley also produces some of the widest variety of crops anywhere in the world, including almonds, peaches, nectarines, grapefruit, walnuts and dates. Cotton is also a major crop. It benefits from the long growing season, the dryness at harvest time and the flat land. About 25 per cent of total US cotton production comes from California.

Agribusiness is not without its problems. The drive to develop mechanisation – especially for harvesting – has led to new plant development and the reduced need to employ the braceros (migrant workers from Mexico). The impact on the production of tomatoes provides an illuminating example. A machine to harvest tomatoes was developed that lifted the vines and shook them. But since not all of the tomatoes ripened at the same time, the process had to be repeated several times. The preferred solution was to make all the tomatoes to ripen at the same time. This was achieved by trimming young plants to encourage branches to multiply and to bear fruit that ripened together. This solved the picking problem, but at the same time created a packing problem. The traditional boxes held 25 kilogrammes but the machine picked the tomatoes faster than they could be packed. So along came the 'bulk bin', which was much larger. However, now there was a new problem. Tomatoes raining down into the big bin became squashed. As an engineer explained, 'We had to design a tomato to fit the bin: a tomato with a rougher skin; easier to handle, and, to save space, not so round: in fact, a tough, oblong tomato!'

The machine halved the costs of harvesting tomatoes and similar successes have been achieved with melons, strawberries and sweet potatoes, among others. However, the cost of these technical developments requires huge capital investment. This has gradually driven out the small farmer. Land has been amalgamated to make steadily larger farm units, which are increasingly run by large companies with the financial backing to afford the new technology. About 7 per cent of the farms in the Central Valley occupy 80 per cent of the arable land, and farms of over 1000 hectares are common.

These intensive farming practices have come under much criticism, not only in the USA but also in the EU. Sustaining the agribusiness in the Central Valley of California has depleted the underlying AQUIFER and lowered the water levels of many lakes. It has required the diversion of large amounts of water in the Colorado River, to the extent that the river is dry before it reaches its mouth in northern Mexico. Over time the amount of labour employed on the farms has declined, replaced by ever more sophisticated machinery. In addition, the widespread practice of factory farming is regarded by many people as being inhumane. The large-scale use of pesticides and other chemicals are considered a real biological threat, polluting water sources and posing a threat to ecosystems. Since the latter part of the twentieth century, there is a move towards making agriculture less intensive and more ORGANIC.

Journal of Agribusiness:
www.agecon.uga.edu/~jab
The history of intensive farming, from
Palomar University:
http://anthro.palomar.edu/subsistence/
sub_5.htm

SEE ALSO

extensive farming, organic farming

INVASIVE SPECIES

The species found in a particular ecosystem coexist in a delicate balance that has been wrought over thousands of years. Everything within that ecosystem survives because its needs are met. For example, plants need to get enough sunlight and moisture to grow and reproduce, and animals need to get enough food (and avoid being eaten themselves) to produce future generations. Limiting factors such as food and light help to control how much a particular species can grow within an ecosystem. The 'checks and balances' in the system usually keep any individual species from taking over.

At times, however, a new species is introduced – intentionally or accidentally – into an ecosystem that upsets this balance. When this happens, a species that does not have a natural predator can grow out of control. As a result, other species get crowded out if they cannot compete for what they need. In the past, natural barriers such as mountains and oceans helped to limit how often a species could settle into a 'foreign' ecosystem. Modern transportation has changed all this. Foreign species can now 'hitch a ride' much

more easily in shipments or on transportation vessels themselves. More often than not, the introduction of an invasive species is done intentionally by humans, without full knowledge of the potential consequences. For example, European starlings were introduced as an exotic species in Central Park in New York City in the nineteenth century. They are now widespread in the USA, crowding out other species of birds. In the UK the aggressive grey squirrel has wiped out the red squirrel from all but a few parts of the country.

Kudzu, an exotic plant originating in Japan, was introduced to the USA in 1876 at the Centennial Exposition in Philadelphia, Pennsylvania. Countries were invited to build exhibits to celebrate the 100th birthday of the

Case study

Zebra mussels

Many US lakes and rivers are now plagued with a foreign killer – zebra mussels (*Dreissena polymorpha*). Although mussels seem harmless enough, they have the power to wreak havoc upon native bivalve species and interfere with water flow and water quality.

Native to Russia, Poland and the Balkans, zebra mussels spread to most of Western Europe by the early twentieth century by attaching themselves to boats travelling down canals. First discovered in North America in the late 1980s, zebra mussels are suspected of hitching a ride to the New World in ballast water of European trade boats bound for the Great Lakes. From the Great Lakes, the zebra mussels have spread throughout the USA and Canada by attaching themselves to boats or barges. Since zebra mussels can stay alive for

Zebra mussels found in a lake in the Upper Peninsula of Michigan, nearly 100 kilometres away from the Great Lakes. Photograph by Joseph Kerski

several days out of water, they have even spread by attaching themselves to boats carried on trailers across country.

Zebra mussels harm native mussels by attaching to them and impeding their feeding, growth, locomotion, respiration and reproduction. As a result, marked decreases in native mussel population have been recorded following the introduction of zebra mussels. Zebra mussels also compete with native zooplankton for food, which has affected the clarity of the water. This in turn allows sunlight to reach new depths, fuelling an increase in aquatic plant growth.

Beyond the direct impact on biodiversity within an ecosystem, zebra mussels become a nuisance by blocking water supply pipes that service power plants, industry and public water supply. They can also impede boat mobility by adding to boat weight and by infiltrating engines, and can precipitate steel and concrete corrosion and deterioration of dock pilings. In all, the US Coast Guard estimates that the damage caused by the introduction of zebra mussels costs US $5 billion per year.

USA. The Japanese Government constructed a beautiful garden filled with plants from its country. The large leaves and sweet-smelling blooms of kudzu captured the imagination of American gardeners, who used the plant for ornamental purposes. Florida nursery operators Charles and Lillie Pleas discovered that animals would eat the plant and promoted its use for forage in the 1920s. By 1972 kudzu had completely engulfed entire areas of native trees and shrubs, and was declared a weed by the US Department of Agriculture.

Whether the new species is introduced accidentally or on purpose, once it becomes established it can be very hard to eradicate. Many agricultural pests are introduced from other ecosystems, and lead to an increased use of polluting chemicals to try to control them. Attempts to introduce a predator species often compound the problem. In 1935 cane toads were introduced to Australia to control greyback beetles. Not only were the toads unsuccessful in controlling the beetles, but they also became an invasive species themselves, since they had no natural predator in their new land.

Further reading

Raven, P., and Berg, L., 2004, *Environment*, 4th edn, John Wiley and Sons.

Nebel, B. J., and Wright, R. T., 2000, *Environmental Science*, 7th edn, Prentice-Hall Inc.

WEBSITES

US Geological Survey of non-indigenous aquatic species – zebra mussels:
http://nas.er.usgs.gov/zebra.mussel
The story of kudzu:
www.cptr.ua.edu/kudzu
Invasive Species Specialist Group – global invasive species database:
www.issg.org/database
Global Invasive Species Programme:
http://www.gisp.org

SEE ALSO

biodiversity

IRRIGATION

Irrigation is the artificial watering of the land. Present-day irrigation schemes exist on a huge range of scales and many have been in use for centuries. Small-scale low-technology methods, like the shaduf (which lifts water from a river into a set of drainage ditches), have been a feature of the rural Indian landscape for generations. The more sophisticated Archimedes screw has been employed here too. While hard labour is necessary to construct these systems, all are relatively simple to operate and maintain. By making use of local materials, these methods are particularly suitable for remote areas. No outside help is required and, even more beneficial, these systems are non-polluting and sustainable.

Rice feeds more people in the world today than any other cereal crop. Much of it is produced as wet paddy (padi) rice, irrigated by the great rivers of Asia, such as the Ganges, which flows across much of northern India. This method of rice growing is very labour intensive. Each field is surrounded by embankments or bunds, built to keep in the monsoon rains and the irrigation water. Canals take irrigation water from the river to the fields. In many ways this remains a very old, traditional system, but several companies now sell much more sophisticated equipment to richer farmers.

Some of the most sophisticated new methods are the systems employed in the USA, which has a long history of irrigation. The first project established in the Great Plains cereal belt was in 1870. It involved taking water from the South Platte and Cache la Poudre Rivers in northern Colorado. Today there are several large-scale projects in the region, such as the Colorado-Big Thompson Project. This involves diverting water from the western side of the Rockies, where there is a surplus, to the east side of the mountains into the South Platte River basin, in the rain shadow of the mountains. A quarter of a million hectares of agricultural land is currently irrigated, producing high yields of cereals and other crops.

While not all schemes in this region are large (many areas rely on small local wells) it is the large-scale systems that have changed the landscape most significantly. Sprinklers operating on a central pivot system (a long arm with sprinkler outlets all along it propelled around the pivot by either electrici-

Case study

Watering the USA

California's ever growing irrigation demands have led to the development of the most ambitious schemes in the whole of the USA. Perhaps the best-known scheme is the Central Valley Project. Here, water is taken from the wetter north of the state, in the Sacramento Valley, to the much drier San Joaquin and Tulare basins. There the temperature is perfect for agricultural production – all that is needed is the water supply. The water is transferred by canal between the two basins. As this supply has been unable to keep pace with demand, so even grander schemes have been proposed that involve the movement of water on a continental scale.

In the North American continent as a whole the Canadian North West has a huge water surplus. The Mackenzie basin, for example, discharges 9710 cubic metres into the Arctic Ocean every second. At the same time the American South West, primarily California and Arizona, is a region where demand is always likely to exceed supply. In the future water could be transported south using a system of canals, river basins, old riverbeds and new manmade tunnels. It might also be possible to route some of this water to the Great Plains of the USA to enhance agricultural production there. Mexico could even benefit from the extension of such a scheme.

The result could be a 'water grid' for the entire continent, akin to the electricity grid. However, the enormous costs of such a project are the key reason why it is not already in operation. It will take some time before the demand is sufficient for the cost to be acceptable. The engineering works would be massive; the USA already has some spectacular water schemes, but this would significantly supersede anything currently in existence.

ty or a hydraulic mechanism) have created numerous lush circular fields in the otherwise dry landscape from the Great Plains of the USA to the Arabian Desert.

There are a number of environmental consequences associated with overirrigation. If water use exceeds natural recharge, as is the case in parts of the Middle East, the resource is unsustainable. In central Arizona some land has subsided by 20 metres since 1950, due to rapid **GROUNDWATER** withdrawal to irrigate the desert's crops, golf courses and lawns. This has caused cracks in foundations of buildings and in roadways, and has led to deep erosional gullies. In such circumstances, either demand needs to be reduced or means found to recharge irrigation supplies artificially. Overirrigation can also lead to rises in the water table, which can in turn promote **SALINISATION**, leading to soil infertility.

Some people believe that future wars will be fought over key resources. Access to irrigation water could be one of these. In 2004 countries with access to River Nile water discussed – and argued – over the percentages of the water to which they were entitled. In such extreme situations, as demand increases with rising populations and improving standards of living,

pressure is placed on existing sources. People will have to become ever more ingenious to provide sufficient irrigation water where it is required – perhaps California's water really will be coming from Canada.

Further reading

White, C. L., Foscue, E. J. and McKnight, T. L., 1979, *Regional Geography of Anglo America*, Prentice-Hall.

Middleton, N., 1988, *Atlas of Environmental Issues*, Oxford University Press.

Reisner, M., 1987, *Cadillac Desert: The American West and its Disappearing Water*, Penguin

WEBSITES

California Irrigation Institute:

www.caii.org

Open University:

http://geography.ou.edu/research/

britishirrigation.html

SEE ALSO

desalination, water supply

LAKES

A lake is a broad term for any body of water collected in a depression on the Earth's surface. Large lakes, if also saline, may be referred to as seas, such as the Caspian Sea and the Dead Sea, both in western Asia. Most lakes are fresh water and there are numerous ways in which they can form. The size and depth of lakes vary greatly, as do the uses to which people put these resources. Some lakes are totally artificial, usually in conjunction with dam construction.

Lakes are an important part of the **HYDROLOGICAL CYCLE**, acting as a store of surface water in the system and a source of water for plant and animal life in times of drought. Some regions of the world take their whole character from the fact that they are based around a lake, for instance the Great Lakes of North America, the Lake District of north-west England, the Highlands of Scotland and the Italian Lakes region.

Glaciation is responsible for many of the lakes in North America and Europe. The Great Lakes were formed when ice sheets melted and filled depressions that the very same ice sheets had earlier eroded. Moraine, a widespread glacial deposit, leaves an uneven hummocky land surface, resulting in the formation of many lakes. Finland's landscape includes thousands of these, as does north-west Canada. The larger lakes in the English Lake District are ribbon lakes, caused by overdeepening of valleys by powerful glaciers. Famous names include Windermere ('mere' is Viking for 'lake'), Coniston Water, Ullswater and Derwentwater. The Lake District's smaller lakes are mostly tarns, or corrie lakes, excavated at the sources of glaciers. Wales is the location of the UK's largest such lake, in Cwm Idwal in Snowdonia.

Lakes are created in other ways. Crater Lake, Oregon, is the result of a massive volcanic eruption forming a deep depression – a caldera – which is now filled with water. Rivers create arc-shaped oxbow lakes. Some of the best examples are found in the lower reaches of the Mississippi River in the USA. East Africa's huge lakes, such as Lake Victoria, are caused by

The world's largest lakes

Name	Location	Surface area (km^2)
Caspian Sea	Asia	371 800
Lake Superior	Canada/USA	82 350
Lake Victoria	East Africa	68 000
Lake Huron	Canada/USA	59 600
Lake Michigan	USA	58 000
Aral Sea	Kazakhstan/ Uzbekistan	36 000
Lake Tanganyika	Central Africa	33 000
Great Bear Lake	Canada	31 800
Lake Baikal	Russia (Siberia)	30 500
Lake Malawi/Nyasa	East Africa	29 600

Case study

The disappearing Aral Sea

The Aral Sea was one of the world's largest lakes (66,300 cubic kilometres in 1960), however, it had shrunk by almost 50 per cent to 36,000 cubic kilometres by the 1990s. The reason for this dramatic reduction in size is the diversion of water from the rivers that feed it, in order to irrigate millions of hectares of cotton fields. As the lake has shrunk, with a corresponding decrease of its depth by 17 metres in places, it has become more saline, with a disastrous effect on wildlife. Fish stocks have been reduced and a once thriving fishing industry has been decimated. Abandoned fishing vessels lie like beached whales on the dried up lake bed, which has itself become a dust bowl. Wind-blown dust contaminates land for hundreds of kilometres around the lake and there is some evidence to suggest that there have been adverse effects on human health.

tectonic Earth movements forming a rift valley as the plates move away from each other.

Human use of fresh water from lakes has led to significant environmental degradation. In some cases entire lakes have dried up as rivers feeding them have been overexploited. For example, as a result of Los Angeles' search for fresh drinking water in the nineteenth century, and an increasing need for irrigation, huge amounts of water were taken from Owens Lake, California. Having had a depth of over 7 metres in 1912, by 1930 it had completely dried up. The dusty sediment left behind then became a hazard to human health as it was blown around by the wind.

Further reading

Philip's Geography Dictionary, 1995, Reed International Books Limited.

Paterson, J. H., 2000, *North America*, 9th edn, Oxford University Press.

WEBSITES

UK Tourist Information (for information on the Lake District):

www.touristinformationuk.com

US Geological Survey Aral Sea Images:

www.glsc.usgs.gov

Council of Great Lakes Governors:

www.cglg.org

Great Lakes of North America:

www.glna.org

US Geological Survey:

http://edcwww.cr.usgs.gov/earthshots/slow/Aral/Aral

SEE ALSO

reservoirs

LANDFILL

A landfill site is a hole in the ground where waste, discarded either by individuals or organisations, is dumped. The waste subsequently decays by physical, chemical and biochemical processes, which may eventually make it harmless.

Landfill has become the main method of waste disposal in the UK and a number of

incentives were introduced in the 1990s to encourage more people to use this method. This included the landfill tax credit scheme in October 1996, which allows landfill site operators to donate up to 6.5 per cent of their landfill tax liability to environmental projects in return for a 90 per cent tax credit.

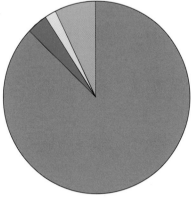

Number of disposal licences	
Landfill	4196
Incineration	212
Treatment	122
Other, including recycling	366

Methods of waste-disposal in the UK in the late 1990s.
Source: K. Byrne, 1997, *Environmental Science*, Stanley Thornes

By the late 1990s landfill accounted for the disposal of 85 per cent of all waste, and 4,196 disposal licences had been granted to landfill operations.

The earliest landfills were simply piles of waste dumped either in huge mounds, often in depressions, or placed in abandoned quarries or mine shafts. However, these early sites often had problems, especially those in saucer-like hollows, since water tended to accumulate in them. This water often became contaminated, known as leachate, and therefore it posed hazards for people, animals and plants.

The water passing through landfill waste assists the micro-organisms by biochemically breaking down the organic matter in the waste, and the subsequent decomposition leads to the creation of a mixture of gases, known as landfill gas (LFG). LFG has a methane content of between 40 and 65 per cent, with the remainder being carbon dioxide, hydrogen sulphide and other gases. The high methane content of LFG, and its ability to migrate upwards towards the surface, therefore poses a risk from fires and explosions.

Another difficulty with early landfill sites was that the untreated rubbish frequently attracted birds, rats, flies and mosquitoes, since these disease-carrying creatures found both food and shelter at the landfill.

To meet these problems, modern landfills are made of both natural and synthetic materials and are designed and operated to contain waste in an environmentally safe manner. The site of the landfill is planned to minimise both the creation and migration of leachate and LFG. Liquid waste is therefore no longer put into landfill, and the moisture content of the material being dumped is carefully monitored. To prevent any percolation, a moisture-barrier layer and other pipes are also present at the base of the landfill. Surface water infiltration is minimised by locating the site away from flood-risk areas, and providing a layer of soil, foam, plastic or other material that covers the waste at the end of each working day at the landfill site.

The main advantage of landfill is that it is a relatively cheap method of waste disposal, and the methane can be collected and used for heating or to generate electricity. But set against this are the problems of finding suitable sites, especially near large urban areas. The land on the rural–urban fringe may be quite expensive, due to its demand for other uses, so this may add to the cost of waste disposal. In addition, since landfill sites generate fumes the local residents may complain if their adjoining land is earmarked for landfill. However, under EU Landfill Directives, the site operators are financially liable for the environmental impacts of the sites. Landfill operators therefore pay the costs of monitoring and controlling sites until they are no longer considered a risk.

Nevertheless, there is still one major problem: LFG is a significant greenhouse gas, and surveys have shown that methane is around 60 times more powerful at contributing to global warming than carbon dioxide. Landfill currently accounts for around 20 per cent of the UKs methane gas emissions, so with this in mind the Department of the Environment announced in 1995 that it would reduce the amount of waste going to landfill by 10 per cent by 2005, and encourage more people to recycle household waste.

WEBSITES

Environment Protection Agency:

www.epa.gov/lmop

Howstuffworks:

http://people.howstuffworks.com/landfill.htm

HM Customs and Excise:

www.hmce.gov.uk/business/othertaxes/landfill-tax.htm

SEE ALSO

solid waste management

LANDSCAPE ECOLOGY

Landscape may connote a rural scene, or a painting of the same, or it may imply a decorative, human-constructed biotic community, such as a garden. But the term has been taken up by ecologists to describe the mosaic of habitat patches and their spatial relationships. Landscape ecology is the 'study of the pattern out of an aeroplane window' or 'the bird's eye view'.

Nature is complex at all scales of observations; a kaleidoscopic pattern across space and time. Ecologists have often sought to simplify nature by defining the concepts of community-types, habitat-types, life zones and other ecological units, realising that these abstractions were only rough approximations of the wondrous and exuberant (even messy) complexity of nature. Landscape ecology is emerging to focus on these environmental patterns and processes and uses concepts from the discipline of **GEOGRAPHY** such as scale and space. Landscape ecology seeks to understand patterns at the complex 'real-world' level where organisms live, in a 'crazy-quilt' of patches of suitable habitat within a matrix of unsuitable or sub-optimal habitat.

Landscape ecology is an interdisciplinary science framework that examines the relationship between spatial patterns of landscape characteristics and conditions of and risks to ecological resources, including forests, rangelands, wetlands, rivers, streams, lakes and urban environmental settings. Local, regional, national and global economies depend upon both goods and services from the environment.

Forests provide materials for many different types of paper products; agricultural systems provide food for the world **POPULATION; CONTINENTAL SHELF** waters provide an abundant food resource; **WETLANDS** and **RIVERS** provide numerous recreational opportunities. Moreover, these resources provide services that sustain ecological goods, jobs and human well-being. Forests help reduce flooding risks by intercepting rain and causing water to trickle down to the **EARTH**'s surface, as well as slowing the speed of surface flow. As a result of gentle impact with the Earth's surface, forested areas are less likely to lose soil. Wetlands and forested areas along streams and rivers (the riparian zone) help reduce the loss of soil off agricultural areas and reduce the loading of excess nutrients (for example, nitrogen and phosphorus) to streams and rivers, which helps sustain a healthy fish population. These interdisciplinary concerns are the focus of landscape ecologists.

Historically, environmental programmes have focused on managing relatively local scale problems. And although it is important to manage the amount of pollution coming from factories, sewage treatment plants and other facilities, many of our environmental problems today result from the cumulative impact of humans across large areas. Studying *all* of the impacts over a large area is what landscape ecology is all about.

Landscapes are dynamic in space and time. Patches develop and disappear. Landscape ecology thus deals with **ecological succession**. Landscape ecology is also concerned with patterns and processes in the physical Earth that underlie and interact with biotic communities. Landscape ecology is a quantitative science and one that is highly dependent on graphical methods to convey its results. The development of computer-assisted mapping technologies (**GEOGRAPHIC INFORMATION SYSTEMS (GIS)**) has been a great boon to landscape ecology, as well as to ecology in general.

An important aspect of landscape ecology as it is evolving is the importance of humans to the environmental pattern. In particular, humans are a frequent cause of *fragmentation* of habitats. The size of a particular habitat patch may have an important influence on the dynamics of species populations within the patch. The smaller the patch, the larger the relative perimeter ('edge') per area. Fragmentation will tend to favour 'edge' specialists and impact negatively on species adapted to patch interiors (whether dense forest or open grassland).

Because of its emphasis on dynamics in the ecosphere over space and time, landscape ecology is also becoming a major focus of conservation ecology. Landscapes, by definition, include a diversity of habitats and hence a diversity of species. Further, because patches recur over the landscape, the landscape includes a degree of redundancy that may allow colonisation of one patch from another, should local extinction occur. Landscape ecology provides another lens through which we can see, understand and ultimately protect our environment.

WEBSITES

Primer of principles and vocabulary of landscape ecology:
www.innovativegis.com/products/fragstatsarc/aboutlc.htm
Landscape ecology information from the US Environmental Protection Agency:
www.epa.gov/nerlesd1/land-sci/intro.htm
The UK's International Association for Landscape Ecology:
www.iale.org.uk/index.html

LANDSLIDES

People have long come to Hebgen Lake in south-western Montana USA to camp, boat and fish in its spectacular scenery. Just 16 kilometres north-west of Yellowstone National Park, people were doing exactly that during a midsummer night in 1959 when a 7.5 magnitude earthquake struck. The strong shaking lasted less than a minute, but sent the mountainside adjacent to the lake crashing into the Madison River Canyon in a huge landslide, hurling 10-metre waves and filling the canyon 1.6 kilometres wide with debris, including dolomite boulders as big as houses. The volume of debris was estimated at between 28 and 33 million cubic metres. The slide caused 28 fatalities and $13 million in damages to property, highways and timber. The campers still lie buried under the debris, as most of the bodies could never be recovered. The landslide debris blocked the Madison River, which began filling up the valley and was named, appropriately, Earthquake Lake. Soon, engineering crews began to dig through the debris, in places up to 100 metres high, because they were worried that a catastrophic collapse of the new dam could send the entire lake in a flash flood to communities downstream.

Landslides involve the movement of soil or rock down a slope, the result of the force of gravity. They include rock falls, deep failure of slopes, slides under the oceans and shallow debris flows. Landslides can be caused by erosion by rivers, glaciers, or ocean waves; rock and soil slopes weakened through saturation by snowmelt or heavy rains; earthquakes and volcanic eruptions; and excess weight from the accumulation of rain or snow. Not all landslides move as rapidly as the landslide after the Hebgen Lake earthquake. Slow movement of landslides and gradual creep of rock and soil downslope can be as destructive as rapid landslides; however, they are not catastrophic, and tend to cause economic damage rather than loss of life. Landslides in populated areas can cause huge loss of life and property. Economic losses due to landslides in Japan alone account for US $4 billion annually.

Debris flows, also referred to as mudslides, mudflows, earth flows or debris avalanches, are debris-laden flows of water, often the consistency of wet concrete. Debris flows move so rapidly down steep slopes (up to speeds of 35 kilometres per hour or more) that they are sudden and unexpected, destroying property and taking

lives. In July 1994 a severe wildfire swept Storm King Mountain, west of Glenwood Springs, Colorado, USA, denuding the slopes of vegetation. Heavy rains on the mountain that September resulted in numerous debris flows, one of which blocked Interstate Highway 70 and threatened to dam the Colorado River.

The world's largest historic landslide was the one that triggered the eruption of Mount St Helens on 18 May 1980 in the Cascade Range of south-western Washington, USA. This landslide was 24 kilometres long, containing 2.8 cubic kilometres of material, which buried 60 square kilometres of the valley of the North Fork Toutle River. The valley was buried with hummocky-surfaced, poorly sorted debris, ranging in size from clay to blocks of volcanic rocks, with individual volumes as large as several thousand cubic metres. Even worse in terms of human life was the 1970 landslide after an earthquake under Huascaran Mountain in the Andes of Peru. The overhanging peak of the mountain collapsed, burying 18,000 people in several towns and villages. A similar event had killed 4000 people in the area only eight years before.

Hundreds of small landslides occurred across the UK during the sustained wet weather of 2000 and 2001, when waterlogged slopes became unstable. No rail service operated between Darlington and York for more than two months while the lines were being cleared of landslide debris. In the UK, coastal areas are particularly vulnerable to landslides, because of waves continually pounding the seaside cliffs, coupled with rising sea levels possibly due to global **CLIMATE CHANGE**. The BBC reported in early 2003 that Lyme Regis on the Dorset coast has been placed on 'landslip watch' following a landslide caused by recent wet weather. Lyme Regis is one of the most active landslip zones in Europe. West Dorset District Council plans to spend nearly £30 million on stabilisation schemes for the town.

All landslides affect their local environment – and in some cases the environment for hundreds of kilometres downstream. Landslides can block river channels, changing the sediments, plants, fish and animal life in the entire river basin. In major landslides, all the soil down to bedrock is carried downslope, taking all the trees and other vegetation with it. Because no soil is left for new plants to grow on, the bare tracks of landslides can remain visible

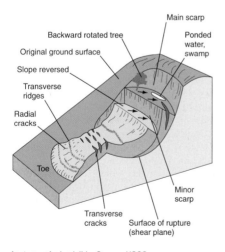

Anatomy of a landslide. Source: USGS

for hundreds of years. Not all the impact is negative. For example, although there is general agreement that landslides result in short-term deterioration of mountain-stream fish habitat, there is a growing feeling among ecologists that landslides may increase the quality of fish habitat in the long term by breaking up stream flow. For example, landslides may deliver large boulders to a stream, thus forming downstream pools that provide quality fish habitat.

WEBSITES

US Geological Survey Landslide Information Center:
http://landslides.usgs.gov
American Red Cross landslides information:
www.redcross.org/services/disaster/
0,1082,0_588_,00.html
Photographs of the Hebgen Lake earthquake and landslide:
http://neic.usgs.gov/neis/eq_depot/usa/
1959_08_18_pics.html
Landslide overview map of the USA:
http://landslides.usgs.gov/html_files/
landslides/nationalmap/national.html

Research on the effect of landslides on the environment:
http://landslides.usgs.gov/html_files/pubs/
hongkong/HongKongJuly21.pdf
Landslide information from the Geological Society of London:
www.geolsoc.org.uk/pdfs/Landsldes.pdf

SEE ALSO

climate change, earthquakes

LEACHING

When rain falls on the ground it is mildly acidic. As it soaks through the soil it dissolves some of the soil's chemicals and transfers them downwards. This is the process of leaching.

Many areas of the world have soils that are formed by the process of leaching – these soils are called podsols. They occur mostly in the damper temperate regions of the world where rainfall totals are high. Vast stretches of land in Canada, Russia and Scandinavia come into this climatic zone. Much of this area is covered by coniferous forest; the leaf litter on the ground, in the form of needles, is itself quite

Case study

Podsols in Eurasia

The word podsol comes from the Russian meaning 'ash-like' because of the grey colour of its upper layers. A podsol is a distinctive soil usually associated with the process of leaching and is characteristic of the northern coniferous forest zones of the world. In terms of latitude they extend across Europe and Asia between 50° north and the Arctic Circle (66.7° north).

The ways in which people can use podsols in this region are limited. As the soils tend to be poor quality, much of the land is used for grazing livestock, mainly sheep. If the soils can be enriched, then cereals may be grown. In the 1960s, when the Soviet Government attempted to expand the cultivable area of the USSR, some podsols in the warmer parts of the belt were ploughed and fed with both chemical and natural fertilisers. The Steppe grassland farming areas spread northwards into much more difficult territory. This expansion proved quite unsuccessful. The area was simply too marginal: soils were too poor and hungry, and also too acidic to maintain any decent yields for more than a few years. Overuse of chemical fertilisers weakened soil structure and constant harvesting took too much organic matter from the soil.

It was soon clear, therefore, that this agricultural expansion programme was a failure, even though the Soviet Government did not want to admit defeat, economic or otherwise. Podsols can be used for agriculture but only within the bounds of their limited potential. On the North German Plains these difficult soils are used for army practice ranges. In much of Eastern and Western Europe it is the natural conifers that are exploited more intensively, to provide for the ever-growing market for paper and softwood furniture products.

acidic and this increases the acidity of the rainwater, promoting the process of leaching.

As the acidic water drains through the soil it dissolves any soluble material with which it comes into contact. The resulting organic solution is known as leachate. These dissolved minerals are important soil nutrients on which plants rely, so, as they are leached, the soil becomes less fertile. As soil formation is a slow process, replacing leached nutrients is also very slow. People can of course add nutrients by spreading fertiliser to help reverse the process of leaching. However, this too may have a detrimental effect on the environment, because it often leads to the increased dependence on fertilisers and pesticides and an unsustainable system. In addition, air pollution and resulting acid rains since the mid twentieth century have increased the rate and extent of the acidic leaching of soils.

As the soil water continues its journey downwards into the lower layers (or horizons) of the soil, its movement tends to slow down as the soil becomes more compacted. The water may well have become saturated with the dissolved chemicals from above and some precipitate out to be deposited within the lower soil layer. Iron oxides can accumulate to form a hard reddish-brown iron-rich layer, known as an iron pan or hard pan.

The presence of a hard pan can cause problems for farmers. Ploughing cannot always break through the hard pan, so only the upper layers of soil are usable. This is particularly the case for poorer farmers with more primitive equipment. Moreover, the hard pan, if continuous, makes the soil impermeable, causing soils to become waterlogged. Concentrations of iron in the soil also reduce its fertility.

WEBSITES

PupilVision.com:
www.pupilvision.com/uppersixth/soiltypes.htm
CST Slovakia:
www.ecosystems.sk/pages/soils.html

SEE ALSO

soils

LEAN-BURN ENGINES

When buying a car, many people simply look at the colour of the vehicle's exterior, the amount of leg room, the standard of the upholstery and the size of the boot. But what really matters, and sadly what very rarely features in the advertising, is the car's safety record, its fuel efficiency and whether or not it will produce vast amounts of exhaust fumes. Since over 600 million vehicles were on the road by the end of the twentieth century, even a modest amount of vehicle pollution reduction would benefit the environment.

Lean-burn engines are designed to operate with a very lean air/fuel ratio. Most modern petrol engines are controlled to run at a chemically correct (stoichiometric) air/fuel ratio (about 14.7:1) to make the three-way catalyst operate at high efficiency and thereby reduce exhaust emissions. Lean-burn engines mix more air with the fuel when full power is not needed, resulting in better fuel economy. Air/fuel ratio in lean-burn engines can be as high as 22:1. When full power is needed, such as during acceleration or hill climbing, a lean-burn engine reverts to a stoichiometric (14.7:1) ratio or richer.

However, a very lean mixture of air and petrol will not ignite as easily as a stoichiometric mixture when a spark is introduced. Several methods can be employed to achieve lean burn, including high temperature, high turbulence and stratification (high concentration of fuel vapour near the spark plug). Therefore, lean-burn engines are often designed with high intake swirl to increase turbulence.

Lean-burn engines became popular during the 1990s, but it was increasingly found that the vehicles, especially those fitted with a pre-chamber system, had high emission rates of nitrogen oxides. As lean-burn engines operate by burning fuel with more air than usual, nitrogen compounds in the fuel are converted with even greater efficiency into oxides of nitrogen. Catalytic converters reduce nitrous oxide emissions in conventional engines, but the exhaust from lean-burn engines requires more vigorous scrubbing.

One approach to removing these nitrous oxides involves first capturing it on some absorbent material and then converting it into a harmless nitrogen gas or re-injecting it into the engine. However the substances that absorb the nitrous oxides also take up sulphur oxides generated from the impurities of sulphur that the fuel contains. The build-up of sulphur oxides on the absorbent material therefore reduces its capacity to take up the nitrous oxides.

Standard catalytic converters use metals such as platinum and rhodium to remove the nitrous oxides, but scientists at the Catholic University of Leuven in Belgium have developed agents based on zeolites that work under lean-burn conditions. Zeolites are mineral-like substances composed mostly of aluminium, silicon and oxygen. Some occur naturally, but others are synthetic, and for many years have been used in the petrochemicals industry for separating and altering the components of crude oil.

One of the most important aspects of the zeolite system is not affected by the presence of sulphur oxides, and scientists are now looking into how they could incorporate this technology into the domestic vehicle.

WEBSITES

ADB.org:

www.adb.org/vehicle-emissions/General/inuse-conversions-1.asp

Energy Technology Research Institute:

www.tokyo-gas.co.jp/techno/eti/ene98/english/13e.html

SEE ALSO

catalytic converters, nitrogen oxides, vehicle exhausts

LEVEES AND EMBANKMENTS

A natural levee is a raised riverbank, formed when floods deposit sediments. When the flood subsides the sediment remains, leaving the levee as the highest part of the floodplain. An embankment is the artificial equivalent of a levee (although in the USA embankments are usually referred to as levees).

When a river floods, its sediment load is deposited all over the floodplain. In order to carry its load a river needs energy. When it has overtopped its banks the river water becomes much shallower and the rate of flow – the energy – drops dramatically, so the river deposits some of its heaviest and largest sediment. In this way sediment builds up on the banks, often remarkably quickly. Finer sediment is then trapped by the larger stones and the gaps are filled in. The levees become increasingly solid and secure.

Some levees grow so large that the flood-water cannot return to the main channel; it has to soak away into the flood plain instead, which is often a slow process. When this happens the river may even end up changing course, as has happened with the River Huang He (Hwang Ho) in northern China.

Levees can be extremely useful as natural flood prevention. By building up the banks, levees increase channel capacity. If the channel can hold more water then flooding may well be avoided for some considerable time. Eventually, however, this process acts against itself. If the river is contained by the levees any deposition that occurs will take place within the channel. In this way the level of the riverbed is also raised. As a consequence channel capacity is consistently reduced and over time flooding becomes increasingly likely once again. Moreover, if flooding does occur it can be even more disastrous because it will flow over the floodplain from an increased height, giving it greater energy to do damage.

The River Wey, a tributary of the River Thames in England, flows through Guildford, the county town of Surrey. Its small levees have been enlarged as part of a flood prevention scheme to protect the town centre. Guildford is located in a gap cut by the river through the North Downs, a range of chalk hills. The town centre is cramped, so that the limited flat flood-plain has had to be developed, even though it was at risk from flooding. Upstream from the town centre lies an area of rough sheep grazing. If river levels are rising and it seems the town centre is at risk, the animals are removed to safety and their pasture is allowed to flood. Built into the natural levees is a series of small sluice gates that can be opened manually to allow rising water on to this low-value part of the floodplain, thus saving the economically more valuable areas downstream.

The best-known levees and embankments in the world are those along the River Mississippi, which drains one third of the USA. These embankments, part natural, part artificial, reach heights of almost 16 metres and in places have been reinforced and enlarged sufficiently to carry roads. Near the mouth of the Mississippi, at New Orleans, the levees are above sea level, but most of the surrounding land upon which New Orleans sits has sunk to below sea level, a situation that seems ripe for disaster. Until 1927 the flood

Case study

The Bangladesh Flood Action Plan

Bangladesh faces enormous problems from flooding – probably more than any other nation on the Earth. The majority of its large and rapidly growing population lives on floodplains at or even below sea level. Heavy monsoon rains often combine with tropical cyclones to bring misery to thousands of people year after year.

Some 4,000 kilometres of embankments have been built to try to prevent flooding. However, poor maintenance over the years has meant that they have been unable to withstand high flows and they have frequently been breached. Embankments have also been deliberately breached by farmers wishing to make use of water for farming or fisheries. Once weakened in this way, the banks are unlikely to cope with high river flows. A further problem is that when flooding occurs as a result of heavy rainfall, water is unable to drain away into the rivers, because of their high banks. Waterlogged areas can lead to health problems and the spread of disease.

In an attempt to improve the situation in the future, the World Bank has coordinated a major project called the Flood Action Plan (FAP). Initially, during the early 1990s, the FAP involved a huge amount of scientific research rather than direct construction. This is because the river system in Bangladesh is extremely complex and dynamic, with the braided rivers frequently changing their course. Without a full understanding of the processes at work, the building of further constructions could make the situation worse.

Various pilot projects have now been undertaken to try to balance the needs of those affected by flooding, both positively and negatively. More embankments have been constructed, primarily to channel water into the major rivers and also to increase protection of major cities, such as the capital Dhaka.

prevention policy in the Mississippi Valley was 'hold by levees'. The 217 deaths and 700,000 evacuations during the 1927 flood changed this policy to a much more multipurpose approach, after which it was once again considered that the flow of the Mississippi was controlled. Over 3,000 kilometres of levees existed along the main channel and tributaries, but they proved insufficient to hold back the 1993 floodwaters. Levels at St Louis, Missouri, reached an all-time high. Thus the most flood-prone rivers levees and embankments can only be used as part of a multifaceted flood prevention strategy.

Further reading

Nickels, S., 1992, *The Big Muddy (A Canoe Journey Down the Mississippi)*, Oriole Press.

WEBSITES

Louisiana Agricultural Center:
www.lsuagcenter.com/Communications/
pdfs_bak/levees.PDF

Lower Mississippi River Conservation Committee:
www.lmrcc.org
Mississippi River resource page:
www.caleuche.com/River/Mississippi.htm
Commonwealth Network of Information Technology for Development:
www.comnet.mt/bangladesh/MasRvr.htm

SEE ALSO
flooding

LIFE ZONE

Life zone refers to elevational bands of life on mountains, characterised by vegetation (and the biota in general) differing from that at lower and higher elevations. Moving up a mountain in western North America, for example, one moves from shrublands (oakbrush, for example), through coniferous woodland, to coniferous forest, through a zone of wind-twisted 'Krummholz' near treeline, to alpine tundra. A similar zonation occurs on mountains worldwide.

C.H. Merriam, who pioneered the life zone concept in North America during the late nineteenth century, noticed that moving up a mountain was – in terms of climate and vegetation – very much like moving from the south toward the North Pole. The Lower Sonoran zone was one of hot desert, with cacti, creosote bushes and desert bighorn sheep. Grasslands, **CHAPARRAL** and woodlands characterise the Upper Sonoran zone. Pine trees grew in the transition zone above that. Douglas fir and Engleman spruce grew in the Canadian zone above transition. The zone dominated by fir trees was named the Hudsonian zone, observed to have the same structure as the taiga of Canada, Siberia and Scandinavia. The alpine **TUNDRA** on top had similar characteristics to Arctic tundra.

The figure below illustrates the different zones. Note also that the **DESERT** zones extend further up the mountains on the south sides. In the northern hemisphere, north-facing slopes are more often in shadow than south-facing slopes, leading to differences in soil moisture, type and size of plants, and animal habitat. The opposite would be true in the southern hemisphere, where north-facing slopes tend to be drier.

In 1889, an expedition funded by the newly created US Biological Survey arrived in Flagstaff, Arizona. Among the members of that expedition were C. Hart Merriam and Vernon Bailey. They quickly discovered 20 new species of mammals, including the Mexican grey wolf and the now-extinct Merriam's elk. While exploring terrain that ranged from desert canyons to the 3,840-metre summit of the San Francisco Peaks, Merriam and Bailey took notes on variations in the vegetation, corresponding with

The San Francisco peaks and the life zone concept

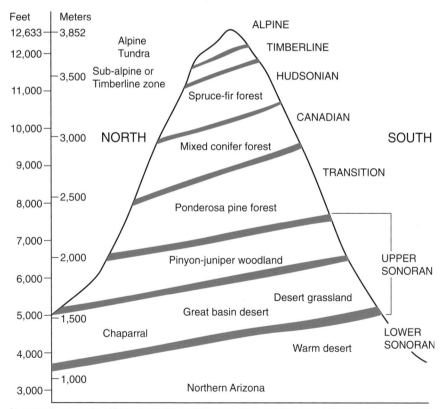

Diagrammatic cross-section of Peaks, showing Merriam's life zone terminology (right) and modern vegetation zone names (left). Zones slope downward on the cooler, moister north slope and upward on the hotter, drier south slope. Drawing by D.M. Zahnie

Florence Merriam Bailey

Born in upstate New York in 1863, Florence Merriam Bailey became a pioneer naturalist, dedicating her life to observing and protecting bird life and recording the wonders of the natural world. She became one of the foremost female writers of her era and travelled for 50 years studying birds. Florence's brother was life zone scientist Clinton Hart Merriam, who eventually became the first chief of the US Biological Survey. Until Bailey's time, most naturalists studied birds only as 'skins' in private or the few public collections, rather than in the field as complete life forms as she did. She preceded Ludlow Griscom in calling for the use of binoculars instead of shotguns when birding.

By 1885, she began to write articles focussing on protecting birds. She was horrified by the fashion trend that not only used feathers, but entire birds to decorate women's hats. Five million birds a year were killed to supply this fashion. Bailey set up the Smith College Audubon Society. First organising bird walks, she eventually involved the students in a campaign to open the public's eyes to this slaughter. She sent out 10,000 circulars by enlisting the help of over 100 students, a third of the college, and wrote articles of protest to newspapers.

When she was 26, Florence collected and developed the series of articles she had written for the *Audubon Magazine* into her first book, *Birds Through an Opera Glass*, published in 1889. Refusing to assume a male *nom de plume*, as was common for women writers at that time, her independent ideas were evident in her writing.

In December 1899 she married Vernon Bailey, who was a pioneering naturalist with the Biological Survey in his own right. They made extensive trips into the American West together, where little was known about its flora and fauna; he focussed on mammals and she on birds. Altogether she published about 100 articles, mostly for ornithological magazines, and ten books, including the *Handbook of Birds of the Western United States* (1902) and *Birds of New Mexico* (1928).

Florence's life work was dedicated to educating people about the value of bird life. Finally a law was passed that made the interstate shipping of birds illegal – the first step in halting the slaughter and decreasing the number of victims, especially among seabirds, such as pelicans and grebes. Eventually change in the law, shifting styles, and continued education meant that birds were no longer killed for hat decoration. A variety of California mountain chickadee was named *Parus gambeli baileyae* in her honour in 1908. She died in Washington, DC in 1948.

changing temperature and precipitation. Meticulously recording the elevation and precipitation at which different plants and animals occurred, Merriam and Bailey mapped out seven basic life zones. For many years, Merriam had felt that temperature exerted the main control over the distribution of plants and animals. By 1889, Merriam was convinced of this and began to categorise the entire North American continent into life zones, which were largely based on temperature. He eventually wrote over 500 articles, most of which were about biology, environmental studies, and Native American culture.

While the life-zone theory is simple by today's standards, it marked the beginning of the sciences of biogeography, modern **ECOLOGY** and **CONSERVATION BIOLOGY**. In focusing on structure and ecological equivalence, the life zone concept foreshadowed the concept of the **BIOME**.

Although the life zone concept was originally elaborated for ecological patterns in mountainous terrain, it is readily applicable to any environment in which gradients of physical factors are steep and simple so that the biotic community exhibits predictable spatial change. Another example is different zones in lakes (lentic biome) or the ocean (pelagic biome).

WEBSITES

A history of Merriam's life zone system and the unique ecology of the Colorado Plateau:

www.cpluhna.nau.edu/Biota/merriam.htm

Merriam's life zones and photographs:

www.radford.edu/~swoodwar/CLASSES/

GEOG235/lifezone/merriam.html

Range types of North America, organised by

life zone, with photographs:

www.tarleton.edu/~range/

Merriam's life and philosophy and the

scientific spirit of his times:

www.aphis.usda.gov/ws/nwrc/hx/

merriam.html

More about Florence Merriam Bailey's life:

www.outlawwomen.com/

FlorenceMerriamBailey.htm

MANGROVE FORESTS

Described as a 'special kind of rainforest', mangroves create one of the most unusual environments in the world. They are smelly places, a result of the masses of sodden rotting vegetation, and, moreover, they are eerie places because of the odd shapes of their highly unusual root systems. Mangrove trees are specially adapted to colonise areas of mud either on the coast or within tidal estuaries. Each tree is supported by a mass of 'prop' roots extending a metre or more above the mud surface. The root systems of neighbouring trees intertwine and overlap, so they are very effective at trapping mud from rivers and sand from the sea. Over time the coastline stabilises and extends seawards.

The watery environment and complexity of the root system in mangrove forests mean that people cannot venture into these areas without difficulty. Mangrove vegetation is adapted to areas flooded daily by the salt water of the rising tide. Trees are therefore halophytic – adapted to a salty environment. Mangroves often develop along a shore where the presence of a coral reef creates a lower energy environment. Sea grass then grows, which in turn succeeds to mangrove swamp.

Mangrove forests exist across a broad area of the Tropics, though their distribution is not contiguous with that of coastal rainforests. They are better developed in some areas than others. The world's most extensive mangrove forest is the Sundurbans of the Ganges Delta along the border of India and Bangladesh. There are also extensive mangroves on the Malay Peninsula, and in Borneo and New Guinea. In contrast, the mouths of the world's two greatest rainforest rivers, the Amazon and the Congo, have only limited mangrove development because of the stronger flows in these larger rivers. Asian and Pacific mangroves are more complex ecosystems. They have around 40 forest species, while African and American examples exhibit only eight or so each.

Another unusual adaptation of mangroves is that they disperse their seeds by water. This means they have reached many remote tropical islands and, because suitable environmental conditions have occurred, they have taken root and created a new ecosystem. They are even found as far south as the harbour of Auckland, New Zealand. Depending on the species, either the fruit or the seed floats for a considerable period of time. Some seeds actually germinate on the tree before shedding, so that they are ready to take root on landing at a suitable location. Roots already sprouted can attach the young plant to the soil immediately.

Mangrove forests form a complex ecosystem; other water-based species rely entirely upon them. Crabs, shrimps and oysters find both food and protection from storms. In fact the whole coastline is protected from the massive potential erosive effect of storms as the root structure helps to break the force of storm-force waves. Particularly unusual species of amphibious fish thrive in mangrove forests. For example, mudskippers feature pelvic fins that also act as legs, by which they push themselves along the shallow sea floor. Even some higher species, including primates such as the proboscis monkey of Borneo, have become adapted to this environment.

People's use of mangrove forests is limited since such a waterborne environment is not so natural to us. Nevertheless, rich fish catches are attractive. Sadly, the potential of this less common form of rainforest is being severely limited by felling, as people seek out materials such as firewood and bark tannin. In the Philippines, Ecuador and India fish farming is expanding and demanding space along these coastlines at the expense of the mangrove environment.

Case study

Mangroves versus shrimps

The ever-increasing demand for protein in India, now the second most populous nation in the world at 1.1 billion people, is destroying valuable mangrove environments within the Bay of Bengal. Fish farming has long been recognised as a huge potential protein resource in less economically developed countries, but India has taken such exploitation further than most. At several locations along its eastern coastline India has developed huge shrimp farms. The mangrove environment is not compatible with this activity. Therefore the 'solution' has been to cut down the mangroves and then develop these coastal industries. Commercial activities have proved profitable; not only are regional supplies of protein increased but exports are possible, bringing in important hard currency.

However, such benefits do not come without costs. Mangroves play an important part in protecting these low-lying areas from the impacts of huge storms and cyclones. They absorb the energy of waves and high winds that move onshore, so protecting settlements and economic activities behind. There is some evidence that the removal of mangroves in parts of India have increased the scale of recent cyclone disasters.

Further reading

RSPB, 1998, *Ecosystems and Human Activity*, Collins Educational.

Wolfe, A. and Prance, G., 1998, *Rainforests of the World*, Crown Publishers.

Middleton, N., 1988, *Atlas of Environmental Issues*, Oxford University Press.

WEBSITES

University of Liege, Belgium:
www.ulg.ac.be/oceanbio/co2/mangroves.htm
Nasa's 'Visible Earth':
http://visibleearth.nasa.gov/egi-bin/viewrecord3457

MICROCLIMATES

Weather can be studied at a variety of scales – from global to local. The scale used on the typical television weather report is called the synoptic scale, about the size of a continent. Local weather reports focus on the mesoscale, allowing for the description of weather phenomena from the size of an individual thunderstorm to the broader regional impact of a typical low-pressure system. What happens in your own neighbourhood is considered at the microscale, and long-term averages and even some extremes are useful to consider at this scale.

Microclimates reflect local conditions that affect air movement, heat transfer or moisture flux, such as local topography, land cover, water bodies or obstacles to air movement such as buildings. Local human-caused sources of heating, such as those found in cities, may also contribute to microclimate variation. Therefore, microclimatic variations affect the environment directly by altering the heat and moisture, and indirectly by altering the land cover and land use in a certain place.

One important microclimatic effect on the environment is the temperature inversion. In an inversion, heavier cold air becomes trapped in a valley, beneath a layer of relatively warm air. Cold air is denser and cannot rise above the warmer air. Inversions typically occur in sheltered locales like mountain valleys, and are particularly noticed during the winter. In the absence of wind, winter inversions may last for days or weeks, and in cities such as Salt Lake City, Denver and Los Angeles, in the USA, pollutants are trapped within the surface layer of cooler air, creating smog.

While mesoscale temperatures might remain above freezing, local radiation conditions and topography might deliver particularly cold air down a slope where it accumulates in a depression and kills plants. Microclimate affects the land use of an area; for example, due to the susceptibility of fruit trees to frost damage, orchards are usually planted along breezier hillsides, rather than in valley bottoms or in hollows. Such situations render the bottoms of sinkholes in Florida unsuitable for citrus growing, while the sinkhole slopes and adjoining ridges can be planted with oranges and grapefruit. In the Champagne region in France, made famous by its expensive sparkling wine, grapes are grown on the warmer south-facing slopes, well away from the valley floors where cold air accumulates during the spring. In gardens, the microclimate is important when considering whether to cover tender plants when temperatures are expected to be near the freezing mark. A low spot or hollow in your garden might suffer from such cold air drainage and plants there need to be hardier or protected.

Perhaps the most widespread example of microclimate is the 'urban heat island'. Urban areas are typically several degrees warmer than the surrounding countryside. In 1965 the researcher Chandler found that, under clear skies and light winds, temperatures in the suburbs of London during the spring fell to a minimum of 5°C, whereas in central London they only dropped to 11°C. Buildings and concrete surfaces usually release, absorb and reflect more heat than natural vegetation. The relative absence of water and plants in urban areas means that less energy is used for evapotranspiration by plants, and more energy is available for heating. Urban areas block some strong winds, which would help disperse the heat and bring in cooler air from the surroundings. In addition, solar radiation is reflected by glass buildings and windows, and the emission of pollutants from cars and heavy industry act as condensation nuclei, leading to the formation of cloud and smog, which can trap radiation. In some cases, a pollution dome can also build up around a city, which ironically can filter some of the incoming solar radiation, reducing some of the heat island effect.

Local winds are another example of microclimates. These can occur at the coast in response to the different thermal capacities of land and sea. During the day the land warms up and air starts to rise. This pulls in colder air from the sea to form a cooling sea breeze. In some parts of the world such a local wind is called a 'doctor' on account of its cooling nature during very hot weather. The opposite happens at night when the colder land surface results in sinking air, which then transfers seaward as a cool land breeze. These land breezes and sea breezes might result in fog

Urban core of Denver, Colorado, as seen from the foothills of the Rocky Mountains. Photograph by Joseph Kerski

where the warm, moist air contacts the cold land surface. In valleys, the movement of cold dense air downslope results in a local wind called a katabatic wind. This is common under clear calm conditions early in the morning. During the day the reverse happens, as warmer air rises up the valley to form an anabatic wind.

The **RAIN SHADOW** effect in mountainous areas is another example of a microclimate. Forests also have their own distinctive microclimates. Trees act as windbreaks, and their leaves and branches also filter out incoming solar radiation and trap outgoing radiation. As a result, forested areas tend to have less temperature extremes. They also have higher levels of humidity because the absence of wind reduces the rate of evaporation. The differences between forested areas and their surroundings vary depending on the season – whether the trees are in leaf – and the type of vegetation – whether deciduous or evergreen.

WEBSITES

UK Met Office microclimate information: www.met-office.gov.uk/education/curriculum/leaflets/microclimates.html
Research on the effect of the urban heat island over Nairobi, Kenya: www.ihdp.uni-bonn.de/ihdw02/summaries/word/Opijah%20text.doc
Urban heat island information from Lawrence Berkeley Laboratory: http://eetd.lbl.gov/HeatIsland/LEARN

NASA Urban Heat Island Study of Salt Lake City:

http://science.msfc.nasa.gov/newhome/headlines/essd21jul98_1.htm

Urban heat island and mitigation strategies from a city planner's perspective:

www.asu.edu/caed/proceedings99/ESTES/ESTES.HTM

Urban heat island information from Geography.about.com:

http://geography.about.com/library/weekly/aa121500a.htm

SEE ALSO

climate

MIGRATION

Migration involves the movement of people from place to place, usually permanently, and over a significant distance. It is difficult to define migration precisely, but most agree that it must be a move of some consequence for an extended period to time. Although migration often involves crossing a political boundary from one country to another, it also includes movement within a country, for example, people moving from New York to Miami. This movement of people over time has led to a social, linguistic and nationalistic mixing of a significant proportion of the world's population. Advances in transportation and communication have allowed more people to move from one area to another.

Migration has affected the environment in many ways. For example, invasive weeds, plants and animals have been carried from one continent to another via ships. In New Zealand rodents from ships threatened native species that previously knew no natural ground predators, including the national bird and symbol of the land, the kiwi. The introduction of rabbits to Australia has had serious effects upon the environment. As people have moved into new areas, so the environment has been put under increasing pressure for housing, services, resources and provision for waste disposal. Those areas suffering severe out-migration have also suffered. Land becomes overgrown and may be mismanaged because it is the younger, active adults who are most likely to have moved away.

Thus migration affects two populations – people in the area of origin, and the population at the destination. Migration is not simply movement from place to place by individuals or families; it can include:

- primitive migration
- group or mass migration
- free-individual migration
- restricted migration
- impelled or forced migration.

Primitive migration is usually associated with groups of people who move from one area to another as a response to the deterioration of the physical environment. It is generally related to the hunting and gathering of food, where the natural environment is not productive enough to support the number of people that must subsist from the land. Sometimes this happens very quickly as a response to drought, or more slowly over time as a response to an ever-expanding population relative to limited natural resources. This type of migration is often used to describe the migration patterns of Native American groups in the USA as Europeans arrived and migrated westward.

Group or mass migration usually involves the movement of a social group larger than a family, such as a clan or tribe. History is replete with instances where entire societies left their original place of habitation and laid claim to or invaded other areas. In some instances armies were involved, and after the invasion the native population was then either displaced or assimilated.

Free-individual migration is a characteristic of much of the international migration that has taken place since the seventeenth century, especially to areas such as Australia, New Zealand and the Americas. Much of the migration to the USA from Europe was the result of individuals or families acting on their own individual initiative to move from their place of origin. 'Nation of immigrants' is a phrase often applied to the USA because of free-individual migration. During the Irish potato famine of the mid 1800s, millions of Irish people emigrated, mostly to the USA. Large numbers of people are still on the move: the late twentieth century saw many South Asian and African immigrants to Europe, for example.

Restricted migration has gradually become

Case study

The 1994 Rwanda genocide and migration

Background information

Tutsi and Hutu are two groups of people in Rwanda who speak the same language and share the same culture. Some identify the Tutsi as 'pastoralists' and the Hutu as 'agriculturalists', the Tutsi as 'patrons' and the Hutu as 'clients', or the Tutsi as 'rulers' and the Hutu as 'ruled', but these stereotypes and generalisations do not tell the complex story of these two groups of people.

Rwanda's President Habyarimana and Burundi's President Ntaryamira were killed in 1994 when a missile, believed to have been launched by Hutu military rulers, shot down their plane. The killings of the Tutsi by the Hutu began half an hour later. The Tutsis were separated from the Hutus and were beaten to death with machetes. More than 800,000 were slaughtered over a period of three months. This represented three-quarters of Rwanda's total Tutsi population. This 'ethnic cleansing' was carried out with extraordinary cruelty. People were burnt alive, thrown dead or alive into pits and often forced to kill their own friends or relatives. The survivors were tracked down all over the country, even into hospitals and churches, where some of the worst slaughters occurred. Women were tortured and raped. In July 1994 the Rwandan Patriotic Front captured the capital city, Kigali. The Government collapsed and a ceasefire was declared. Around 2 million Hutus fled the country to the Congo, one of the largest sudden migrations of recent decades.

more significant than free migration. Since the beginning of the twentieth century laws have been enacted to restrict the migration of people between countries. These laws can involve a total ban on immigration, or they may permit migration for people of certain social or economic classes or from specific countries. Immigration to the USA has been restricted since 1921, when both the types and numbers of migrants were restricted. Similar restrictions exist in other countries such as Canada, Australia and many countries of Europe – countries that originally had very open immigration policies. One result of these restrictive policies has been an increase in illegal migration, especially to the USA. As a result, an increasing number of people are urging the use of the military or National Guard to patrol the USA borders. The USA–Mexico border area has experienced rapid population growth over the past 30 years, all the way from Texas to California. Communities known as *colonias* along the border cope with a host of environmental problems – including poor sanitation, industrial waste from the *maquiladores* (border factories) and air pollution.

When the activating agent in migration is the state or other political or social institutions, that migration is categorised as impelled or forced. If the migration is impelled, the migrant holds some degree of choice, but the migrant has no control in forced migration. Perhaps the most blatant example of forced migration was the slave trade associated with the settlement of North and South America. Unfortunately even today we have the equivalent of slavery in the form of bonded labourers or child labourers. Children have been sold in slave markets in Thailand and the Sudan, and have been used for everything from sex slaves to child labourers. One recent estimate suggests that as many as 10 million youngsters worldwide are involved in child prostitution. In 2000 the Central Intelligence Agency estimated that approximately 50,000 women and children were illegally brought to the USA to work as prostitutes, servants, and sweatshop labourers.

The twentieth century also saw millions of people forced to leave their homelands due to wars and ethnic strife. These refugees are now found living in squalid conditions throughout the world with little hope that they can return to their homeland. Recent estimates are that the number of refugees worldwide varies from 15 to 20 million. Although exact numbers are hard to verify, it has been estimated that since the beginning of the twentieth century more than 100 million people have become refugees.

WEBSITES

International Organization for Migration:

www.iom.int

United Nations High Commissioner for

Refugees:

www.unhcr.ch

Reports from the BBC on the Rwanda

genocide:

http://news.bbc.co.uk/1/hi/in_depth/

africa/2004/rwanda/default.stm

MINING RECLAMATION

In modern societies people use a vast quantity of minerals over their lifetime. According to the USA's Minerals Information Institute, every American born will demand 259,880 kilograms of coal, 630 kilograms of copper, 13,303 kilograms of salt, 2,205 kilograms of bauxite (aluminum ore), 14,879 kilograms of iron ore, and an amazing 697,000 kilograms of stone, sand and gravel. Since all these minerals need to be extracted from the ground, mining has been an omnipresent activity ever since the Industrial Revolution began. The need for minerals has not abated, despite the move away from heavy industry towards the information age. Most of the electronic products we use, such as computers and mobile telephones, are made from a variety of minerals.

When minerals are taken out of the ground, the land and the environment are disturbed. Two main types of mines exist – surface mines (sometimes referred to as opencast) and deep mines. In surface mines minerals are extracted near the surface over a broad area, but in a deep mine shafts take workers to underground caverns where the minerals are removed. Surface mines disturb the surface environment to a greater degree than deep mines. However, both can cause other environmental problems associated with waste, including spoil heaps and toxic chemicals. Abandoned mine sites pose health, safety and environmental hazards. Every year, people fall victim to the hazards of mines, such as falling into open shafts or becoming lost in adits (mine tunnels). Thousands of sites have the potential to contaminate surface water, groundwater or air quality, sometimes because of the substances used in extracting the minerals. For example, the toxic substance mercury was used in gold mining for many decades.

Another problem is acid mine drainage, caused by water flowing from surface mining, deep mining or coal refuse piles; it is typically highly acidic with elevated levels of dissolved metals. This highly toxic water results from complex geochemical and microbial reactions that occur when water comes into contact with pyrite (iron-based minerals). The resulting iron hydroxide and sulphuric acid contaminates local water supplies, kills fish, amphibians and wildlife, and alters riparian plants (those that grow along streams). This is a particular problem in the Appalachian Mountains of the eastern USA, where coal was extensively mined for over 100 years.

Today, in many countries such as the UK and the USA, those who operate mines are required to reclaim the surface during and after mining is completed, returning the land to other useful purposes. Reclaimed land is often quite attractive to wildlife and human uses. Reclamation removes dangerous health and safety hazards such as cliffs, mine buildings, tailings, pits and shaft openings, improves the environment, and restores resources to make them available for economic development, recreation and other uses.

In the UK old quarries have been used for a variety of purposes. Some are simply left to allow plants and animals to inhabit the area naturally. Others are used as landfill sites and then grassed over for farming. Some are used for recreation activities, often making use of the natural lakes that form in the bottom of the quarries. An enormous quarry south-east of London was converted into the Bluewater Shopping Mall.

During the mid-1970s the US Congress found that over 600,000 hectares of land had been directly disturbed by coal mining and over 18,400 kilometres of streams had been polluted by sedimentation or acidity from surface or underground mines. In response to the problem associated with inadequate reclamation practices, Congress enacted the Surface Mining Control and Reclamation Act of 1977. A main purpose of this act was to promote the reclamation of mined areas that, in their unreclaimed condition, substantially degraded the quality of the environment, prevented or damaged the beneficial use of land or water resources, or endangered the health or safety of the public. Title IV of this Act established a fund to be used for the reclamation and restoration of areas affected by past mining. The fund is derived from a reclamation fee of 35 cents per tonne of clean coal produced by surface coal mining, 15 cents per tonne of clean coal produced by underground mining, and 10 cents per tonne for lignite coal. The fund is supplied with money from the active operators who are mining coal today.

WEBSITES

Mine reclamation resources from the Minerals Information Institute:
www.mii.org/recl.html
National Association of Abandoned Mine Land Programs:
www.onenet.net/%7Enaamlp
US Office of Surface Mining:
www.osmre.gov
British Land Reclamation Society:
www.blrs.org
The Eden Project in Cornwall, UK (where an old china clay quarry has been used for the creation of giant greenhouse biomes):
www.edenproject.com

SEE ALSO

coal

MINING WASTE

Mining can cause great harm to the environment. The disruption of the surface disturbs habitats, vegetation and drainage patterns, the 'tailings' or solid waste products can cause siltation of rivers and lakes or contain hazardous chemicals, and the liquids used in the extraction of minerals can themselves be hazardous. Gold miners, for example, have for years used substances such as cyanide and mercury; if these liquids contaminate local water supplies, the impact is unthinkable.

Unfortunately, if there is a huge demand for valuable metal ores and minerals that lie beneath the surface then the environment often suffers as a result. Immediately following the Second World War in the UK, when heavy manufacturing was still vibrant, economic arguments won the day. But in the 1960s a change in thinking took place, especially over the issue of what to do with mining waste. A major influence in changing people's minds was the tragic accident on 21 October 1966 at Aberfan in South Wales.

WEBSITES

International Institute for Environment and Development:
www.iied.org/mmsd/finalreport/index.html
National Institute for Occupational Safety

Case study

The Aberfan Disaster, South Wales

On 21 October 1966 the small mining village of Aberfan in the South Wales valleys made news headlines all over the world as a spoil tip of mine waste slid down the valley sides killing 144 people, including 116 children and 5 teachers in Pant Glas Junior School. The children and their teachers had just finished their morning assembly at which they had sung the hymn 'All Things Bright and Beautiful'. A few minutes later the spoil demolished their school, killing around half of the pupils.

Waste from the Merthyr Vale Colliery had been piled up on the hillside because there was no room on the valley floor – the valley land was needed for housing, shops and the local school. The spoil tips covered a natural spring that, in the damp autumn of 1966, became active again, turning the coal slag into liquid slurry.

On the morning of the disaster, it was sunny on the mountain but foggy in the village, with visibility down to about 45 metres. The tipping gang up on the hillside saw the slide start at around 9.15 a.m., but they could not raise the alarm because their telephone cable had been stolen. Down in the village, nobody saw anything, but everybody heard the noise as the tip slid down the hillside, demolishing a farmhouse and then engulfing the Pant Glas Junior School and 20 terraced houses nearby.

continued over

An inquiry was held, lasting for 76 days – at the time, it was the longest inquiry ever in British history – and it called 136 witnesses, ranging from schoolboys to professors of geology. As 300 exhibits were examined and 2,500,000 words were heard, it emerged that there had long been local worries over the stability of the tip. The National Coal Board had no tipping policy at all. Lord Robens, the National Coal Board chairman, appeared dramatically in the final days of the tribunal to give evidence, and admitted that the Coal Board had been at fault.

The Inquiry's report made clear that it was a tale 'not of wickedness but of ignorance, ineptitude and a failure of communications'. No one faced criminal proceedings, but various individuals, as well as others who were cleared, had to live with the disaster on their consciences for the rest of their lives. As far as the National Coal Board was concerned, it subsequently took greater care over where it placed the spoil tips, and steps were taken to improve the drainage and stability of slopes where the spoil heaps were placed.

and Health's Mining Resources:
www.cdc.gov/niosh/mining/default.htm
Environment Protection Agency:
www.epa.gov/epaoswer/osw/mission.htm
Nuffield Institute at Oxford University:
www.nuff.ox.ac.uk/politics/aberfan/home.htm
BBC:
www.bbc.co.uk/wales/walesonair/database/aberfan.shtml

SEE ALSO

coal

THE MISSISSIPPI

The Mississippi-Missouri is the fourth largest river in the world, being exceeded only by the **NILE**, Amazon and Yangtze. Today's version of the name, Mississippi, comes from the Ojibway Indian word *messipi*, meaning 'big river'. From its headwaters in Minnesota to its mouth in the Gulf of Mexico, it is 4083 kilometres long. Some 82,000 cubic metres of water pour into the Gulf of Mexico every second, along with 400 million cubic metres of sand, mud and gravel. Along its route the Mississippi flows through ten of the USA's 50 states – Minnesota, Wisconsin, Iowa, Illinois, Missouri, Kentucky, Tennessee, Arkansas, Mississippi and Louisiana – but it drains water from all or part of 31 states plus two Canadian provinces, Ontario and Manitoba. Over 12 million people live in the counties immediately bordering the river.

The Mississippi Valley played an important part in the early history and peopling of the USA. Some of the earliest European explorations took place along it. In 1672 Louis Jolliet, a French Canadian, and Jacques

Mississippi characters

Black Hawk

Chief of the Sauk tribe from 1788 to 1832, Black Hawk was tricked into signing away tribal lands on the eastern side of the river in 1804. Despite the huge risks, he later led his people back across the Mississippi to reclaim their lands. His peace emissaries were shot and he was faced with militia who fired, despite his white flag. In this way Mississippi lands were taken and kept by settlers.

Henry Miller Shreve

One of the most important steamboatmen in the river's history, Shreve was Master of the Mississippi and Superintendent of Western River Improvement between 1827 and 1841. He pioneered several technological innovations on the early steamboats, including double decks for passengers, and the first high-pressure engine. It was his first steamboat, the *Enterprise*, which proved in 1814 just how effective steam power was in navigating the strong current of the Mississippi. Shreve remains a giant among the USA's river men, having changed inland river travel more than anyone else in the first half of the nineteenth century.

Marquette, a Frenchman, aimed to see whether the great river flowed westwards into the Gulf of California, which would enable the French to reach the Pacific and the Far East. For the first seven days of sailing down the Mississippi they saw no one, but on the eighth day human footprints led to an encounter with the Peoria Indians. Later, downstream of the Arkansas confluence, they met other tribes who already traded with the Spanish, at which point they gave up their quest. Later, in 1682, La Salle led the first expedition to explore the full length of the river and he claimed the whole valley for France, though it was not until 1821 that Henry Rowe Schoolcraft identified the true source of the Mississippi.

The Mississippi Valley is hugely important for wildlife. It is a major flyway for migrating birds because it is so easily recognisable from above. Around 40 per cent of the North American continent's duck, goose, swan and eagle populations use it in this way. Some of the nation's most precious wetlands lie along the river too. However, 90 square kilometres of this habitat are lost each year along the coast of Louisiana alone. The Mississippi River Basin Alliance is a coalition of over 150 local grassroots organisations along the whole length of the river that was formed to bring people together to save the Mississippi.

The waters of the Mississippi River carry dissolved contaminants and bacteria that originate from a variety of municipal, agricultural and industrial sources. About 2 per cent of the average discharge of the Mississippi River

Mississippi facts and figures

- The Mississippi provides over 4 million people with their water supply.
- The navigability of the river boosts its contribution to the economy: 29 locks and dams control navigation on the Upper Mississippi alone.
- The river transports 472 million tonnes of cargo annually, including 46 per cent of the USA's grain exports.
- $7 billion worth of agricultural and forest products are produced within the valley itself each year, but manufacturing is worth much more at $29 billion.
- Tourism generates a large percentage of income today. Visitors from abroad create 53,000 jobs and $2.6 billion every year. More people visit the Upper Mississippi River National Wildlife Refuge (3.5 million/year) than Yellowstone National Park.

comes from municipal and industrial point sources. As most of the basin is cultivated as farmland, the major environmental concern in the river is from pesticides and fertilisers from farming. The distribution of contaminants along the Mississippi River depends on the nature and location of their sources, the degree of wastewater treatment, the stability of the contaminants, and their dilution by receiving waters. Fecal coliform bacteria derived from human and animal wastes survive only briefly in river water, but their average concentrations exceed the maximum

Mississippi flood damage

Damage	1927 flood	1993 flood
Area flooded	70,000 square kilometres	40,400 square kilometres
River volume	70,000 cubic metres/second	28,000 cubic metres/second
Deaths	246	47
Displaced people	700,000	74,000
Financial loss	$347 million (= $4.4 billion in 1993 $ value)	>$7.5 billion
Structural damage – number of buildings destroyed or damaged	137,000	47,650

contaminant level (MCL) established by the US Environmental Protection Agency (2000 per litre for recreational use) in much of the Mississippi River because of incomplete wastewater treatment. Biodegradable detergents, primarily derived from domestic sewage, exist in high concentrations in the Mississippi River in the St Louis area. Even caffeine, a stimulant chemical in coffee and soft drinks, is found in the river.

The Mississippi is renowned for its flooding. In the last hundred years two particularly catastrophic floods occurred, in 1927 and 1993. Prior to the 1993 flood huge engineering works had been implemented to reduce the flood hazard and to make the river more navigable. The river had been dammed, its banks were lined with concrete and artificial levees were constructed to stop the water overtopping its banks. However, following a summer of heavy rain the huge volume of water caused the levees to be breached in several places and large areas of floodplain – including settlements, roads and farmland – were inundated. Some people felt that the artificial nature of the river channel contributed to the disaster by preventing smaller floods occurring in the upper reaches that would have prevented the build-up of water further downstream.

WEBSITES

Great River.Com:

www.greatriver.com

Mississippi River Parkway Commission:

www.mississippiriverinfo.com

Oldmanriver.com:

www.oldmanriver.com

Mississippi River Basin Alliance:

www.mrba.org

Upper Mississippi River Conservation Committee:

www.mississippi-river.com/umrcc

Mississippi River Museum:

www.mississippirivermuseum.com

US Geological Survey:

http://water.usgs.gov/pubs/circ/circ1133

Public Broadcasting Service with an archive collection on flooding:

www.pbs.org/amex/flood

SEE ALSO

flooding

MORTALITY

Graveyard in front of St Martin's Church, oldest church in England, Canterbury. Photograph by Joseph Kerski

Mortality, or data describing death, is an important component of population change and hence human impact on the environment. Death rates provide valuable information about the nature of a particular society. The most common means of measuring deaths is by the crude death rate. This is the number of deaths in one year per 1000 population:

Crude death rate =
$$\frac{\text{Number of deaths in one year}}{\text{Mid-year population}} \times 1000$$

Like the crude birth rate, the crude death rate is fairly easy to obtain with a minimum amount of data. Since it is based on the total population from an area, it does not take into account the age or sex differences within that population, and therefore it is referred to as a 'crude' measure. The death rate is especially affected by the age structure of a population. All things being equal, a population with a large proportion of young people is going to have a lower death rate than a population made up of many elderly people.

Because the age structure differs among countries, assessing the true death rate between countries is difficult. This is immediately apparent when comparing the small differences between the death rate of developed countries and that of the world's developing countries. In 2003 the world crude death rate and the death rate for the USA were the same at 9 per thousand, but most developing countries also had a rate of less than 20. The primary reason is that the USA has a much older population than most of the countries in the rest of the world, especially the poorer developing countries that make up a large proportion of the world's population.

In an attempt to account for the effect of age structure, other more refined measures of mortality have been developed. The infant mortality rate is such a measure. It measures the number of deaths to infants under one year of age relative to the total number of births:

Infant mortality rate =
$$\frac{\text{Deaths of infants under one year old}}{\text{Annual number of births}} \times 1000$$

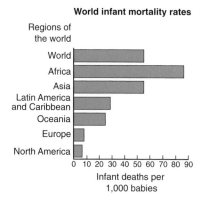

World infant mortality rates

Regions of the world

World
Africa
Asia
Latin America and Caribbean
Oceania
Europe
North America

0 10 20 30 40 50 60 70 80 90

Infant deaths per 1,000 babies

A look at infant mortality rates throughout the world reveals some interesting patterns.

The overall infant mortality for the world was 55 in 2003. That means that for every 1000 babies born in a given year, 55 of them die before their first birthday. The area of the world with the highest infant mortality rates is Africa. The African nation Mozambique has the highest rate in the world: 201. Europe and North America have the lowest rates, with the Scandinavian countries amongst the lowest in the world. Iceland, with a rate of 2.7, is the country with the lowest rate in the world. The disparity between the highest and lowest rates is striking, with Mozambique having a rate more than 70 times that of Iceland. Although

Latin America and Oceania have moderate rates, some individual countries like Haiti, with a rate of 80, stand out. Iceland's excellent healthcare facilities, nutrition, and standard of living contribute to its low infant mortality rate. In contrast, countries with high infant mortality rates have been plagued by infectious diseases such as AIDS and cholera, and also by civil wars and unrest.

The infant mortality rate for the USA is relatively low at 6.9, but 33 other countries have lower rates than the USA, including Cuba with a rate of 6. Almost all the other wealthy nations in the world have rates lower than the USA. Thus some people claim that the USA still has a way to go before it provides infants with the type of medical care they deserve.

Mortality rates are the result of the interaction of a variety of variables or factors. The last century has seen a major change in the role of diseases in the overall death rate. This change has been the result of the 'epidemiologic transition' that has taken place in the past hundred years. The primary causes of death a hundred years ago were infectious diseases such as typhoid, cholera, tuberculosis, malaria and Bubonic plague. Today, degenerative diseases such as heart disease, stroke, cancer and diabetes play a much larger role, especially in the wealthier countries. Race and socioeconomic factors also seem to play a significant role in death rates. Throughout the world, people of colour have higher death rates than whites. Socioeconomic status is also important: poor people in general have higher death rates, as a result of a combination of poor diets and insufficient medical care.

Life expectancy has also increased worldwide, from about 30 years in 1900 to around 66 years in 2000. Worldwide, however, there are still great variations, with some African nations still having very low life expectancies. A person born today in Mozambique can only expect to live to age 34, and a person born in Zimbabwe can only expect to see age 41. The worldwide AIDS epidemic has played a significant role in reducing life expectancies in many parts of Africa and Asia, although wars and other violence have played a role as well.

Recently, researchers at the Worldwatch Institute say rising death rates are slowing world population growth for the first time since famine in China claimed 30 million lives in 1959–61. They say that about two thirds of

Case study

The great influenza epidemic of 1918–19

The year 1918 has gone: a year momentous as the termination of the most cruel war in the annals of the human race; a year which marked the end, at least for a time, of man's destruction of man; unfortunately a year in which developed a most fatal infectious disease causing the death of hundreds of thousands of human beings. Medical science for four and one-half years devoted itself to putting men on the firing line and keeping them there. Now it must turn with its whole might to combating the greatest enemy of all – infectious disease.

So wrote the *Journal of the American Medical Association* in its final edition of 1918. Deep within the trenches of the First World War, men lived through conditions that seemed could not be any worse. Then, across the globe, something erupted that seemed as benign as the common cold. The influenza of that season, however, was far more than a cold. In the two years that this pandemic ravaged the Earth, one-fifth of the world's population was infected. The influenza was most deadly for people in their prime, from ages 20 to 40. It infected 28 per cent of all Americans, and an estimated 675,000 Americans died of influenza during the pandemic – ten times as many as in the world war. The influenza virus mortality rate was 2.5 per cent, compared to previous influenza epidemics, which were less than 0.1 per cent. The influenza spread following the path of its human carriers, along trade routes and shipping lines. Outbreaks swept through North America, Europe, Asia, Africa, Brazil and the South Pacific.

In some areas of the USA, the nursing shortage was so acute that the Red Cross asked local businesses to allow workers to have the day off if they volunteered in the hospitals at night. Emergency hospitals were created to take in the patients from the USA and those arriving sick from overseas. Schools in the UK were closed. Even President Woodrow Wilson suffered from influenza in early 1919 while negotiating the crucial Treaty of Versailles to end the world war. Public health officials distributed gauze masks to be worn. Stores could not hold sales and some towns required a signed certificate to enter – railways would not accept passengers without them. Those who ignored the influenza ordinances had to pay steep fines enforced by extra officers. Besides the lack of healthcare workers and medical supplies, there was a shortage of coffins, morticians and gravediggers.

Funerals were limited to 15 minutes. The conditions in 1918 were not so far removed from the Black Death in the era of the Bubonic plague of the Middle Ages. The influenza pandemic happened less than 100 years ago. More recent outbreaks of disease, such as SARS in 2002, have led people to ask: could such a pandemic happen again?

the slowing is explained by falling birth rates, but one third is caused by an increase in the death rate. The Institute says the two regions where death rates are either rising already, or are likely to do so, are sub-Saharan Africa and the Indian sub-continent. This is due to three specific threats: the AIDS epidemic, the depletion of aquifers, and shrinking cropland area per person.

The mortality rate is affected by changes in the environment. Probably the strongest link between the two is safe and adequate drinking water. When water is scarce or polluted, infectious diseases are difficult to contain, and the mortality rate is high. Death rates declined in many countries when sanitation facilities and municipal water facilities were constructed. Just as mortality is affected by the environment, so too is the environment affected by the mortality rate, but in a more subtle and complex way. If the mortality rate is low and a country is wealthy, the country can afford to spend money on such things as reducing emissions and setting aside land for open space. Some of the most polluted, congested cities are in developing countries because here resources are lacking for such positive endeavours. However, the wealthier countries also consume more, placing enormous pressure on environments that may be far from the source of consumption. For example, people may consume a great deal of water for municipal use, requiring enormous reservoirs to be constructed hundreds of kilometres away. Their electrical requirements may require coal or oil to be mined, transported and burned in another country or another region of their own country, with environmental hazards present at each step along the way.

WEBSITES

Weekly reports about death rates from various diseases in the USA from the Center of Disease Control:

www.cdc.gov/mmwr

National Statistics UK:

www.statistics.gov.uk

Information about the countries around the world:

www.who.int/en

World Population Data Sheet:

www.prb.org/pdf/WorldPopulation DS03_Eng.pdf

The 1918–19 influenza pandemic:

www.stanford.edu/group/virus/uda

SEE ALSO

fertility, population growth

MOUNTAINS

The greatest objects of Nature are, methinks, the most pleasing to behold; and next to the great Concave of the Heavens, and those boundless Regions where the Stars inhabit, there is nothing that I look upon with more pleasure than the wide Sea and the Mountains of the Earth. There is something august and stately in the Air of these things, that inspires the mind with great thoughts and passions; We do naturally, upon such occasions, think of God and his greatness: and whatsoever hath but the shadow and appearance of Infinite, as all things have that are too big for our comprehension, they fill and overbear the mind with their Excess, and cast it into a pleasing kind of stupor and admiration. (Thomas Burnet, *The Sacred Theory of the Earth*, 1684)

Mountains, with their awesome landscapes, snow-capped peaks and deeply dissected valleys, have inspired writers and artists throughout the ages. They provide some of the most magnificent natural environments on the Earth, yet despite their grandness they are environmentally fragile. Easily damaged, their sensitive ecosystems take many years to recover.

Most mountains are concentrated in linear belts around the world, coinciding with the major tectonic plate boundaries. The Rockies and Andes stretch for some 12,000 kilometres north to south through the western side of the Americas. On the other side of the world, the Himalayas form a west to east mountain range, stretching for some 4,000 kilometres. The world's highest mountain peaks – Mount Everest at 8,850 metres and K2 at 8,611 metres – are found within the Himalayas. The Alps in southern Europe are another great mountain range, stretching across several countries including France, Italy, Austria and Switzerland. All these mountain ranges can be classified as being active, rising centimetre by centimetre as vast tectonic plates converge. Elsewhere in the world, mountain ranges such as the Urals in Eurasia and the Appalachians in North America, while no longer active, bear testament to great tectonic forces of the past. Some mountains appear to be more isolated in their extent, such as those resulting from volcanic activity: Mount Mayon in the Philippines and Mount Fuji in Japan are good examples.

Mountain regions face a number of environmental issues, most of which are linked to human activity. **LANDSLIDES** and avalanches are common, due to the steep slopes and the frequently heavy rainfall or snowfall. They are often triggered by human activity, such as excavating the base of slopes for road building or quarrying, deforestation or simply skiing and snowboarding. Not only do landslides and avalanches cause damage to property and loss of life, they disrupt ecologies by ripping up trees and exposing the soil to subsequent erosion.

DEFORESTATION is a major issue in mountain regions. In the foothills of the Himalayas, trees are chopped down to be used for firewood and some scientists believe that this may be partly responsible for the severe flooding witnessed in Bangladesh in recent years. Deprived of the forest cover, the monsoon rains readily wash soil into river courses, which then become clogged and unable to

Case study

The impact of global warming on the Alps

Scientists believe that global warming is leading to an altitudinal rise in the snowline in the European Alps. With warmer winter temperatures, snowfall below about 1000 metres has become less reliable in recent years and scientists believe that this trend will continue into the future. Already many of the lower slopes in Italy, Austria and Germany have been affected, and internationally famous resorts, such as Kitzbühel in Austria (760 metres above sea level), face economic disaster.

Apart from the impact on local skiing economies, there will be a change to the alpine ecosystems. Vegetation zones will shift up the mountainsides at such a rate that some species of flora and fauna may fail to adapt quickly enough to the changing conditions. Scientists also suggest that landslides and avalanches will become more common as temperatures rise. Frost weathering will also become exacerbated as temperatures fluctuate more frequently either side of freezing and this, too, could promote slope failure.

hold as much water as they used to. As a consequence, flooding takes place.

Tourism is a major industry in mountain regions. It brings massive economic benefits and employs many thousands of people. However, the sheer volume of people who wish to ski, snowboard, walk and cycle have caused environmental degradation in many 'honeypot' sites. The French Alps is an extremely popular tourist destination for both winter and summer sports. There are many attractive resorts, with their characteristic wooden chalets, cable cars and ski runs. While tourism certainly brings some advantages, it has its costs too. Deforestation has taken place to make way for ski developments and the often severely eroded and dissected ski slopes and mountain bike trails bear witness to heavy and inappropriate use.

WEBSITES

UN Environment Programme:

www.unep.org

Manitoba University, California :

www.umanitoba.ca/institutes/

natural_resources/mountain/book

UNEP Mountain Environments:

www.unep-wcmc.org/mountains/

mountain_watch/pdfs/

mountainEnvironments.pdf

University of East Anglia, UK:

www.cru.uea.ac.uk/tourism/position_perry.pdf

SEE ALSO

deforestation, landslides

MULCH

Mulch is organic material such as cocoa pods, manure, leaf mould and bark chippings, which is used to retain moisture, enrich the soil and prevent erosion. Mulch mimics the conditions in a natural forest environment where tree roots are anchored in a rich, well-aerated soil blanketed by a layer of leaves and rotting organic materials that replenish soil nutrients. Gardeners especially in urban areas use mulches to mimic the more natural environment and to improve plant health.

As well as fertilising the soil mulch also has a positive impact on reducing soil erosion. The organic nature of the mulch helps soils hold more water and reduces evaporation. The mulch also helps to bind the mineral components of soil, reducing their susceptibility to the action of wind and rain. Spread over the soil surface, mulch also reduces the likelihood of weeds developing, as it blocks out light. The mulch acts as an insulating blanket that keeps soils relatively warm in winter and cool in summer. In addition, mulches add organic material to soils and inhibit some plant diseases.

However, if incorrectly applied – if it is too deep or composed of the wrong plant material – then mulch can cause damage to trees and other plants. 'Mulch volcanoes' is a new term used to describe the way some mulches are piled up against the base of trees – often to the tree's detriment. The practice can lead to root rot because there is excess moisture in the root zone. These mulches encourage pests and diseases, including rodents who like living in the mulch and chewing the bark.

Case study

The big mulch at the Eden Project, UK

In January 2002 the Eden Project's 'green team' carried out the biggest mulching experiment in the history of greenhouses. They covered the beds within the enormous greenhouse 'bubble' that comprises the Humid Tropic Biome with a lush covering of recycled organic matter in order to help soil and plant health. Over the forthcoming months and years, it is expected that this mulch will improve soil structure, water retention and nutrient levels. Scientists at the Eden Project plan to analyse and compare the effects of a variety of different mulch recipes that have been spread over the beds, in order to determine which mulch works most effectively in this unique environment.

There are two types of mulch – organic and inorganic. Inorganic mulches, such as rubber chippings and lava rock, do not contain organic material and therefore do not decompose to add nutrients to the soil. As a result, however, they do last a long time. Organic mulches, such as wood chips, pine needles, wood bark and peat, decompose gradually, adding nutrients to the soil. These mulches therefore need periodic replenishment.

WEBSITES

Eden Project:

www.edenproject.com

International Society of Arboriculture:

www.isa-uki.org/pages/Mulch.htm

SEE ALSO

organic farming

MUTUALISM AND NEUTRALISM

Mutualism and neutralism are common forms of **SYMBIOSIS** that operate in the environment. The terms describe vastly different things.

Mutualism

Mutualism is a 'win–win' symbiosis, which is 'good for' both species involved. Some authors and many amateur naturalists use 'mutualism' and 'symbiosis' as synonyms, but this leaves no general heading for all of the modes of biological 'togetherness'.

Examples of mutualism are legion. These examples were collected with relish by early evolutionists as intricate examples of the evolutionary process. This occurred before it was generally appreciated that the biosphere as a whole and every particular bit of it exemplifies the evolutionary process. Even earlier, the Natural Theologians of the eighteenth and early nineteenth centuries catalogued examples of mutualism as evidence of the intricacy and cleverness of divine design.

Mutualism is common in pollination relationships. A plant supplies energy in the form of rich nectar 'in exchange' for the transport of its pollen by an insect or by birds or bats. Mutualistic seed dispersal also is common. Nutrient-rich fruits with tiny seeds (raspberries and figs, for example) are often adaptations to dispersal by birds or mammals. The

Ants and acacias

The Bullhorn acacia is a shrub or small tree of the pea family that is found in Central America. At the tips of the leaflets in this acacia are small glands called Beltian bodies – named after Thomas Belt, a nineteenth-century naturalist who explored Nicaragua – that produce a sugary substance. This sweetness attracts ants of the genus *Pseudomyrmex*. The ants live inside the thorns and feed on the product of the Beltian bodies. In return, they do what tropical ants do best: they bite. They bite insects that alight on the tree, they bite plants that grow too close and any humans who may bungle into the tree. In short, they protect the plant from other, detrimental forms of symbiosis, in particular from competition by other plants and predation by 'uninvited' insects.

Beltian bodies in Bullhorn Acacia.

Source: Michael Clayton, University of Wisconsin

fruit attracts the mobile animal and rewards it enough to attract it again. The seeds themselves – the next generation – are carried away from the parent plant, lowering the probability of intraspecific COMPETITION.

Neutralism

Neutralism is also a form of symbiosis, but unlike mutualism, it represents biological independence. Neutralism is the most common type of interspecific interaction. Neither population affects the other. Any interactions that do occur are indirect or incidental. One example is tarantulas living in a desert and the cacti living in close proximity.

One could argue that neutralism is not really symbiosis, 'life together', but rather, 'life apart'. It would simplify things to ignore neutralism, but considering it makes us think about the nature of environmental communities. The fact is that examples of neutralism are difficult to come by and impossible to prove. However, in a scientific and logical context, one cannot prove a negative, and thus it cannot be asserted positively that there is no relation whatsoever between any two co-occurring species. The symbiotic relationship between a fruit-eating bat and an earthworm may be highly indirect, but it probably is there, perhaps mediated by the fruiting tree that feeds the bat (and whose seeds the bat disperses). The relationship may be tiny, and perhaps it is negligible, but at the level of cells, minute things, even seemingly negligible things, can make a difference, where ecologists have the patience to search for patterns.

Therefore, it may well be that true neutralism does not in fact exist.

WEBSITES

List of biological community interactions, including neutralism:
www.csuchico.edu/~pmaslin/fbiol/CmmInter.html

The concepts of symbiosis are sometimes applied to the study of sociology, for example, in this paper about the study of peace and human interactions:
www.peace2.uit.no/hefp/contributions/papers/Dalland_Oystein_12E-2.pdf

NATIONAL PARKS

View from Boulder Pass at Glacier National Park, Montana. Photograph from US National Park Service

Despite the rapid advance of urban growth, the environment has one fairly safe breathing space – national parks. National parks and other protected areas make important contributions to human society and the environment by conserving the natural and cultural heritage for the enjoyment of people and aiding ecological balance as human populations increase. National parks have been designated across the world, including in the USA and the UK, South Africa (Kruger), Tanzania (Serengeti) and Kenya (Tsavo).

In the USA the establishment of Yellowstone National Park by Act of Congress on 1 March

1872, for the first time signified that public lands were to be set aside and administered by the federal government 'for the benefit and enjoyment of the people'. The National Park Service Act signed by President Wilson in 1916 (also referred to as the Park Service's Organic Act) created a park system that has become the world's largest, both in the number of parklands (384) and in total land area (337,000 square kilometres). There are 21 types of parklands in the USA's National Park System, including parks, monuments, preserves, seashores, lakeshores, historic parks, battlefields, historic sites, memorials, wild areas, scenic and/or recreational rivers, parkways, trails, recreation areas and scientific reserves. Units are generally added to the National Park System by an Act of Congress, although the President may proclaim national monuments on federal land for inclusion in the system.

Pressure on the parks

Dr Michael Frome, in his remarks before the 50th annual banquet of Olympic National Park Associates in 1998, claimed that many national parks were being turned into theme parks. He claimed that national parks are 'overused, misused, polluted, inadequately protected, and unmercifully exploited commercially and politically. Clearly, we the people need to redefine and reassert the rightful role of national parks in the fabric of contemporary high-tech, materialist-driven society. We need to rescue the national parks from being reduced to popcorn playgrounds.'

With the boom in recreation and tourism during the last half of the twentieth century, people flocked to national parks in ever-increasing numbers. The UK's Countryside Agency placed the number of annual visitors at 110 million; one estimate of the number of visitors to USA national parks was 60 million annually. According to the National Parks Conservation Association, national parks in the USA are under assault from air pollution, development, poor funding and other woes. This organisation includes several of the 'crown jewels' of the US park system, including Yellowstone, Glacier Bay and Mojave, on its annual list of 'America's 10 Most Endangered Parks'. For example, Big Bend National Park in Texas is threatened by water diversion from the Rio Grande and trans-border air pollution from Mexico. The report says domestic power plants have lowered air quality in the Great Smoky Mountains National Park on the North Carolina–Tennessee border. Everglades National Park faces an oil-drilling proposal at its northern border in the Big Cypress National Preserve.

Increased traffic congestion, erosion from trails and litter are just a few of the environmental concerns in national parks. Parks even face noise pressure. The US Federal Aviation Administration presented at least seven proposals to expand or construct major airports near national park units across the country in 1999. An air tour overflights provision was incorporated into law during 1998, requiring the Federal Aviation Administration to cooperate with the National Park Service in developing management plans in parks where overflights could occur. This legislation, however, does not regulate the operations of large commercial airports just outside parks, or the large jet overflights that pose a menace to park wildlife and natural tranquillity.

Parks are responding first by restricting vehicle traffic or banning it altogether by forcing tourists to take shuttle buses, walk, or bicycle. For decades, the best view of the 740-metre-high Yosemite Falls in Yosemite National Park was from an idling tour bus or parked cars in a crowded carpark 'perfumed by diesel exhaust', according to an article by the Environmental News Service. During 2003 the carpark was removed and turned into a picnic area, trails to the foot of the falls were rerouted around sensitive stream beds, and vehicles and buses exiled to a permanent parking area several hundred metres from the old one. The aim of the work is that the view will belong exclusively to pedestrians for the first time in more than a generation. 'We can handle the 3-and-a-half to four million visitors [a year]', said Yosemite National Park spokesman, Scott Gediman. 'We just can't handle them in their cars.'

In England and Wales national parks were first established following the National Parks and Access to the Countryside Act (1949). Originally ten national parks, covering an area of 13,000 square kilometres, were designated, selected because of their outstanding natural beauty, and their ecological, archaeological, geological and recreational value. The broad aim of the national parks is to retain current land uses and employment opportunities but in such a way as to minimise their impact on the natural environment. Additionally, national parks are to be used and enjoyed by people for a variety of recreational activities. Since the original act, two more areas in England and Wales have been granted similar status: the Norfolk Broads and the New Forest (which became an officially titled national park in 2004). A third, the South Downs, is due to be added. There are now two national parks in Scotland: Loch Lomond and the Trossachs and the Cairngorms.

In the UK national parks are not publicly owned land, unlike in the USA. However, many national parks in the UK do have large areas that are accessible to the public, and national parks sometimes negotiate such access agreements with landowners. The National Trust, a private charity, owns about 12 per cent of the Peak District National Park, the oldest national park, and more than 25 per cent of the Lake District, as well as many other large areas of other parks. Although the National Trust is independent of the national parks, most of the land it owns is open for public access. In these parks, resources are available to promote and manage tourism, special funds may be available to landowners, and certain restrictions apply on many types of development.

Maintaining national parks requires substantial human and financial resources, particularly difficult for developing countries. To recoup these costs, many countries promote tourism in national parks. Such a move not only recognises the desire of people everywhere to seek solitude and contact with nature, but also offers them a chance to be acquainted with the natural heritage that they will hand on to future generations. Tourism can serve as a self-financing mechanism and may help as a tool for conservation. This can only happen, however, if the level, type and management of tourism are appropriate and, in particular, the 'carrying capacity' of the area is respected.

Over 20,000 protected areas have been established around the world, covering more than 5 per cent of the globe, or an area roughly equivalent to twice the size of India. Because most of these areas were established after people were living in them, only 1470 of these areas are national parks in the Yellowstone model. The relationships between people and land have often been ignored and even destroyed by well-intentioned but insensitive resource conservation and management initiatives. Community participation and equity are necessary components in decision-making processes, together with mutual respect among cultures.

WEBSITES

National parks in the UK:
www.naturenet.net/status/npark.html
UK Council for National Parks:
www.cnp.org.uk
US National Park Service:
www.nps.gov
Dr Frome's national parks versus theme parks speech:
www.wildwilderness.org/docs/frome.htm
National Parks Conservation Association:
www.npca.org
Research article on protected areas for the twenty-first century:
www.fao.org/docrep/v2900E/v2900e03.htm

NATURAL GAS

Natural gas is one of the cleanest, safest and most useful of all modern-day energy sources. But natural gas is certainly not new – the gas that is extracted from underground is millions and millions of years old. What is new are the methods for obtaining this energy supply, bringing the gas to the surface, and safely putting it to use.

For thousands of years the natural gas seeping out from the Earth's surface was not harnessed by humans. On occasions, lightning strikes would ignite the gas, creating fires that puzzled early civilisations, and led to myths and superstition. One of the most famous fires was on Mount Parnassus in Greece, where in 1000 BC a goat herdsman came across a flame rising up from a fissure in the rock. Unable to explain where the fire came from, the Greeks believed the 'burning spring' to be of divine origin. A temple was subsequently built around the flame, housing a priestess, known as the Oracle of Delphi, who gave prophecies that she claimed were inspired by the flame.

Around 500 BC the Chinese discovered ways of harnessing the energy of natural gas. On finding places where gas was seeping to the surface, they developed simple pipelines, made out of bamboo shoots, that transported the gas to areas where it was used to boil sea-water, so that the salt could be separated and the water could be drunk.

The commercial use of natural gas began in Britain from around 1785, with natural gas being produced by burning coal to light houses and streets, especially in rapidly growing industrial cities. Throughout most of the nineteenth century natural gas was used as a source of light, largely because there were no pipelines so it was too difficult and dangerous to transport the gas very far, for instance into homes to be used for heating or cooking. This started to change in the late nineteenth century, following the invention in 1885 by Robert Bunsen of a device known as a Bunsen burner, which mixed natural gas with air in the right proportions, creating a flame that could be safely used for cooking and heating.

In the second half of the twentieth century welding techniques, pipe rolling and metallurgical advances meant that safe and reliable pipelines could be built. Since then, natural gas has increased in usage, to heat homes, in manufacturing industries, in electricity generating plants and even to power buses.

In terms of its chemistry, natural gas is a very uninteresting gas – it is colourless, shapeless and odourless in its pure form. Its most useful characteristic is that it is a highly combustible mixture of hydrocarbon gases, including methane, ethane, propane, butane and pentane. When burned, it gives off a great deal of energy. Unlike other **FOSSIL FUELS**, natural gas is clean burning and emits lower levels of potentially harmful by-products into the atmosphere.

Natural gas has the same origins as other forms of petroleum, and it is derived from the action of heat on organic matter in sediments. Most of the commercially exploited reserves of natural gas occur where the source rock has been heated to temperatures in excess of

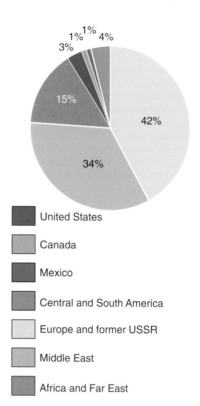

World Gas Reserves – January 1st 2000 (source – International Energy Annual)

160°C. The global reserves of natural gas currently far exceed those of crude oil. Currently it is estimated that the global natural gas reserves stand at around 146 billion cubic metres. Of these reserves, 34 per cent are located in the Middle East, with a further 42 per cent in Europe and Russia. Only 3 per cent occur in the USA, yet the USA is amongst the greatest users of this type of energy, once again showing that the location of peak demand rarely coincides with areas of greatest supply.

WEBSITES

Most of the major oil and gas companies have websites with further information, for example:

www.naturalgas.org/overview/overview.asp

www.bg-group.com/index.html

SEE ALSO

fossil fuels

NICHE

In the tangled web of relationships in the Earth's environment, each species has evolved a particular role. This is its niche. The niche (sometimes referred to as the ecological niche) is the sum of the species' resource needs – the 'ecological space' occupied by a species population. If a species' habitat is its 'residence' or its 'address', then its niche is its 'profession' or its 'lifestyle'. A given animal or plant lives in a particular place, is active at particular times and eats particular things, and these factors define its niche. The Earth's environment is divided into millions of ecological niches, each of which represents a potential 'home' for life.

The niche concept was first stated formally in 1913 by English ecologist named Charles Elton, but long before Elton people had observed that each species has a particular place in the 'scheme of nature'. More recently, ecologists have sought to quantify the niche concept; Yale ecologist G.E. Hutchinson has stated that the niche should be considered as an 'n-dimensional hypervolume'.

Imagine a cube. A cube is a three-dimensional volume. Suppose each axis of the cube (length, breadth, height) represents some dimension of the species' resource needs. Let height represent moisture, for example, breadth represent temperature and length represent soil nitrates. A given plant species has a genetically influenced range of tolerance for each of these factors. Each factor is a dimension of the species' niche or lifestyle.

We describe our daily spatial world in terms of three dimensions. Sometimes we throw in a fourth dimension – time. To deal with more than three dimensions we need a metaphor more complex than a mere volume – we need a hypervolume. But how many dimensions do we need to describe a species' niche? Our everyday, commonsense graph is fully occupied with soil moisture, temperature and soil nitrate. We have no room for day length, carbon-dioxide demand, soil acidity, soil particle size, calcium, potassium or phosphorus, let alone the host of other resources that our plant needs to survive. Its niche has a large, unknown number (n) of dimensions. The shape that describes its niche is not three-dimensional, but n-dimensional. It is not just a commonsense space, but a hyperspace (*hyper*,

from Greek meaning 'more than normal'). This niche concept has proved invaluable when looking closely at the structure of communities and the evolution of that structure.

Under laboratory conditions, ecologists can study a species' niche in isolation from

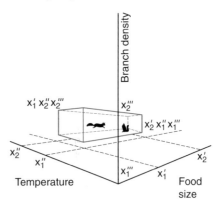

The niche of a tree squirrel in three dimensions (from G E Hutchinson, *An Introduction to Population Ecology*, Yale University Press, 1978)

other species, and determine its tolerance for as many of those n dimensions as they have the patience or funds to study. A picture can then be drawn (and it must be a mental picture, of course) of the fundamental or potential (pre-competitive) niche of the species. The n dimensions can, of course, be easily handled and drawn by a computer.

In the real world, however, the species occupies a niche that is determined partly by its genetic and physiological tolerances and partly by its biotic environment. Various competing species may occupy parts of some or all of the same resource axes. These competing species 'cramp the style' of the species under study, making the actual, so-called 'realised' niche – the post-competitive niche we can observe in the real world – more restrictive than the potential niche. In the figure, we might consider the outer volume as the fundamental (pre-competitive) niche of the squirrel and the inner volume as its realised niche.

The niche is thus a property of the species – a product of its evolutionary history in its environments and the sum of its ecological tolerances and resource demands.

We sometimes read about an 'empty-niche'. In a sense, this is an absurdity, because species define niches. If there is no species, there is no niche. What is really meant by an empty niche is, that a community seems to have resources that are not being utilised and hence there is 'room' for other species. The Shasta ground sloth of the American South West ate creosote bush. Since this beast became extinct some 10,000 years ago, no other species has come along that can stomach the creosote bush. Hence, there is an underutilised resource in the south-western deserts – room for an 'ecological entrepreneur'. The same would be true for a consumer with a taste for ferns, which are a largely unexploited resource in most biotic communities.

That no two species occupy the same niche is a 'law' of ecology. No two species have the same niche because no two species have the same limits of tolerance. No two species have the same resource needs because no two species have the same evolutionary history and genetic program. However, the more similar two species are (that is, the more similar their niches), the more they will compete. When one species achieves the slightest reproductive advantage over the other, it will drive the other to local extinction in a process known as **COMPETITION**.

Sometimes, when species with closely similar niches are present in a community, they adjust to the presence of each other. The adjustment may be a local shift in resource exploitation or habitat selection, or it may be a shift in physical structure (see **SYMBIOSIS**, and **COMPETITION** and **CONVERGENT EVOLUTION**).

The Earth's environment is constantly changing over time. Therefore, some niches are destroyed, driving the species that occupied them to extinction as they lose their ecological home. For example, current rising sea level, which could be caused by human influence on the climate, may destroy the niche for many wetland species as these areas are flooded.

WEBSITES

Physical Geography Net's explanation of the niche concept:
http://www.physicalgeography.net/fundamentals/9g.html
Explanation of ecological niche:
www.purchon.com/ecology/niche.htm

Ecological niche explanation from the BBC's Evolution pages:
www.bbc.co.uk/education/darwin/exfiles/ec_ext_eco.htm
Examples of the niche concept in the news and in research:
http://ecological-niche.wikiverse.org

THE NILE

Egypt is the gift of the River Nile (Herodotus)

Egypt has the longest known history of any African country. The Nile allowed this part of the world to develop as one of the early cradles of civilisation. The river used to flood annually in the summer, and dependably so, spreading successive thin layers of fertile silt, or alluvium, over the valley floor.

It was not only the river's regime that allowed this important civilisation to develop, but also certain discoveries. Cultivation of wheat and barley, derivatives from wild grasses, plus technology geared towards construction of irrigation systems were part of the basis of this early nation. An excess of farm production led to urban development, which in turn fostered trade, hence economic stability and leisure. Arts, crafts, political systems and academic development – writing, arithmetic, geometry and astronomy – were the consequences.

It is easy to see how people in the past were so directly dependent on the waters of the Nile. The modern Egyptian economy also depends on the Nile, and 96 per cent of the present population lives in the Nile Valley and Nile Delta. However, without strict management of its waters, the river would remain strictly seasonal in flow. Control and storage have therefore become essential for the organisation of a modern-day economy.

Dam building began early on the Nile. In 1902 the first dam was constructed at Aswan. Sixty-eight years later the Aswan High Dam, 6.4 kilometres upstream, was constructed. It holds back 25 times as much water as the original 1902 dam and so allows year-round storage much more efficiently. Consistent flow throughout the year is therefore possible. Lake Nasser, the lake formed behind the dam, was named after the past president of Egypt. It is around 5000 square kilometres in area – depending on the exact time of year. One major consequence of the creation of this lake was that 50,000 people had to be displaced from their homes and land.

However, on the positive side, this massive supply of water has been used for the irrigation of large areas of land that were once desert. More agriculture is therefore possible, including the cultivation of a greater variety of crops. Water-hungry varieties, such as rice, are now produced on a scale for export, adding considerably to Egypt's balance of payments. Cotton is another major crop. While not as thirsty a plant as rice, cotton has been boosting the Egyptian economy for a longer time. Long stapled varieties are grown; developed in the USA, some varieties have longer fibres within the cotton boll, ultimately creating a higher quality fabric. The majority of this product also goes for export.

Today, farmland in Egypt is under major pressure from alternative land uses, which also rely on the Nile as a crucial resource. The Aswan High Dam, and others along the river, are also a source of hydroelectric power, and thus they are a source of power for industry and a rise in the standard of living. Egypt's human development index (HDI) has now risen to 0.614 (1.0 is the maximum possible under this system), making it 109th in the world and considerably above the average for less economically developed countries. This is largely a consequence of the sensible use of the Nile resource.

The construction of the Aswan High Dam has had a number of environmental consequences. Prior to the construction of the dam, Wadi Allaqi upstream from the lake was a typical arid wadi – a dry valley for most of the year, with the occasional flash flood. However, since the creation of Lake Nasser the water table has risen such that the lowest 50 metres of the wadi is now permanently inundated. The ecology of the whole area has changed, with previously marginal areas now cultivable and thriving agriculturally. However, downstream in the delta region the impact has been more negative. The reduction in silt deposition has made some areas more vulnerable to erosion and the intrusion of salt water has polluted fresh water. Furthermore, silt deposited in the annual floods sustained the fertility of the soils along the Nile Valley and

in the Nile Delta. Without these annual floods, farmers are required to use artificial fertilisers and pesticides to maintain crop production. Some people believe that a decline in the sardine catch in the Mediterranean is also the result of the dam construction. However, this must be countered by the increase in fish stocks resulting from the creation of the lake, which provides employment for some 7000 people.

WEBSITES

AncientNile:
www.ancientnile.co.uk/nile.html
Gary Cook Photography:
www.garycook.co.uk/Africa/index.html

SEE ALSO

dams, flooding

NITRATES

Nitrates are nutrients derived from nitrogen that are essential for plant growth. Nitrogen is one of the three key elements that soil needs to produce high crop yields. Look at any bag of artificial fertiliser and you will see the letters, N, P and K. These represent nitrogen, phosphorus and potassium, the three key nutrients.

The main sources of nitrogen are from within the soil as well as directly from the air. Nitrogenous plants are able to 'fix' nitrogen, that is, they take this gas from the air and process it into a usable form for themselves and other plant life. Soils naturally acquire nitrates from rotting plant materials.

Left alone, nature manages the nitrogen cycle between soil, air, water and plants perfectly well. Some crops, such as alfalfa, a type of clover, also do this, and so create a relatively natural way of adding nitrogen to the soil in areas of intensive agriculture. However, this is only appropriate in some farming systems (those with livestock, as alfalfa is a useful fodder crop) and in certain climates. When people want to increase the yield in arable farming, natural methods prove insufficient and farmers may resort to using chemical fertilisers rich in nitrates. Other alternatives include manure or silage liquor, the liquid collected from fermenting silage, which is laden with such nutrients.

Although nitrates are vital nutrients, if excess nitrates are applied to the soil they can cause problems. Nitrates are easily soluble in water, so as water passes through the soil some are dissolved and eventually find their way into rivers, ponds and groundwater. High concentrations are not normal and can be extremely damaging. In periods of heavy rainfall overland flows of rainwater can wash excess nitrates straight into rivers. In addition to causing high levels of nitrates in drinking water, this also encourages rapid algal growth, leading to problems of **EUTROPHICATION**.

In the UK the pollution of drinking water by nitrate seepage into the groundwater is a big problem. Areas where agriculture is particularly intensive suffer most. People in East Anglia, Lincolnshire, Nottinghamshire and Staffordshire often consume drinking water with nitrate levels above the World Health Organization safety limit of 50 parts per million. Today almost 2 million people in the UK are affected in this way. In some cases levels reach almost double the permissible limits. The main cause of the problem has been the increasing use of fertilisers to maximise production.

Wide stretches of North America suffer similar problems. The High Plains of the USA and the Canadian Prairies are the most affected areas, since they concentrate on grain production.

Once in the groundwater store, nitrates can be trapped there for 40 years or more. They may not appear in the water supply for that time, or even longer. Nitrate contamination is therefore a long-term problem, as is any solution. Consequences of nitrate ingestion by humans include birth defects, blood poisoning in babies, hypertension in children and, later in life, gastric cancer. Even if we begin to solve the input of these toxins into the system it will remain a problem for the future generation.

If artificial fertilisers are used as recommended, only a small amount of nitrate residue finds its way into groundwater; problems on the scale we are now experiencing would not disappear but would be less of an issue. It is excessive applications of these fertilisers that result in an overload situation, with potentially serious consequences for ourselves and for the ecosystems of which we are a part. The problem of nitrate pollution on a global scale is increasing. As less economically developed countries increase their wealth and consequently their use of nitrogenous fertilisers, the area over which the problem can occur increases. Many poorly educated, even illiterate farmers are using these products. Experience to date has shown that there tends to be a belief that if a little of something is good, then more of it must be better. Fertilisers are therefore overused – and with increasing environmental consequences.

Remedies for high nitrate levels in drinking water are far from simple. One possible strategy is the denitrification of water at purification plants, that is, the removal of the nitrogen compounds. This is expensive and would add significantly to the costs of production. Alternatively, and more cheaply, drinking water supplies with excess nitrates could be mixed with uncontaminated sources. However, in the longer term the best solution will be to change farming practices to make them less reliant on the addition of nitrates to the soil.

WEBSITES

US Geological Survey publication 'Nutrients in the nation's waters – too much of a good thing?':

http://water.usgs.gov/nawqa/CIRC-1136.html

UK Department for Environment, Food and Rural Affairs:

www.defra.gov.uk

ENGETEC USA Ltd:

www.triwan.com/nitrates.htm

NITROGEN CYCLE

The most common element in the Earth's atmosphere is nitrogen. However, nitrogen must be changed into a more reactive form to allow plants and animals to use it. Living things need nitrogen to make proteins, but they cannot get it directly from the air – nitrogen is too stable to react inside an organism to make new compounds. Plants can use nitrogen when it is in the form of nitrates or ammonium salts. Changing nitrogen into a more reactive substance is called nitrogen fixation.

Nitrogen is the prime ingredient of the 20 different kinds of molecules that link together in long chains as amino acids. Amino acids make proteins, and proteins, in a certain sense via their genetic connection, make life, as least they specify the amazing diversity of life. A corn plant consists of about 1.5 per cent nitrogen and a human being is about 3 per cent. We have twice the nitrogen of a corn plant because our support systems – muscles and bones – are high in nitrogen-rich protein. The support system of corn and other plants, on the other hand, is cellulose, a carbohydrate, and a compound of carbon, hydrogen and oxygen.

Nitrogen is a fairly abundant element on Earth. Nitrogen makes up less than 0.1 per cent of the crust, but it comprises 70 per cent of the atmosphere. However, most organisms cannot deal directly with elemental nitrogen from the atmosphere. Rather, they must have their nitrogen served up with hydrogen or oxygen attached, as ionic ammonium (NH_4^+), nitrate (NO_3^-), or nitrite (NO_2^-). Because of this requirement for ionic (combined) rather than elemental nitrogen, the biogeochemical cycle of nitrogen is remarkably complicated.

The nitrogen cycle is shown in very simplified form in the figure below. Each step in the nitrogen cycle involves some process at the cellular level and some discrete product. The processes are often a highly complex series of biochemical reactions, simplified here to a single word. Many of the processes are accomplished by microbes, especially bacteria. There is no substitute for the work these microbes do. At some of these microbial junctures, several kinds of microbes have evolved, adapted to 'doing their thing' under particular environmental conditions. The energy inputs to the cycle include biological ones (from the Sun) as well as geophysical ones (such as lightning).

The principal reservoir of nitrogen is the atmosphere. Nitrogen is present in the air as a molecule which has two atoms of the element (known as 'diatomic'), that is N_2. The very strong triple bond between the atoms must be broken ('activated') before other atoms can be attached and that takes energy. The energy can come from cosmic rays or even from lightning. Or, the energy may come from the chemical bond energy contained in organic molecules. Once activated, nitrogen combines with hydrogen to make ammonium (NH_4^+), which can carry nitrogen into the synthetic processes of plant cells. About ten times as much nitrogen enters the biosphere through nitrogen fixation by microbes via these chemical bonds as through non-biological means such as the Sun or lightning. In other words, the productivity of the biosphere is utterly dependent upon microbial nitrogen fixation. Perhaps we should seek to better understand the existing microbe-driven nitrogen cycle, appreciate it and care for it in order to protect the environment.

Over the long run, the nitrogen cycle has been in equilibrium, with fixation balanced by denitrification (conversion of nitrate to nitrogen gas), volcanic addition balanced by sedimentation, and release balanced by storage. At the present time, however, industrial humans are adding fixed nitrogen to the

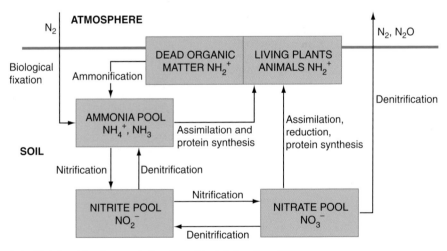

A simplified view of the nitrogen cycle (from PR Ehrlich et al., Ecoscience, 1971)

cycle, effectively speeding it up. One study estimated that the world applies fertilisers containing 90 million tonnes of nitrogen each year. Where does the excess nitrogen go and what difference does it make?

In part, the nitrogen goes into storage as living and dead organisms. To some degree, the denitrifying bacteria respond by redoubling their efforts. In part also, the fixed nitrogen, as ions in solution or as pieces of organisms, flows downhill in the water cycle. The net result of this is not yet known. There are some intriguing possibilities, however. An increase in nitrates makes seawater more acidic. The more acidic the seawater, the more quickly carbon dioxide is released from carbonate rocks to the atmosphere and hydrosphere. Hence, industrial fixation of nitrogen, resulting in nitrous oxide and carbon dioxide, may contribute indirectly to the greenhouse effect. Nitrogen can also be a pollutant when nitrogen compounds are mobilised in the environment. They can leach from fertilised or manured fields, they can be discharged from septic tanks or feedlots, they can volatilise to the air, and they can be emitted from combustion engines. As pollutants, nitrogen compounds can have adverse health effects and contribute to degradation of waters and an overabundance of nutrients (EUTROPHICATION).

The more the nitrogen cycle is understood, the more complex it becomes. The cycle is notoriously difficult to study, partly because the cycle operates on a minute scale, a sort of 'cottage industry' of microbes with immense cumulative results. Microbial nitrogen fixation depends upon an enzyme called nitrogenase, which is needed to break diatomic nitrogen gas (N_2) into elemental nitrogen to allow oxidation of nitrogen.

WEBSITES

Biogeochemistry of the nitrogen cycle from the World Resource Institute:
www.wri.org/wri/wr-98-99/nutrient.htm
Human impact on the nitrogen cycle:
www.esa.org/sbi/sbi_issues/issues_text/issue1.htm
Nitrogen cycle from Physical Geography.net:
www.physicalgeography.net/fundamentals/9s.html

NITROGEN OXIDES

Nitrogen oxides, or NOx, are the generic term for a group of highly reactive gases, all of which contain nitrogen and oxygen in varying amounts. Many of the nitrogen oxides are colourless and odourless, but nitrogen dioxide (NO_2) along with other particles in the air can form a reddish-brown layer of air pollution over many urban areas.

Nitrogen oxides are a major source of pollution and are produced by the combustion of fossil fuels in heating systems, power stations and motor vehicles, as well as from fertiliser application and the burning of crop residues. Nitrogen oxides are also the ingredients in photochemical smog and acid rain. Exposure to this substance can lead to respiratory infections, throat and eye irritations, as well as skin problems. In contrast to carbon dioxide and methane, nitrous oxide is released in small quantities from anthropogenic (i.e. resulting from human activity) sources, but its 100-year global warming potential of 310 makes it a significant contributor to atmospheric warming. In 1990 anthropogenic emissions of nitrous oxide contributed to 8.5 per cent of the total amount of UK greenhouse gas emissions.

In the early 1990s surveys conducted by the Friends of the Earth at 20 sites in British cities showed that emissions of nitrogen dioxide were very close to, or had exceeded, EU safety levels at 40 parts per billion. Five sites in central London, as well as those in Birmingham, Manchester and Cardiff, exceeded these safety levels between December 1991 and January 1992.

Since the 1990s, legislation has been introduced in the UK to reduce emissions, particularly the 1995 Pollution Act. This requires careful monitoring of air quality, as well as steps to reduce the level of these harmful substances in the lower atmosphere.

Controls have also been in place in the USA since the 1970s. The US Environmental Protection Agency (EPA) has required motor vehicle manufacturers to reduce NOx emissions from cars and trucks, and in the 1990s NOx emissions from highway vehicles decreased by more than 5 per cent.

To help reduce acid rain, the EPA devised a two-phased strategy to cut NOx emissions from coal-fired power plants. The first phase, agreed in 1995, aimed to reduce NOx emissions by

Nitrogen dioxide in British cities

Exceeding EU safety levels (>40 part per billion)		Not exceeding EU safety levels (<40 parts per billion)	
City Road, London	59	Sheffield	34.5
Streatham, London	58.5	Glasgow	34
Whitehall, London	57.5	Bristol	33
Queenstown Road, London	48	Bradford	30
Great Ormond Street, London	45.5	Oxford	29.5
Birmingham	44.5	Belfast	22
Manchester	41.3	Derby	21
Cardiff	40		

Source: Daily Telegraph, 24 January 1992, data from Atomic Energy Authority

over 400,000 tonnes per year between 1996 and 1999. The goal of the second phase was to reduce emissions by over 2 million tonnes per year, beginning in the year 2000.

However, with many of the urban areas still sprawling, and car ownership showing no sign of falling, it seems likely that levels of nitrous oxides in the USA will remain high. Perhaps the only answer will be to discourage car usage in central areas. The congestion charge zone, recently introduced in central London, might be hailed by environmentalists in the future as the most successful strategy to combat the emission of these harmful nitrous oxides.

WEBSITES

UK Department of Environment, Food and Rural Affairs:

www.defra.gov.uk/environment/climatechange/cm4913/4913html/28.htm

Environment Protection Agency:

www.epa.gov/air/urbanair/nox

SEE ALSO

pollution, urban sprawl, vehicle exhausts

NOISE POLLUTION

Noise is a feature of everyday life in the modern world. While we may enjoy or seek moments of peace and quiet, it is not long before our tranquillity is broken by some sort of noise. Some low-level noise can be easily tol-erated, indeed, in the case of so-called 'background noise' it is actively encouraged, but there is often a point when the level of noise reaches an unbearable level – this is when noise pollution occurs.

Noise pollution is different to other forms

How is noise measured?

The decibel (dB) is a measure of sound intensity – the magnitude of the fluctuations in air pressure caused by sound waves. The decibel scale is based on a logarithmic scale. This means that each time the decibel level doubles, the actual sound level increases exponentially.

The intensity of noise diminishes with distance. Outside, and in the absence of any close reflecting surface, the effective decibel level diminishes at a rate of 6 dB for each increase in distance by a factor of two. Therefore, a sound measuring 100 dB at 10 metres would be 94 dB at 20 metres, 88 dB at 40 metres, and so on.

Permanent hearing loss is usually a long-term process. So it is impossible to know at exactly what point noise becomes loud enough to cause damage to the ears. However, the US Environmental Protection Agency has established that 70 dB is a safe average for a 24-hour day.

Case study

Noise pollution in Sydney, Australia

In 1991 it was estimated that 1.5 million people living in Sydney were exposed to outdoor traffic noise levels defined as undesirable (between 55 and 65 dB). Of these, 350,000 were also estimated to be experiencing noise levels that were considered as unacceptable (greater than 65 dB).

The New South Wales Road Traffic Noise Taskforce reported in 1994 that road traffic noise had become a major urban environmental problem. This was a result of a combination of factors, including poor land use planning, the creation of transport corridors in close proximity to residential areas without appropriate buffer zones or treatment to buildings, an increase in the level of traffic well above expectations and the reluctance to switch over to public transport. Despite the report, the problems worsened. By 1997 many of the major roads within Sydney had traffic volumes in excess of 30,000 vehicles per day, with some surveys estimating that this high volume of traffic was producing a noise level of around 70 dB.

of pollution. First, the noise may only be temporary, so once the noise stops the environment is free of it. Another difference is that the degree of pollution is not directly proportional to the actual volume of pollution. In this case, the loudness of the sound and the impact of the noise vary depending on time, place and whether the listener has any control over it. Most people would object to the very loud, but very short, sound of a 21-gun salute at a military parade or civic reception. In contrast, the repetitive, yet softer, thump-thump coming from a neighbour's stereo system at 2.00 a.m., will cause more stress and more complaints, even though it is quieter than the gun salute. Globally, persistent noise problems are often associated with mining and construction equipment, as well as transportation – such as freeways and airports. Community action groups have formed in response to transportation noise in particular. These groups have campaigned for rerouting of aeroplane flight paths and motorway noise barriers. Noise barriers have helped reduce sound for nearby residents, but some say that these high walls create visual pollution.

The World Health Organization has suggested that noise can affect human health and well-being in a variety of ways, including annoyance reaction, sleep disturbance, interference with communication, performance effects, effects on social behaviour and hearing loss. Research in more economically developed countries has shown that people experiencing high noise levels, especially close to airport runways or near busy motor-ways or dual carriageways, are more likely to have an increased number of headaches, greater susceptibility to minor accidents, increased reliance on sedatives and sleeping pills, and increased mental hospital admission rates.

WEBSITES

Noise Pollution Clearinghouse:
www.nonoise.org/library.htm
Environmental Protection Agency:
www.epa.nsw.gov.au/soe/97/ch1/15_4.htm

SEE ALSO

pollution

NORTH SEA OIL AND GAS

For many years, the main claim to fame of the North Sea was that it was one of the shallowest and youngest seas in the northern hemisphere. Everything changed in 1959 when an extensive **NATURAL GAS** field was discovered in north-east Holland, near the town of Slochteren. Dutch geologists subsequently found that the geological formation continued under the North Sea, but as soon as the maps had been drawn up, a problem was discovered – who actually owned the bed of the North Sea?

All the European countries with a North Sea coastline had a vested interest, so in 1964 an international agreement was drawn up, with the sea bed being divided up into sectors, based on the length of North Sea coastline that each country possessed. The UK received the largest sector.

For the next few years, geologists and petroleum companies undertook surveys, using echo-sounders and seismic equipment, in search of oil and gas traps in the sea floor strata. In October 1965 the West Sole gasfield was discovered, followed in 1970 by the Forties oilfield. Subsequent explorations found the CRUDE OIL to be light and low in sulphur, making it especially valuable, although the varying depth of the sea, between 80 and 180 metres, caused difficulties in exploiting the oil and gasfields.

The first oil came ashore in 1975, and by 1980 the UK was producing enough oil to satisfy its own demands, and start exporting as well. This gave the local and national economy a boost and helped to reduce the reliance on imported petroleum from OPEC nations, especially in the Middle East. Both oil and gas were brought ashore by pipelines, built directly to terminals on the east coast of England (such as Teesside) and Scotland (such as Flotta in the Orkneys). Oil tankers of between 30 and 100,000 tonnes were also used, loading from floating buoys moored above oilfields that were considered to either be too small in area, or to have too short a lifespan to justify the construction of a pipeline.

Scottish towns such as Peterhead and Aberdeen have greatly benefited from the development of North Sea oil and gas. For every job created on the oil rigs, another three or four are needed on land, such as constructing the rigs, supplying and maintaining the equipment, and building the oil and gas terminals. In all, around 100,000 jobs have been created in the UK by North Sea oil and gas. Huge improvements have taken place to the infrastructure, and the isolated, resource-poor islands of the Orkneys and Shetlands have also benefited from the oil boom. Sullom Voe on the Shetlands is now served by oil pipelines from the Brent, Murchison and Ninian fields, and has had a new airport and a large oil terminal.

Despite the obvious economic benefits associated with North Sea oil and gas, there have been concerns from environmentalists about the onshore developments in remote and often sensitive parts of the Scottish coastline. As with the exploitation of oil from Alaska, there is always an environmental price to pay when developing a resource in a previously unspoilt area.

Additionally, there has been some social conflict between the tightly-knit traditional communities and the migrant workers who have arrived there, and the UK and Denmark have clashed over the rights for explorations in a disputed area in between the Shetlands and the Faroes. However, all these problems pale into insignificance compared with the awful events of one evening in July 1988 when the world's worst offshore oil disaster took place in the North Sea – the Piper Alpha disaster.

WEBSITES

Detailed information, plus diagrams of the Brent oilfield:

www.schoolscience.co.uk/content/4/
chemistry/findoils/findoilch6pg3.html

Case study

The Piper Alpha disaster

On the evening of 6 July 1988 a huge fire broke out on the Piper Alpha oil and gas platform, 176 kilometres north-east of Aberdeen. There was no time to send out an SOS message and those aboard the stricken rig had to jump hundreds of metres from the platform into the cold sea – as one survivor later said 'It was either jump and try to swim, or fry and die'.

An RAF rescue helicopter from Lossiemouth could not approach nearer than 1.5 kilometres because of the heat, and flames could be seen over 100 kilometres away. By sheer coincidence, a specialised oilfield firefighting vessel was nearby, but although it directed all of its pumps and jets on to the platform, the flames were uncontrollable and the rig was totally destroyed.

In all, 167 people were killed, and in the subsequent inquiry, led by Lord Cullen, the platform's operators, Occidental, were found guilty of inadequate safety procedures.

Institute of Petroleum, including maps of the various fields:

www.energyinst.org.uk/education/coryton/page2.htm

For a Norwegian perspective, plus other historic information:

www.explorenorth.com/library/weekly/aa091500a.htm

SEE ALSO

fossil fuels, natural gas, oil

NORTHERN SPOTTED OWL

Perhaps no other battle between industry and environmental preservation is more famous than that of the northern spotted owl. The northern spotted owl (*Strix occidentalis caurina*) depends on old-growth forests for its hunting and mating territory. The towering trees of the north-western USA and south-western Canada, some more than 500 years old, form a high, protective canopy, providing habitat for animals that are rarely found anywhere else in the world. These owls need to roost in old-growth forests because these forests offer cool, damp conditions, with plenty of holes and cavities in which to roost. These trees also harbour rodents such as the red tree vole – some of the owls' main prey. Northern spotted owls do not build nests; instead they find naturally occurring sites like crevices and ledges of cliff faces, or tree cavities – cavities that are often found in old-growth trees.

Amid 60-metre-tall, 2.5-metre-diameter Sitka spruce and western hemlock, this dark brown bird flies silently, hunting only at night in search of small mammals and even other birds. The brown-eyed owl calls through the trees with a low hoot that resembles a soft, dog-like bark. An individual spotted owl needs more than 1200 hectares of old growth to survive, due to its scarce food supply. Because the spotted owl can live only in an old-growth forest environment, it is considered an 'indicator species'. This means that the health of the spotted owl population indicates the health of the old-growth forest ecosystem.

The fibrous, grainy structure of old-growth timber is what makes this wood so valuable to lumber companies. Most of the timber is on US Forest Service public land, much of which can be leased to lumber companies for logging. More than 80 per cent of old-growth forests, from northern California to British Columbia, have now been cut down – a boon for lumber companies, but a drastic loss of habitat for the northern spotted owl. When an old-growth forest has been clear-cut for logging, and converted into a less biologically diverse second-growth forest, the owl's habitat is destroyed and its numbers dwindle. One estimate has placed the number of remaining owls at only 4720. In 1990 the bird was officially listed as 'threatened'. The listing immediately limited logging in the remaining stands of old-growth forest. This resulted in a dilemma – and a confrontation – that has been popularly portrayed as 'jobs versus owls'.

During the late 1980s news of the owl's plight spread. Eventually, concern for the owl caught the attention of the media and mainstream environmental groups. Clashes erupted between environmentalists and loggers, and the northern spotted owl practically became the mascot for the environmental movement in the USA. Extremist groups even tied themselves to trees marked for logging and damaged logging equipment.

The controversy led to the filing of dozens of lawsuits, involving the US Forest Service, the US Fish and Wildlife Service, the Audubon Society, private industry and other groups. It raised deep societal issues of how public land should be managed, and the balance between economics and preservation. In 1994 the US Government set aside millions of hectares of forests as reserves for the spotted owl and other species known to dwell amid old-growth trees. Called the Northwest Forest Plan, it sharply limited a once-booming logging economy.

The US Fish and Wildlife Service and the Weyerhaeuser Lumber Company agreed to a habitat conservation plan in 1995. Under the plan, Weyerhaeuser maintains 'dispersal habitat' – areas of forest that are large enough to sustain groups of spotted owls and close enough to one another to allow movement of the owls among the forested areas. In between these areas, Weyerhaeuser will have access to enough timber to maintain its required production levels.

Because spotted owls use a wide range of forest types, managers have had difficulty developing a simple description of owl habitat

Northern spotted owl. US Fish and Wildlife Service

that can be applied to all areas. This has led to debate over how much habitat is still available for spotted owls. As a result of the controversies, more is known about the distribution and abundance of the northern spotted owl than about any other owl in the world, but the status of the species is still hotly debated. At the same time, barred owls may in places have become a greater factor in the spotted owl's decline than habitat destruction from logging. Barred owls are larger and more aggressive than spotted owls, pushing the spotted owls from their nesting areas.

Northern spotted owls do not nest exclusively in old-growth forests. Some have been reported as nesting in less mature, managed forest areas as well. Is the small population of owls that nest in younger forests sufficient to ensure survival of the species as a whole? Do these owls actually choose the younger forest sites, or were they forced into them because their old nesting areas were destroyed? How successfully the owls can adapt over time to these new surroundings is under close research.

WEBSITES

Spotted owl information from the American Museum of Natural History:
www.amnh.org/nationalcenter/Endangered/owl/owl.html
Historical data on the spotted owl timeline:
www.lib.duke.edu/forest/usfscoll/policy/northern_spotted_owl/timeline.html
Description of the northern spotted owl:
www.birdweb.org/birdweb/species.asp?id=248
NASA's Earth Observatory resources on the spotted owl:
http://earthobservatory.nasa.gov/Study/SpottedOwls/
Species at risk's owl Information:
www.speciesatrisk.gc.ca/search/speciesDetails_e.cfm?SpeciesID=33

NUCLEAR POWER

Ask anyone what they think about nuclear power, and you will get a very firm view – either supporting its development, or firmly against it. Those in favour will probably point to the relatively low cost of producing energy in a nuclear reactor, while those against will point to the fears about safety and the issues regarding the disposal of radioactive waste. But while the costs and effects are clear in people's minds, the mechanisms by which nuclear energy is generated are not well understood.

In simple terms, there are two sources of nuclear power – nuclear fusion and nuclear fission – but only one is possible on the Earth. Nuclear fusion takes place in the Sun and occurs when two atoms fuse together under extremely high temperatures and pressures to create a single atom. Nuclear fission was discovered in Berlin in 1938 by Otto Hahn and Fritz Strassman. They found that when an atom, usually uranium, is bombarded with neutrons, the nuclei of the uranium absorb the neutron. But the atom splits, releasing energy and more neutrons (daughter elements), which in turn split more uranium atoms and create a nuclear fission chain reaction.

Energy is generated in a nuclear power station by harnessing this heat from the fission process to create steam, which in turn rotates a turbine attached to a generator. Within the nuclear reactor, a graphite moderator encases the fuel rods – usually made of uranium – in order to slow down the neutrons so that they can be absorbed more easily, while the reactor is encased by a thick layer of concrete so that harmful radiation cannot escape.

Unlike other power stations, there is no combustion; instead, heat is transmitted to the boilers by a sealed coolant fluid that circulates between the boilers and the hot reactor core. Many different substances are used as the coolant – carbon dioxide in a Magnox plant, helium gas in advanced gas-cooled reactors, and water in the aptly named pressurised water reactors.

There are several arguments in favour of nuclear power. First, one tonne of uranium contains as much energy as 2 million tonnes of coal. Second, less uranium is needed compared with the amount of coal at coal-fired plants – just 50 tonnes of uranium are needed each year, compared with 540 tonnes of coal per hour at coal-burning power stations. Moreover, there are plentiful supplies of uranium in politically stable countries, making it not so finite a resource as oil, coal or natural gas. The power produced from a nuclear plant is also cheaper compared with electricity from coal-fired power stations. Based on 1990 prices, this was £2.32 per unit, compared with £2.51 from coal-fired sources.

But these production costs have to be set against safety fears and other long-term dangers. Some would say these are the real costs of nuclear power and cite examples of radioactive leaks and accidents such as the one at CHERNOBYL in April 1986, which contaminated land in the former USSR and killed hundreds of people. Some scientists have also shown that people living near nuclear power stations are more likely to develop leukaemia and other cancers. In the aftermath of September 11th, many also fear that a rogue state, or terrorist group, could attack a reactor and cause widespread devastation.

There are also huge start-up costs to be taken into account. It costs £3 billion to construct a 1000-megawatt power station, and can take up to seven years to build. A gas-fired power station with the same production capacity would only cost £400 million and can be completed in two years. Therefore the nuclear power stations take many years to pay for themselves. Closing down, or decommissioning, obsolete plants is also extremely expensive – currently estimated at about £20 billion – and there is no way to guarantee the safe and permanent disposal of high-level nuclear waste, such as spent fuel rods.

Many also question whether it is worth taking these risks when there are many renewable and safer alternatives, such as reduced demand through recycling, solar power, wind power, hydro power, wave power, natural gas, biomass conversion and others described in this book.

WEBSITES:

How a nuclear power plant works:
http://people.howstuffworks.com/nuclear-power.htm

Issues associated with nuclear power:
www.nuc.umr.edu/~ans/QA.html

SEE ALSO

Chernobyl

NUCLEAR WASTE

Nuclear waste is one of the most toxic substances ever produced by human activity, and it is something that requires extremely careful

Disposing of nuclear waste in the UK and USA

The UK and the USA have at times approached nuclear waste in similar ways. NIREX, the UK's nuclear waste disposal company, considered drilling a waste depository, deep underground for high and intermediate-level waste produced at the Sellafield plant in Cumbria. Local people and environmental groups protested that rainwater, percolating through the rocks, could flush out waste and potentially expose people to radiation doses up to 10,000 times the legal limit. After a study by Her Majesty's Inspectorate of Pollution, the plan was shelved.

The US Department of Energy began studying Yucca Mountain, Nevada, in 1978 to determine its suitability for storing nuclear waste. The site is 160 kilometres north-west of Las Vegas on the Nevada Test Site, a military reserve. The area had a nuclear history before the mountain was investigated: during the 1940s and 1950s the area was used to test nuclear bombs. Approximately 70,000 tonnes of high-level nuclear waste currently stored at 131 sites around the country would be shipped to Yucca Mountain if the plan is approved. Over 25 years after the studies began, tribal and state governments and concerned citizens are bringing court cases in attempts to block the storage of the nuclear waste at the site.

handling. It is also a highly sensitive issue for the world's politicians. The disposal of radioactive waste from nuclear power plants and nuclear missiles has been as politically intense an issue as the opening of the plants and the use of the missiles themselves.

Radioactive waste can be any form of solid, liquid or gaseous waste that contains radionuclides. There are three broad categories of nuclear waste, all of which require careful disposal:

- High-level nuclear waste: such as spent fuel rods.
- Intermediate nuclear waste: including fuel cladding and the components from the reactor.

Case study

Depleted uranium in the USA

Depleted uranium is the radioactive by-product of the uranium enrichment process, and is around 60 per cent as radioactive as naturally occurring uranium.

If uranium is used as a fuel in the types of nuclear reactors that are common in the USA, the uranium must be enriched. To enrich uranium, a process called gaseous diffusion was developed during the 1940s. This creates two products: enriched uranium hexafluoride and depleted uranium hexafluoride. The depleted uranium decay chain includes hazardous radioactive thorium, radium, radon and lead. It also has a half-life of 4.5 billion years. Currently, the USA has in excess of 5 million kilograms of depleted uranium waste material.

■ Low-level nuclear waste: including protective clothing.

The high-level waste is highly radioactive material from the core of the nuclear reactor or nuclear weapon. This waste includes uranium, plutonium and other highly radioactive products from the reprocessing of spent nuclear fuel. They all release large amounts of ionising radiation and contain elements that decay slowly and remain radioactive for hundreds if not thousands of years. High-level waste must be handled by remote control from behind protective shielding to protect workers. There are, as yet, no ways that guarantee its safe and permanent disposal.

In contrast, low-level nuclear waste includes material used to handle the highly radioactive parts of nuclear reactors, for example cooling water pipes and radiation suits and waste from medical procedures involving radioactive treatments or X-rays. It is comparatively easy to dispose of low-level waste. Storing the low-level waste for up to 50 years should allow most of the radioactive isotopes to decay, by which point the waste can be disposed of as normal refuse.

All high-level radioactive waste must be stored over a long-term period in secure areas so that the dangerous waste cannot escape into the outside environment by any foreseeable accident, through malevolent action, or by geological activity. At the present time, areas being evaluated for the storage of this type of nuclear waste include areas of the sea floor and large, stable geological formations on land. When choosing which locations are best for this toxic waste, scientists have to consider that this waste will take at least 20,000 years to decay. It was only 3000 years ago that the Egyptian Empire was at its peak, so the timescale here is over six times the span of both modern and ancient history. So much could happen in this time!

WEBSITES

NIREX:

www.nirex.co.uk

US Nuclear Waste Technical Review Board:

www.nwtrb.gov

Yucca Mountain Project:

www.ocrwm.doe.gov/ymp/about/index.shtml

SEE ALSO

nuclear power

OCEAN CURRENTS

An ocean current is a huge flow of water within an ocean. These currents may be close to the surface or very deep indeed, hugging the ocean floor. Hundreds of kilometres wide and thousands of kilometres long, their behaviour affects several global systems, most particularly our weather. The warm Gulf Stream transports 55 million cubic metres of water per second – this is 50 times more water than all the world's rivers. Without it, the temperate lands of north-west Europe would have a sub-Arctic climate.

Ocean currents are also very important

economically. For example, the Peruvian Cold Ocean Current is responsible for supplying vast amounts of nutrient-rich waters, which support the massive stocks of anchovies off the west coast of South America. Peru is heavily dependent on these anchovies. In recent years, increases in sea temperature associated with the **EL NIÑO** effect have had a serious impact on the Peruvian fishing industry, as stocks of anchovies have declined. There is also increasing concern about the possible impact of long-term global warming on ocean currents.

The biggest deep-sea current begins in Antarctica. The freezing of the Antarctic sea around the edges of the continent increases the saltiness and therefore the density of the water in the Southern Ocean. This water, being denser than normal seawater, tends to sink. This mass of heavier, very cold water reaches the ocean floor around Antarctica and from there spreads out towards the Equator, bringing a cooling influence to the hottest zone of the world.

The major surface ocean currents are generated by a number of factors. These include the prevailing winds blowing across the surface, the rotation and tilt of the Earth and the uneven distribution of the landmasses. A world map shows how the currents basically follow a huge circular route within each ocean basin. These circuits are called gyres. In the northern hemisphere gyres flow in a clockwise direction; in the southern hemisphere they are anticlockwise. The exception is the Surface Circumpolar Current, which flows continually around Antarctica from west to east. There is simply no land to obstruct or redirect the flow.

A key function of ocean currents within the global system is to distribute the Sun's heat more evenly around the Earth. Tropical areas have a net gain of solar radiation and polar areas a net loss. So some tropical heat needs to be redistributed, because otherwise these regions would simply become hotter and hotter and the poles colder and colder. Ocean currents remove about 20 per cent of this excess tropical heat to the colder parts of the globe. The remaining 80 per cent is transferred by wind.

Ocean currents affect our climates more than most of us realise. New York, at 40° north, is on the same line of latitude as Lisbon, yet the climates of these two cities are very different. This contrast is explained by the fact that the northern part of the eastern seaboard of the USA is affected by the cold Labrador Current, bringing cold water and icebergs south. It was one such iceberg carried by the Labrador Current that sealed the fate of the *Titanic* in 1912. On the other hand, Portugal is barely influenced by the cool Canaries flow, much further offshore, so its temperature remains warmer.

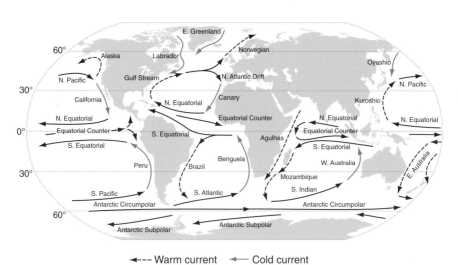

Map of ocean currents of the world. From the University of Southern California

California's climate is also strongly influenced by the flow of one particular ocean current. The cold California Current flowing from the north is the cause of the fogs, and partially of the smogs for which the coast of the Golden State and its major urban areas are renowned. Warm moist air blowing from the Pacific finds itself over an ocean cooled by this flow. Moisture held in the relatively warm air is chilled when it passes over the cooler sea, resulting in the formation of sea fog. This in turn lowers the temperatures of coastal areas, especially around San Francisco. Urban pollutants can then become trapped in the fog, causing unpleasant and dangerous smog.

The climate of the UK would be considerably colder without the positive effects of the Gulf Stream/North Atlantic Drift. Therefore recent suggestions in the press that one result of global warming might possibly be to 'turn off' this ocean current, making north-west Europe actually colder in a warming world, caused consternation. In fact, the North Atlantic Drift has been 'turned off' before – at the end of the Quaternary Ice Age – but it is thought that a contributory factor was the vast quantity of fresh glacial meltwater entering the Atlantic from the ice sheets of North America. Therefore the chance of this ocean current failing once again seems slight.

WEBSITES

The Center for Improved Engineering and Science Education:

www.k12science.org/curriculum/gulfstream

USC's Ocean Currents and Climate:

http://earth.usc.edu/~stott/catalina/oceans.html

SEE ALSO

oceans

OCEANS

The famous science fiction writer Arthur C. Clarke once remarked: 'How inappropriate to call this planet Earth, when clearly it is Ocean'. Oceans cover about two-thirds of the Earth's surface. For many years the oceans represented the last frontier yet to be explored on the Earth. It was only after the Second World War, when great strides in undersea technology took place, that we were really able to discover the wonders of the oceans. We found the answer to the continental drift puzzle in the mid-ocean ridge, where sea-floor spreading and the theory of **PLATE TECTONICS** was proposed and validated. We found hydrothermal vents (undersea hot springs) where new and unusual forms of life were discovered. We drilled into the floor of the ocean, where the Earth's crust is thinnest, and found answers about the structure of the Earth's interior (see **EARTH**).

How were the oceans formed? One theory is that gases trapped in the Earth's core as it was cooled were expelled by volcanoes to form the Earth's atmosphere. These gases included water vapour and carbon dioxide. It is thought that the oceans formed about 4 billion years ago when widespread volcanic activity released hydrogen, and smaller amounts of carbon dioxide, chlorine and nitrogen, which produced a water-vapour atmosphere that also contained carbon dioxide, methane and ammonia. As the Earth cooled, water vapour condensed and fell to the Earth's surface as rain, where it accumulated to form the oceans.

An alternative theory is that comets were responsible for the formation of oceans. However, the idea that comets encouraged life on Earth with water and essential molecular building blocks is hotly debated.

Oceans have an enormous effect on climate and, therefore, on land environments. **OCEAN CURRENTS** are caused by winds that blow over the ocean, causing surface water to drift along with them. Without the moderating influence of one of these currents, the Gulf Stream, for example, much of Europe would be too cold for dense human habitation.

Land near the oceans tends to have a more moderate climate than land further from coasts. Therefore the plants and animals tend to be different as well. One example is the marine west-coast climate, characterised by mild winters and summers, a low annual temperature range, cloudy, humid days, and frequent cyclonic storms with prolonged rain, drizzle and fog. It occurs in such places as the Pacific Coast from Oregon to southern Alaska, in north-west Europe, southern Chile, southeast Australia and New Zealand. These environments support such diverse and unique species as the California redwood and the myrtlewood tree.

In terms of climate change, the oceans may have a strong impact, because of the huge amounts of carbon dioxide stored in them. Microscopic plants extract carbon dioxide, a greenhouse gas, from the atmosphere during photosynthesis. Warmer ocean temperatures could produce increased numbers of these plants, which could reduce carbon dioxide in the atmosphere. This would make the future climate cooler. On the other hand, warmer ocean temperatures could increase the carbon dioxide in the air due to the fact that warmer water cannot dissolve as much carbon dioxide as colder water. This would make the future climate warmer.

WEBSITES

Comet theory of ocean formation:
www.gsfc.nasa.gov/gsfc/spacesci/origins/linearwater/linearwater.htm
Continental drift theory:
http://pubs.usgs.gov/publications/text/developing.html
United Nations Ocean Atlas:
www.oceanatlas.org/index.jsp
Galapagos Islands natural history:
www.galapagosonline.com/nathistory/nathistory.htm
Ocean currents map:
www.physicalgeography.net/fundamentals/8q_1.html

SEE ALSO

climate change, ocean currents

OIL

The fame and fortune of many individuals, as well as the economic growth of some Middle Eastern nations, is all the result of one thing – the remains of living organisms from millions of years ago. That is precisely what crude oil is, and it is this 'black gold' that industrialised nations crave – to the extent that some nations have even gone to war over the control of the oilfields.

Crude oil or petroleum is a complex mixture of hydrocarbons that developed under ocean and sea beds following the decomposition of microscopic marine organisms called phytoplankton.

How crude oil is formed

Oil is formed in several stages. The first involves the build-up of a fine-grained mud, rich in planktonic remains, in low-energy offshore environments. In the second stage, decomposition by anerobic (low oxygen) bacteria transforms the planktonic remains into an amorphous ooze or sludge known as sapropel. In the third stage, an increase in temperature and pressure converts the sapropel into petroleum compounds within layers of rock – usually black shales – called source rocks.

Subsequent compression and compaction causes oil and natural gas to migrate towards the surface into areas of lower pressure. These are known as the reservoir rocks. Gas and oil occupies the pore spaces of more coarsely grained deposits than the source beds. Finally, an oil and gas trap is formed when the further migration of gas and oil is contained by an overlying impermeable layer known as a cap rock.

In some cases the hydrocarbons have migrated up to the surface, with the oil and gas seeping into porous sand – these are known as tar sands. The most famous are the Athabasca Tar Sands in Alberta, Canada. Here asphaltic hydrocarbons, containing an estimated 600 billion barrels, have seeped into quartzitic sands, which since 1966 have been exploited by opencast methods by Canadian companies.

There are global variations in the composition of crude oil. North Sea crude oil is low in sulphur content, while crude oils from the Middle East are all high in sulphur.

The traditional way of recovering oil involves primary recovery methods, with the drilling of a well so that the crude oil surges to the surface. As the enormous pressure is released, the oil spurts to the surface forming oil fountains. Modern rotary drilling rigs, either on land or at sea, have steel towers, called derricks, which support the hollow drill pipe tipped by a drilling bit.

Secondary and tertiary recovery methods

The chemical composition of petroleum		
Element	Crude oil (%)	Natural gas (%)
Carbon	82.2–87.1	65–80
Hydrogen	11.7–14.7	1–25
Sulphur	0.1–5.5	Trace–0.2
Nitrogen	0.1–1.5	1–15
Oxygen	0.1–4.5	0

are employed after an oilfield has passed its peak production. This involves flushing out the residual oil, either by injecting natural gas into the reservoir above the oil, or by flooding with water below the oil. The secondary and tertiary methods are very useful if the residual oil is too viscous, too dilute or too dispersed to surge out by conventional means. They are also very economical, as in Canada, where demands for petroleum have increased and concerns were raised about relying too much on foreign imports. Here secondary methods were employed, at a cost of $7 million, but they boosted output by an additional 40 per cent. This proved more economical than surveying for new fields on the perimeters of glaciers or beneath the permafrost.

While oil offers massive opportunities for economic development and energy production, there are environmental costs involved in both extraction and its use. To develop an oil well, huge tracts of land are required for supply roads and storage facilities. This frequently occurs in pristine wilderness areas, such as Alaska and Siberia, where environmental recovery rates are slow. Transporting the oil to refineries often involves the use of massive ocean-going tankers, which require deep harbour facilities. Over the years several accidents have occurred involving oil tankers, which have resulted in terrible OIL SPILLS. If transported by pipeline, again significant developments are required in order to facilitate construction and maintenance (see POLAR ENVIRONMENTS). In its use, oil-fired power stations and vehicles produce toxic gases, which have been linked with acid rain and global warming. These environmental 'costs' need to be weighed against the economic 'benefits' of further oil exploitation.

WEBSITES

The corporate websites of the world's oil companies contain much information about the current rates of extraction and value of oil. Some also contain background information about oil:

www.chevron.com/learning_center/crude

www.bg-group.com/index.html

US Geological Survey on energy:

http://energy.usgs.gov

SEE ALSO

fossil fuels

OIL SPILLS

Since the late 1800s oil has been one of the world's most commonly used substances – it has been used to heat homes; to lubricate machinery; to make products ranging from asphalt to toys to CD players and computers; to make food, medicine, ink, fertiliser, cosmetics, pesticides and paint; and to generate electricity. The USA uses about 2.6 billion litres of oil every day, while the world uses over 11 billion litres each day.

Since oil is used so frequently and for so many different purposes, it needs to be stored, refined and transported by pipelines, ships and trucks. Occasionally, these transport mechanisms, or the people who operate them, fail. At other times, natural disasters, or malicious acts by countries at war, terrorists, vandals or illegal dumpers result in oil spills.

Crude oil is a foul-smelling viscous (thick) liquid that can quickly contaminate the environment. Since oil is less dense than water, it tends to float on the surface – this is particularly true with the salty water that comprises the oceans. An oil spill spreads out rapidly across the water surface to form a thin layer referred to as an oil slick. As the spreading process continues, the layer becomes thinner and thinner, finally becoming a very thin layer called a sheen, which often looks like a rainbow. Sheens can often be seen on roads after a rainstorm or in rivers or ponds where motor craft are present.

Oil spills can be very harmful to the environment. Marine birds and mammals, fish and shellfish are all affected, as are plants and the wildlife that feeds on plants. Oil destroys the

Case study

The *Exxon Valdez* oil spill

At 9.12 p.m. on 23 March 1989 the *Exxon Valdez* departed from Valdez, Alaska, to take some oil from the Trans-Alaska pipeline to California. After navigating the 300-metre ship through the Valdez Narrows, Pilot Murphy left the vessel and Captain Hazelwood took over the wheelhouse. After the ship encountered icebergs in the shipping lanes, Captain Hazelwood ordered Helmsman Claar to take the ship out of the shipping lanes around the icebergs. Hazelwood then handed over control of the wheel-house to Third Mate Gregory Cousins, with precise instructions to turn back into the shipping lanes when the tanker reached a certain point. At that time, Claar was replaced by Helmsman Robert Kagan. For reasons that remain unclear, Cousins and Kagan failed to make the turn back into the shipping lanes, and the ship ran aground on Bligh Reef at 12.04 a.m. on 24 March 1989. Captain Hazelwood was in his quarters at the time. While the state charged him with operating a vessel while under the influence of alcohol, a jury found him not guilty of that charge. The jury did find him guilty of negligent discharge of oil, a misdemeanour for which he performed community service.

Over 41 million litres of oil were spilled over the next several days into Prince William Sound. Some pointed to the fact that while this was a big spill – enough to fill nine school gymnasiums or 430 class-rooms, it was actually less than 2 per cent of what the USA uses everyday. Nevertheless, after four summers, $2.1 billion and a peak workforce of 10,000 people, nearly 2100 kilometres of shoreline were contaminated. The timing of the spill, the rugged and wild shoreline, and the abundance of wildlife in the region combined to make it an environmental disaster well beyond the scope of most other spills. Sea birds, sea lions, eagles, fish and sea otters were among the animals devastated. In addition to the direct economic impact of the clean-up, additional economic loss was felt by the tourism and sport fishing industries, as well as by those making their living by fishing.

Five years later, the incident was still causing front-page headlines. Local people had yet to be com-pensated, due to lengthy legal procedures, and many people had decided to move away from the blighted fishing communities. Some had even committed suicide.

insulating ability of fur-bearing mammals, such as sea otters, and the water-repelling abilities of a bird's feathers, thus exposing these creatures to the elements. Birds and animals also swallow oil when they try to clean themselves, which can poison them. If an oil spill occurs on land, the local and even the regional aquifers and watersheds can become polluted.

Despite safety regulations and measures in place to prevent spills, almost 14,000 oil spills are reported each year in the USA alone. Although many spills are contained and cleaned up by the party responsible for the spill, some spills require assistance from local and state agencies and, occasionally, the fed-eral government. Under the USA's National Contingency Plan, the Environmental Protection Agency is the lead federal response organisation for oil spills occurring in inland waters, while the US Coast Guard is the lead response agency for spills in coastal waters and in deep water ports. These organisations often call on other agencies, such as the

National Oceanic and Atmospheric Administration (NOAA) and the US Fish and Wildlife Service, for help and information.

When oil spills, it can be cleaned by booms and skimmers (floating barriers), chemical and biological dispersants, burning, washing off with hoses, vacuum trucks, shovels and road equipment. However, none of these methods can truly clean all of the oil, particu-larly if the spill contaminates a river, lake or part of an ocean.

The methods and tools people choose depend on the circumstances of each event. These include the weather, the type and amount of oil spilled, how far away from shore the oil has spilled, whether or not people live in the area, and what kinds of bird and animal habitats are in the area. Different clean-up methods work on different types of beaches and with different kinds of oil. For example, road equipment works very well on sand beaches, but cannot be used in marshes or on beaches with big boulders or cobbles (round-ed stones larger than pebbles, but smaller than boulders). People also may set up stations where they can clean and rehabilitate wildlife. Sometimes people may decide not to respond to a spill, because in some cases responding may even add to the damage from the spill.

In the USA the goal of new federal regulations is to prevent oil spills from happening. People who cause oil spills must now pay severe penalties, and the regulations also call for safer vessel design in the hopes of avoiding future spills. People who respond to oil spills must practise by conducting training drills, and people who manage vessels and facilities that store or transport oil must develop plans explaining how they would respond to a spill, so that they can respond effectively to a spill if they need to.

Because we all use oil, individuals must share the responsibility for reducing the hazards of oil pollution, for instance by not dumping oil or oily waste into the sewer or rubbish. In fact, the amount of oil pollution from run-off from roads and routine ship operations dwarfs the amount spilled each year. Storm sewers typically run directly into rivers, not to treatment facilities. Because we rely on oil, we run the risk of oil spills. Therefore, the best way to reduce the chances of oil spills is to use less oil to begin with; for example, by walking, cycling, sharing cars or using public transportation to go where we need to, and to support research into alternative sources of energy.

WEBSITES

US NOAA oil spills response and restoration:
http://response.restoration.noaa.gov/kids/spills.html

Environment Canada's oil, water and 'chocolate mousse' resource:
www.ec.gc.ca/ee-ue/pub/chocolate/toc_e.asp
Oil spill information from NASA:
http://seawifs.gsfc.nasa.gov/OCEAN_PLANET/HTML/peril_oil_pollution.html
Exxon Valdez oil spill information:
http://www.evostc.state.ak.us/facts
http://library.thinkquest.org/10867/home.shtml
http://response.restoration.noaa.gov/spotlight/spotlight.html
Sea Empress disaster in South Wales:
www.aber.ac.uk/iges/cti-g/STHAZARDS/seaempress/sea.html
Prestige oil spill of Spain:
www.guardian.co.uk/oil/story/0,11319,843559,00.html

SEE ALSO

pollution

ORGANIC FARMING

Do you buy free-range or battery-farmed eggs? Organic or agribusiness-generated fruit and vegetables? Are you more concerned with how your food is produced and the impact on the environment, or do you just buy what is cheapest?

For about 50 years the word organic has been used to describe food grown without artificial fertilisers or pesticides. To avoid using fertilisers and pesticides, crop rotation is encouraged, as well as making the most of natural fertilisers such as farmyard manure and

Case study

Organic ranching in New Mexico, USA

Matt Mitchell became the first organically certified rancher in New Mexico in 1997. Initially he was attracted by the potentially greater profits – organic meat fetches a higher price and fluctuates less. However, it has not been easy. Markets were hard to come by and the larger organic retailers, such as Wild Oats and Whole Foods, did not want to deal with small family ranchers. 'I got into it and then I had to create a market. One guy told me that Whole Foods would buy all the meat we could produce, but it didn't work out that way,' Mitchell says. 'When I first talked to Wild Oats they just kind of shut me out. I tried selling quarter halves and boxed beef at farmers' markets, and that didn't work too well.' Whole Foods asked him to ship his products to the chain's warehouse in Austin, Texas, but Mitchell could not afford it.

continued over

Seven years later he has found markets for his 150-head of organic cattle he runs on his Reunion Ranch Organic Beef operation east of Roy. The food goes mostly to La Montañita Co-op Natural Foods Market, which has stores in Albuquerque, Santa Fe and Taos. Mitchell says it has been a difficult seven years. 'Sometimes you start pulling your hair and wondering why you got into this. But the next day you get a call from someone who says they haven't been able to eat beef for ten years because of their immune system, and that now they can eat meat, and that feels pretty good.'

There are now 100 certified organic farmers in New Mexico, up from zero in 1990. The New Mexico Organic Commodity Commission has been formed to inspect and certify farms that want the organic label. The Federal Government lays down the regulations, but in general cattle may only be fed vegetarian-based food and crops have to be grown without chemical fertilisers and pesticides. The 100 organic producers have about 4000 hectares and account for about $18 million of sold product. Increasingly, fears about the disease BSE (bovine spongiform encephalopathy) are encouraging more farmers to think about going organic.

ensuring that the health of the soil is maintained. Animals are kept in humane ways that minimise the need for medicines and other chemical treatments. One of these practices involves allowing animals free range over a farm, rather than keeping them confined to small cages. Organic production aims to work with natural systems, preserving the existing landscape, encouraging wildlife habitats and avoiding all forms of pollution.

Since 1993 in the EU regulations have governed the inputs and practices that may be used in organic farming and the inspection system that must be in place. Genetically modified ingredients and artificial food additives are never allowed in organic foods, although up to 5 per cent of certain non-organic food ingredients may be used – otherwise products such as bread could never be labelled as organic.

WEBSITES

Organic Farmers and Growers:

www.organicfarmers.uk.com

Soil Association:

www.organicfarmers.uk.com

Department for Environment, Food and Rural Affairs:

www.defra.gov.uk/farm/organic/default.htm

Organic Farming Research Foundation:

www.ofrf.org

California Certified Organic Farmers:

www.ccof.org

SEE ALSO

agriculture

OVERFISHING

About 1 billion people, mainly in poorer countries, rely on fish as their main source of protein. As population continues to grow, this demand will rise in the future. Around 70 per cent of the world's catch is used to satisfy this demand for food, while the rest is used for animal feed, fish oil, margarine and fertilisers. Over 12 million people worldwide earn their living directly from fishing, with 2 million of the world's fishermen found in the Philippines alone.

Overfishing occurs if the amount of fish caught is so great that it fails to leave the minimum amount required to sustain the stocks for the future. For example, two-thirds of the USA's commercial fish resources are currently being overexploited. In the Pacific Ocean stocks of tuna are under threat, and in the North Atlantic cod and haddock are becoming increasing scarce, with the EU enforcing strict quotas on fishing in the North Sea. In Chesapeake Bay catches of oysters halved between the 1960s and the 1980s, and the striped bass catch has declined by over 90 per cent. Pilchard stocks are under threat in the South Atlantic, and the Southern Ocean around Antarctica is already 'fished out' of cod.

Today there are fewer fishing vessels than in the past, but their size has increased dramatically. In line with this, catches have grown, and it is this that has led to overfishing. Fish stocks are not evenly distributed over the oceans. Ninety per cent are in shallower coastal waters, where nutrients washed off the land encourage plankton to grow. The most

Case study

Peru's anchovies

The surface waters of the south-eastern Pacific are particularly rich in fish. This is because it is an area where very cold water from the ocean depths well up to the surface. This water carries a high density of nutrients and oxygen, which then supports a large fish population, especially anchovies.

However, anchovy stocks have recently come under severe stress, with catches some 6 per cent of what they were in the 1960s and 1970s. South America's population has grown relatively rapidly by global standards, and this has led to an increase in the demand for fish. In addition, the El Niño climatic phenomenon, which temporarily cuts off the upwelling of cold water and therefore the nutrients too, has dramatically reduced the anchovies in these waters. Dolphin populations, which rely on the anchovies, are suffering badly. As El Niño events are increasing in frequency, so the chances of the anchovy reserves recovering for Peru are slight.

important fishing nations in the world at the start of the twenty-first century are Russia and Japan, followed by the USA, Canada, China, Indonesia, Thailand, the Philippines, Peru, Chile, Spain and Norway.

Fish shortages can have an affect on market prices and therefore on demand. For example, in the 1960s North Sea cod was perceived as a second-rate fish to haddock, and this was reflected in the cost. Today, the reverse is often true, as cod prices have risen dramatically because of shortages. Much cod now comes from further afield, and the UK is more dependent on Icelandic stocks. During the 1950s and 1960s it was believed that world fish stocks had the potential to provide a vast proportion of the growing world demand for protein. It was simply never thought that supplies would be under such pressure – and certainly not so soon.

One solution to over-fishing is to limit catches. This has been tried in Europe, but it usually leads to feelings of great unfairness due to the uneven distribution of quotas. Additionally, some French and Spanish trawlers have in the past retained small-meshed nets to gain the maximum catch, even though these net many immature fish which should be left to grow and breed the next generation. The desire for short-term profit is thus making the whole industry unsustainable for the future.

There remains great potential in fish, but careful management and strict conservation methods are needed. Perhaps we need to look elsewhere for fish supplies rather than simply counteracting the effects of overfishing. Large-scale fish farming might offer us a solution.

Superfleets

Since the 1970s fleets of long-distance 'supertrawlers', which can stay at sea for months, have radically altered the nature of fishing – and have been blamed for massive overfishing. Having overfished their own waters, countries such as Russia, Japan, North and South Korea, and the USA are now fishing in the open ocean, sometimes even within the 320-kilometre limits of other countries. A recent location for some huge factory trawlers is well off the Atlantic coast of the USA, primarily seeking herring and mackerel. Atlantic cod, haddock, blue-fin tuna and striped bass in these waters are already overfished or recovering from overfishing.

These industrial-scale fishing vessels comprise only a tiny percentage of the world's fishing vessels, yet they are responsible for most of the world's fish catches. They have the highest proportion of by-catch – the unwanted fish and other sea creatures that are simply thrown back dead. One quarter of the world catch is wasted in this way. Supertrawlers employ few people because of their high level of mechanisation and so generate the least revenue for local communities. Indeed, they take the resources from smaller-scale operations and thus reduce incomes.

Further reading

Middleton, N., 1988, *Atlas of Environmental Issues*, Oxford University Press.

Marshall, B. (ed.), 1991, *The Real World*, Houghton Mifflin.

WEBSITES

Nature:
www.nature.com/nsu/030106/030106-13.html
Dolphin Conservation in Peru:
www.delphinschutz.org/acorema-uk.htm
Sierra Club Save the Fish Committee:
www.sierraclub.org/marine/activism/save_the_fish.asp
National Coalition for Marine Conservation:
www.savethefish.org
Greenpeace:
www.greenpeace.org/oceans/
overfishingtext.htm
Response to editorial on overfishing:
www.fishingnj.org/artnyted.htm

OVERGRAZING

Overgrazing is the exhaustion of the land by the continuous grazing of animals. Grazing land or pasture may have different carrying capacities depending upon its fertility and its ability to recover from grazing. However, in each instance of overgrazing too many animals cause the land to become degraded. Degraded means that both the quality and the quantity of the vegetation deteriorates.

Overgrazing usually occurs as a result of increasing the size and numbers of animals grazing in an area. This is usually a response to the need to produce more food for a growing human population. However, in some cases herd sizes have remained the same and overgrazing has resulted from climate change – after a number of years of reduced rainfall, pastureland has failed to recover from grazing activities. The combination of climate change and overgrazing has been a major cause of desertification in the Sahel region, south of the Sahara in Africa.

Overgrazing is a problem in many parts of the world – both developed and developing. Its impact is often greatest in developing countries, where the grazing of animals constitutes the sole food supply for communities. In the developed world few people rely absolutely on food that they grow – most have the luxury of a nearby supermarket.

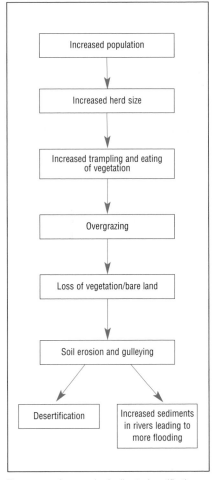

The process of overgrazing leading to desertification

WEBSITES

ManagingWholes, with ecological perspectives on overgrazing:
http://managingwholes.com/overgrazing.htm
IPS News Service, with an article on overgrazing:
www.ipsnews.net/fao_magazine/
environment.shtml

Case studies

Overgrazing in the Himalayas

Rearing animals is an important occupation in large areas of the Himalayas because other forms of agriculture, especially crop growing, is impossible due to the harsh environmental conditions. Some local populations, such as the Gujars and Gaddis, make their living primarily by rearing animals. Animals reared include sheep, goats, cows, mules, horses and yaks. Some animals are reared for milk production, leather, meat or wool. Some animals are more useful as beasts of burden, and for their dung, which is used as manure.

In the Himalayas overgrazing has had detrimental effects on both alpine grasslands and forests, leaving the land bare and prone to erosion. The animals graze in spring when new seedlings of trees, herbs and grasses are growing. The animals therefore prevent regeneration of the vegetation by eating the seedlings and trampling on new growth. Loss of the vegetation ultimately leads to changes in the soils, which become compacted, reducing porosity and aeration. This makes it even more difficult for vegetation to regenerate, so encouraging soil erosion to take place where water flows across the bare ground. Soil erosion is a particularly significant problem in the Himalayas, exacerbated by the very steep mountain slopes.

Along with deforestation, overgrazing in the Himalayas is thought to be one contributor to increased silting and incidents of flooding of rivers such as the Ganges and Brahmaputra. These floods bring great misery to countries such as India and Bangladesh. The governments of these countries would like to see international agreements to control deforestation and overgrazing in order to reduce the flood threat.

China's dust bowl

In April 2001 huge dust storms from northern China blanketed areas in North America from Canada to Arizona with a layer of dust. This followed an earlier storm that blanketed Beijing in March 2001, which came after other storms in 2000. The storms led to reduced visibility, slowing traffic and closing airports.

These dust storms were the result of increasing human pressures on the land in north-west China, exacerbated by a drought that has lasted for three years. In addition, Chinese policy states that all land used for construction must be offset by land reclaimed elsewhere for agriculture. This has led to an increase by 22 per cent of farmland in Inner Mongolia. On this marginal land, overgrazing and over-cultivation has increased soil erosion, especially by wind, leading to the dust storms. The erosion fuels the abandonment of land and migration of people – similar to that which occurred in the Great Plains of the USA during the Dust Bowl years.

Overgrazing is a key cause of the problem. In 1978 controls on herd sizes were removed and livestock populations exploded. China has 127 million cattle, compared to 98 million in the USA, and 279 million sheep, compared to 9 million in the USA. In Gonge county in Quinghai province the estimated carrying capacity of the land is 3.7 million sheep. In 2000 the province held 5.5 million sheep, causing rapid deterioration of the grassland, desertification and the formation of sand dunes. Research suggests that about 2330 square kilometres are turning into desert each year. Undoubtedly, livestock populations need to be reduced to prevent the spread of these dust bowl conditions.

Overgrazing in the Himalayas:
http://library.thinkquest.org/10131/
problems_overgrazing.html?tqskip1=1

SEE ALSO

desertification, flooding, soil erosion and conservation

OZONE

Ozone is a naturally occurring atmospheric gas with concentrations between 12 and 50 kilometres above the surface of the Earth. It is essential to human survival, shielding the Earth from much of the ultraviolet radiation that comes to us from the Sun. Without this

Protecting yourself from ultraviolet radiation

Ultraviolet light is made up of three different types: UVA, UVB and UVC. The latter is effectively blocked from reaching the Earth, leaving only UVA and UVB to concern us. Of the two, UVA is much less dangerous, though it can lead to premature ageing and wrinkling of the skin. By itself this is not a major health concern, but it can increase the damage caused by UVB. UVB rays are the ones responsible for the most damage to skin cells, ranging from a mild sunburn to skin cancer. According to the American Cancer Society, about one out of every seven Americans will contract some form of skin cancer in their lifetime, almost all due to the Sun's rays. More than 9000 deaths in the USA each year can be attributed to skin cancer.

The most effective strategy to reduce the risk of skin cancer is to minimise exposure to UVB radiation. Depending on your skin type, you could be at risk with only a few minutes of exposure. Consequently, skin covering, sunscreen and sunglasses are all needed to provide protection.

Given the importance of the OZONE LAYER for our health and even our survival as a species, scientists, environmental activists and ordinary citizens have been very concerned about its apparent destruction over the past few decades. Ever since the 'hole' in the ozone layer was discovered by scientists of the British Antarctic Survey in the late 1970s, efforts have been underway to reduce the presence of chemicals in the atmosphere that are dangerous to the ozone layer. In particular, great attention has been given to the role of chlorofluorocarbons (better known as CFCS) in the environment. Once used frequently in common products such as aerosol sprays and air conditioners, as well as for industrial purposes, their use has been curtailed dramatically as a result of international agreements. The Montreal Protocol to Reduce Substances that Deplete the Ozone Layer was crafted by the United Nations, hoping to eliminate CFC production and use. In response, Canada reduced its use of CFCs by 95 per cent from 1987 to 2002 through strong laws that include fines up to $1,000,000 and prison terms of up to five years. CFC emissions dropped dramatically by 1993. Recovery of the ozone layer, however, is expected to take 50 to 100 years. Damage to the ozone layer can also be caused by drops of sulphuric acid produced by volcanic eruptions.

blocking, the radiation would threaten the survival of much of life on Earth. Humans would suffer more skin cancer and ailments affecting vision and immune systems. Other plants and animals, and thus the environment, are also affected by ozone depletion. Studies have shown that certain plants grow less well when they are more susceptible to disease, and that populations of phytoplankton – small, usually microscopic plants (such as algae), found in bodies of water and an essential part of the ocean FOOD WEB – are disrupted. If this disruption reduces their productivity, all the species that depend directly or indirectly on the phytoplankton are in jeopardy.

Chemically, ozone (O_3) forms when ultraviolet radiation breaks apart oxygen molecules (O_2) creating two 'free' atoms of oxygen (O). When one of these has a collision with another molecule of oxygen (O_2), ozone is formed.

Further reading

Moran, J. M., and Morgan, M. D., 1997, *Meteorology*, 5th edn, Prentice Hall.

Tarbuck, E. J. and Lutgens, F. K., 2000, *Earth Science*, 9th edn, Prentice-Hall.

WEBSITES

Evergreen Resources, 'Canada's success with ozone proves it can achieve Kyoto targets': www.edie.net/news/Archive/5107.cfm
University of Cambridge Centre for Atmospheric Science, 'The ozone hole tour': www.atm.ch.cam.ac.uk/tour/index.html

SEE ALSO

CFCs, climate change

OZONE LAYER

Ozone is a form of oxygen (O_3), very faintly blue in colour, found in greatest quantities at a height of 20–25 kilometres in the Earth's atmosphere. It is thought to be formed by photochemical changes resulting from the absorption of ultraviolet radiation by oxygen. Indeed, it acts as a filter, protecting the Earth from harmful UV rays.

The Earth's ozone layer was not always present. When the Earth was formed, little free oxygen existed in the atmosphere. As photosynthesis by cyanobacteria and blue-green algae in the early oceans commenced, more and more oxygen was added to the atmosphere. At first, the oxygen attached to iron-based compounds, oxidising them. But as more and more of the exposed metallic surfaces were oxidised, increasing amounts of oxygen filled the atmosphere. As the atmosphere evolved, increasing oxygen levels allowed some of the oxygen to migrate into the stratosphere, where it was turned into ozone by the process described above. When levels approached current amounts, the world was safe from UV radiation, and plants and animals began to colonise the land surfaces of the Earth.

Ozone is a molecule with three oxygen atoms. The bonding of these atoms is weak, so the molecule has a short life. Ozone is created when a normal two-atom oxygen molecule is broken into two separate single atoms by ultraviolet (UV) light, high temperatures or chemical reactions. The individual atoms do not like flying solo – they seek to be bound back with other oxygen atoms. The likelihood of individual atoms finding their original counterpart is not very likely. More likely, they will find a normal two-atom oxygen molecule. When they attach themselves to this molecule, a three-atom oxygen molecule – ozone – is born.

A return to the lower ozone levels of the Earth's early history would mean a return to the seas. Land life and terrestrial ecosystems would not be able to exist. This is why the world became alarmed when a huge hole in the southern hemisphere ozone layer was revealed by satellite during the 1980s. Research revealed that human-made chlorine and bromine compounds, particularly chloro-fluorocarbons (CFCs) casually used at terrestrial levels in household and industrial applications (such as refrigerators and aerosol spray cans) were migrating to the stratosphere. Once there, they entered into a complex chemical reaction that destroyed ozone faster than it was being created.

Under the Montreal Protocol (1987) there have been significant cutbacks in the use of **CFCs**. Consumption worldwide has dropped by 84 per cent, and in the industrialised countries it has dropped by 97 per cent. The ozone holes are expected to 'fill' and recover to reach their pre-1980 level by 2050. The success of the international agreement – preventing an environmental disaster from becoming too serious – has renewed hope among environmentalists: might it lead the way to forging an agreement relating to global warming?

WEBSITES

History of the ozone hole:

www.atm.ch.cam.ac.uk/tour/part1.html

The ozone hole:

www.theozonehole.com

US EPA's ozone resource:

www.epa.gov/ozone

British Antarctic Survey:

www.antarctica.ac.uk/met/jds/ozone

SEE ALSO

CFCs, climate change

PESTICIDES

Pesticides are just one of a vast range of chemicals used in farming. They are commonly thought to be used to eradicate pests such as spiders and rodents, but are also used against ticks, bacteria, birds, fungi, weeds, snails, nematodes, fish and even animals. They are widely used, along with herbicides and other chemicals, in intensive farming. Pesticides often receive a bad press and many have been linked to cancers and other unwanted side effects in humans, as well as being accused of harming the ozone layer and other components of the environment. Another reason for their unpopularity is the drive in some sectors of the population for a return to organic methods of farming.

Over time various pesticides have been approved and later banned from use. Perhaps the most well-known pesticide to fall into this category is dichloro-diphenyl-trichloroethane, commonly known as DDT. DDT was originally prepared in 1873 but was not widely used until the Second World War when its effectiveness as an insecticide was recognised. Paul Muller of Geigy Pharmaceutical in Switzerland received the Nobel Prize in medicine and physiology in 1948 for this discovery. DDT's first use was by the US Army in Naples to stop a typhus epidemic. It was then realised that DDT killed mosquitoes that spread malaria, and DDT began to be used all over the world. It was, on the face of it, very successful. For example, in India deaths from malaria fell from 500,000 in 1960 to 1000 per year in the 1970s.

By the mid 1960s concerns were being expressed that DDT was perhaps not as safe as had been first thought. Research was showing that DDT concentrated itself through food chains so that animals higher up the food chain ended up with very high concentrations of DDT in their bodies. For example, DDT concentrations, measured in parts per million of body weight (ppm) may show the following ranges:

- lake water: 0.00002
- mud at the bottom of a lake: 0.014
- amphipods: 0.140
- trout (and other fish): 6
- herring gulls: 99
- peregrine falcons: 5000

At levels of below 6 ppm DDT is relatively harmless. However, by the time it reaches the concentrations found in birds it becomes toxic, leading to serious health problems or even death.

As a result of these problems of toxicity, and the fact that pests became increasingly resistant to DDT, the chemical was banned in the USA in 1973. It has since been banned in many other countries throughout the world. Other chemicals called organophosphates replaced it, but these were about three times more expensive than DDT – an obvious disadvantage, especially for countries in the developing world. Today these organophos-

How green does your garden grow?

As research develops into the effect of pesticides on the environment, various chemicals may be banned from use. In 2003 10 per cent of all garden pesticide products were removed from sale in the UK, bringing the UK in line with a EU-wide directive on pesticide regulations. However, the ban has created one huge headache of how to dispose of the unused chemicals – they cannot be tipped down drains or sinks, or put in dustbins to go to landfill sites. David Bowe, Labour Member of the European Parliament said, 'We must work to develop a chemicals policy that strikes the right balance between the enormous benefits that chemicals can bring to every aspect of our lives and the need to protect public health and the environment'.

In Canada, Quebec pledged to ban the use of most non-farm pesticides by 2005 in the entire province, following the ban of 30 noxious pesticides on public lands in 2002. This follows a 2001 Canadian Supreme Court decision to allow cities to ban pesticides in residential areas. Fines will range from $500 to $3000.

phates are also being banned because research shows they have unwelcome side effects for humans.

It is easy to be critical of chemicals such as DDT and their supporters and users. But it must be remembered that, in spite of the now-proven environmental problems, the chemical brought great benefits and saved millions of lives from malaria and typhus. It does however serve as a powerful reminder that any chemical being considered for use in the environment should be fully tested in order to establish whether any problems may result from its use.

Pesticide residues in foods are now being extensively researched. The most recent results of the Pesticide Residues Committee in the UK showed that 50 per cent of lettuce samples and 27 per cent of apples contained pesticide residues. Worryingly, some of the chemicals detected had in fact been banned.

WEBSITES

Report by the Pesticide Residues Committee in the UK:

www.pesticides.gov.uk/index-ns.htm

US Environmental Protection Agency:

www.epa.gov/pesticides

Beyond Pesticides (campaign group):

www.beyondpesticides.org/main.html

SEE ALSO

organic farming

PHOSPHORUS CYCLE

Phosphorus is only about one-tenth as abundant in living matter as nitrogen, but it is essential to the structure of three biologically universal molecules – DNA, RNA and ATP. DNA and RNA are nucleic acids, compounds responsible for information storage and information retrieval (such as for genes) in living systems. ATP (adenosine triphosphate) is the 'energy currency' of cells (see **PRODUCTION/RESPIRATION RATIO**). Phosphorus figures prominently in some other biological structures as well, including the mineral component of bone.

The biological cycle of phosphorus is rather simple, as shown in the figure below. The terrestrial (earth) cycle and the aquatic (water) cycle are linked via dissolved phosphate that flows downhill with water. Phosphate eventually ends up in storage in deep ocean sediments, and remains there until long-term physical restructuring of the globe raises the sediment to the surface as phosphate rock.

The total amount of phosphorus in the biosphere (organisms living and dead, plus soil phosphates, plus phosphates dissolved in water) is about 3.1×10^{11} metric tonnes. This is approximately 0.007 per cent of the amount sequestered in sediments. Uplifted sediments provide around 1.9×10^{7} metric tonnes of phosphorus annually to the biosphere, an amount roughly equalled by losses to the ocean bottom. The amount of phosphorus released by excretion, death and decay (2×10^{8} tonnes on land, 1.2×10^{9} tonnes in the sea) is more or less balanced by uptake.

There are six major elements in the **BIOSPHERE** – carbon, hydrogen, nitrogen, oxygen, phosphorus and silicon (see **BIOGEOCHEMICAL CYCLE**). Of these, the phosphorus cycle is peculiar. The cycles of water, carbon and nitrogen each involves some gaseous form: water vapour (and oxygen), carbon dioxide and nitrogen. The phosphorus cycle does not.

Phosphorus has an atomic weight of 31, much heavier than other common elements.

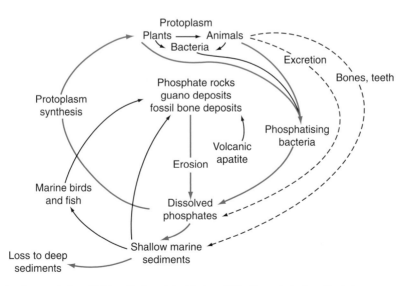

The phosphorus cycle (from EP Odum, *Fundamentals of Ecology*, 1953, with permission from Elsevier)

The dead zone

It sounds like the title of a horror movie, but the 'dead zone' refers to the biogeochemical cycles of nitrogen and phosphorus. The Mississippi River system in the USA drains one of the most productive agricultural regions on Earth. Each summer, an area of the Gulf of Mexico off the coast of Louisiana becomes a low oxygen (hypoxic) 'dead zone', encompassing some 36,000 km^2. This results from the interaction of several factors, some of them human-caused, or 'anthropogenic.'

The fresh waters of the Mississippi are less dense than the salt water of the Gulf of Mexico. In the calm summer, fresh and saltwater remain stratified. With stratification, there is little mixing of oxygen-rich surface waters with saltwater beneath.

The freshwater is laden with nutrients, especially nitrates and phosphates, runoff from agricultural fields, as well as municipal sewage and industrial wastes. It is estimated that the 'corn-belt' states of Iowa and Illinois (which comprise only 9 per cent of the watershed) contribute over a third of the total nitrogen load. This nutrient-rich water stimulates the growth of phytoplankton (algae). Aquatic consumers eat the algae and defecate in the water. This organic matter sinks through the water column and is decomposed. Decomposition (respiration) removes oxygen from the water, making it uninhabitable for aerobic life – hence the term 'dead zone'. The dead zone persists until the autumn 'hurricane season' returns to stir the surface waters and disrupt the stratification, mixing freshwater with saltwater and increasing oxygen levels.

The dead zone off the Mississippi delta is not alone. The United Nations Environmental Programme (UNEP) estimates that 150 dead zones exist worldwide, in such places as the Baltic Sea, the Kattegat, the Black Sea and the northern Adriatic Sea. These dead zones threaten already fragile fish stocks and the millions of people who depend upon fisheries for food and livelihoods. UNEP advises international action to reduce the amounts of fertiliser and sewage runoff by planting forests along rivers to 'soak up' excess nitrogen, improving 'precision agriculture' with GLOBAL POSITIONING SYSTEMS to waste less fertiliser, improving sewage treatment, more widespread use of technologies that remove nitrogen compounds from vehicle fumes, and using alternative energy sources that are not based on burning fossil fuels. These actions can and do work. An agreement between the European countries that share the Rhine River has reduced by half the levels of nitrogen being discharged and has cut the amount of nitrogen entering the North Sea by 37 per cent.

A metallic element, phosphorus reacts readily with oxygen to make a compound too heavy 'to get off the ground'. About the only way to get phosphorus 'airborne' is to get it up via winged organisms such as seagulls or dragonflies. These organisms snatch phosphorus from the water in the form of aquatic invertebrates or fish and deposit it on land in the form of droppings. In certain places, like the coast of Peru, this nutrient-rich accumulation of dung (termed *guano*) is mined for fertiliser about as fast as the birds can deposit it, something like 10^5 tonnes per year.

Humans are also a factor in the global phosphorus budget. Nearly 100 million tonnes of phosphate rock are mined annually worldwide, which represents about 20 million tonnes of phosphorus. This is equal to the amount of phosphorus made available to the biosphere by natural weathering of rocks. Thus, as in the carbon and nitrogen cycles, industrial humankind has assumed a role in the Earth's phosphorus economy of geophysical magnitude.

Phosphorus-rich run-off and organic waste in the water can produce a rapid spurt of growth of aquatic producers. These increase their demand on other resources, squeezing out other organisms. For example, an increase of phosphorus in a temperate-zone lake causes an increase in algal growth. Cyanobacteria respond more quickly to enrichment than do green algae. The water becomes cloudy, decreasing the availability of sunlight to producers. The blue-greens are toxic to some consumers and are poorly utilised. They die unused and float to the surface. The surface scum prevents oxygen exchange and aerobic life can no longer persist. The anaerobic microbes that come to dominate do not produce good, clean wastes such as water and carbon dioxide; they produce smelly by-products such as methane and alcohols. What was once a balanced and efficient ecosystem has been disturbed. This process of nutrient-glut is called **EUTROPHICATION**, literally, 'well fed'. Eutrophication is a natural process in the development of **LAKE** (lentic) ecosystems, but it is hastened by cultural processes.

WEBSITES

The US non-profit Environmental Literacy Council provides information on the phosphorus cycle with links to numerous other ecological processes and concepts: www.enviroliteracy.org/article.php/480.php Dead zone information: www.peopleandplanet.net/doc.php?id=2183 Phosphorus cycle from Virginia Tech University: soils1.cses.vt.edu/ch/biol_4684/Cycles/Pcycle.html LennTech's Phosphorus cycle information: www.lenntech.com/phosphorus-cycle.htm

PHOTOSYNTHESIS

Starflower
Source: Joseph Kerski

Photosynthesis is the process whereby green plants, algae and some bacteria harness the energy of the Sun, using carbon dioxide and hydrogen, and fix it in the chemical bonds of carbohydrate. Oxygen is released in the process of photosynthesis, making all life possible.

Photosynthesis and respiration are complementary processes that run most of the **BIOSPHERE**. Processes occurring in cells and subcellular organelles run biogeochemical cycles on a global scale.

Photosynthesis involves combining carbon dioxide and the hydrogen from water to make a molecule of carbohydrate, specifically the carbohydrate *glucose*, a simple six-carbon sugar ($C_6H_{12}O_6$). The hydrogen from water is needed to 'fix' or establish carbon dioxide as carbohydrate. Breaking water apart takes energy and so does building new bonds between carbon dioxide and hydrogen. This energy comes from the Sun.

None of this happens spontaneously, of course. We are surrounded by an atmosphere containing carbon dioxide and water vapour, shot through all day with the energy of sunlight, yet it never rains glucose. Something more is needed – the DNA-coded structure of the chlorophyll molecule.

The production of carbohydrate occurs in cellular organelles called chloroplasts, and involves two distinct but linked processes. The light reactions take place within chloroplasts in stacks of membrane-bound structures called grana; the dark reactions (so-called because they can take place in the absence of light) are localised in the intervening stroma of the chloroplasts.

The light-dependent reactions of photosynthesis involve pigment systems. A pigment is a molecule that absorbs light. The common green pigments in higher plants are chlorophyll *a* and chlorophyll *b*. Pigments are complicated molecules; for example, the empirical formula for chlorophyll *a* is $C_{33}H_{71}O_5N_4Mg$.

Pigment molecules are 'excited' by light. In human skin, pigment that becomes too 'excited' by the sun results in a sunburn. The energy absorbed pushes an electron in the pigment to a higher energy level. An analogy clarifies what we mean by energy level. Imagine a ball, a set of stairs and a well-trained footballer. The ball starts at the bottom of the stairs, a position of no potential energy, its 'ground state'. The footballer invests some energy (an appropriate amount, neither too much nor too little) to lift the ball to the top of the stairs. The potential energy of the ball is increased by the expenditure of biological (chemical bond) energy by the footballer. A ball on the top step (like a rock perched on a ledge) is not very stable, so it rolls off. As the ball bounces down the stairs, it does some work, dissipating its energy. We might install a lever on each step, for example, to turn a ratchet to turn a generator. Then, the energy of the falling ball could be harnessed to provide a little electricity. An electron in chlorophyll is 'kicked upstairs' by solar energy. As it returns to its ground state – bouncing down the steps – it can do work.

Some plant species are much more efficient than others in converting carbon dioxide and water into oxygen and organic matter. They may have become efficient due to the low atmospheric concentration of carbon dioxide for the past several hundred million years. These more efficient plants, which include corn, sorghum and a few weed species, have a unique photosynthetic system that permits carbon dioxide to be converted into organic matter at a faster rate than with the more common photosynthetic system of other plant species. The presence of this efficient system and a long period of seasonal growth are important reasons why corn, for example, produces higher yields of organic matter (biomass) per hectare of land area per year, relative to most other plant species.

White visible light is comprised of different wavelengths that correspond to Red, Orange, Yellow, Green, Blue, Indigo and Violet. The light that excites chlorophyll is in the blue and red range of the spectrum of visible light. The green light between is reflected unused. We appreciate this in the green, verdant beauty of the English countryside and other green landscapes around the world.

As winter approaches, some plants and trees begin to shut down their photosynthesis. The green chlorophyll disappears from the leaves. As the bright green fades away, we begin to see yellow, orange, red and other colours. Small amounts of these colours have been in the leaves all along, but we do not see them in the summer because they are obscured by the green chlorophyll. In some trees, such as maples, glucose is trapped in the leaves after photosynthesis stops. Sunlight and the cool nights of autumn turn this glucose into a red colour. The brown colour of trees such as oaks is made from wastes left in the leaves. It is the combination of all these things that make the beautiful colours of autumn leaves.

Human activities can influence photosynthesis and therefore have an adverse effect on the environment. Since paper mills opened in 1973 on the world's deepest freshwater lake, Lake Baikal, in Russia, air pollution has dried out the trees in the region's taiga over an area of about 5,000 square kilometres. Some 61 per cent of the taiga around the southern basin has been weakened by the impact of air pollution, with an associated sharp decline in photosynthesis. In any land that is overgrazed and deforested, photosynthesis is by definition reduced. Along coral reefs, water pollution reduces the sunlight, and therefore the amount of photosynthesis, that can occur in such reefs.

WEBSITES

Illustrated tutorial on photosynthesis from the US Public Broadcasting System:
www.pbs.org/wgbh/nova/methuselah/photosynthesis.html
Comprehensive photosynthesis and related processes information:
www.emc.maricopa.edu/faculty/farabee/BIOBK/BioBookPS.html
Photosynthesis Center at Arizona State University:
http://photoscience.la.asu.edu/photosyn/

PLATE TECTONICS

Look at a world map and you will see that South America and Africa might 'fit together' if the Atlantic Ocean was not between them. In the early twentieth century Alfred Wegener, a German geologist, noticed the same thing – that the continents look much like jigsaw puzzle pieces that could fit together. He proposed that the continents were once part of a single continent he named Pangaea – Greek for 'all lands.' Wegener's theory started a geologic controversy that lasted for half a century and eventually developed into the theory of plate tectonics, which is accepted today.

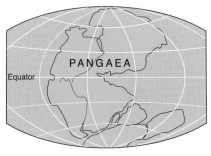

Permian
225 million years ago

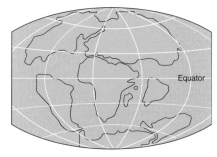

Cretaceous
65 million years ago

Triassic
200 million years ago

Present day

According to Wegener's continental drift theory, the supercontinent Pangaea began to break up about 225–200 million years ago, eventually fragmenting into the continents as we know them today. Image from USGS, not copyrighted

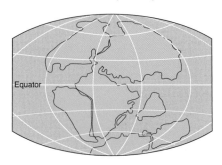

Jurassic
135 million years ago

The problem with Wegener's theory was that there was no acceptable mechanism that would explain the motion of the continents – known as continental drift. It was not until people were able to explore the ocean floor that Wegener, who died in 1930, was vindicated.

The concept of thermal convection is the key to understanding the theory of plate tectonics and supplies the mechanism that was missing from the theory of continental drift. Until thermal convection, it was unknown how

the giant plates could move around on the mantle of the Earth. Convection provides the engine that moves the plates. Thermal convection occurs when a substance is heated and its density decreases, causing it to rise to the surface; then, as it cools, it sinks again. This theory was championed by Arthur Holmes during the 1930s. Holmes's idea did not gain many supporters until the 1960s when major ocean floor features, such as mid-oceanic ridges, sea floor spreading, island arcs and oceanic trenches near continental margins, were discovered.

The continents are part of huge crustal plates at whose boundaries thermal convection acts like a conveyer belt, recycling the Earth's topmost layer. Convection currents beneath the plates move the plates in different directions. The source of the heat driving the convection currents is radioactive decay deep within the Earth's mantle.

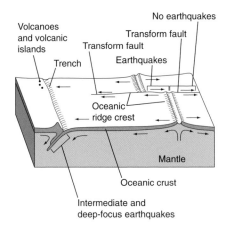

Sea-floor spreading. (After G. Gross, *Oceanography*, 3rd ed., Charles E. Merrill, 1976)

Four types of plate boundaries exist and each behaves differently:

■ Divergent boundaries: new crust is generated as the plates pull away from each other. The Mid-Atlantic Ridge is an example.
■ Convergent boundaries: crust is destroyed as one plate dives under another, or subducts. Convergent boundaries are responsible for deep ocean trenches such as the Peru-Chile Trench, and tall mountains such as the Andes. At some convergent boundaries, continental crustal plates collide and crumple to form massive mountain ranges. The Himalayan range, which is rising at a rate of 5 centimetres a year is a good example. Here there is no subduction of crust – no magma is created and no volcanoes are formed.
■ Transform boundary: crust is neither produced nor destroyed because the plates slide horizontally past each other. The San Andreas Fault Zone in California is probably the best known example of this type of boundary.
■ Broad tectonically active belts: boundaries are not well defined and the effects of plate interaction are unclear. One of these zones marks the Mediterranean–Alpine region between the Eurasian and African Plates, within which several smaller fragments of plates (microplates) have been recognised.

Because plate-boundary zones involve at least two large plates and one or more microplates caught up between them, they tend to have complicated geological structures and earthquake patterns.

Seismic activity at plate boundaries increases the frequency of earthquakes and volcanic activity. Earthquakes disrupt the local environment where they occur, while volcanoes can affect the environment on a global scale through airborne ash and subsequent effects on weather and climate.

As continents have moved on the plates that carry them, the direction of the ocean currents have changed, as well as the air masses that affect climate the world over. In addition, as the continents have joined together and separated over time with oceans in between, plant and animal migrations have been opened up and then closed. This has had a profound influence on today's environments and **biomes**. Why are certain animals, such as elephants, found only in Africa and Asia? The answer lies partly in plate tectonics. As India broke away from Africa 20 million years ago, it very likely ferried some unsuspecting elephants (along with many other organisms) northwards to Asia. The Asian and African elephants have slight physical variations, but they are clearly cut from the same genetic mould.

Plate tectonics have affected people and environments in other ways. The physical breakdown and chemical weathering of volcanic rocks have formed some of the most fertile soils on Earth. Most of the metallic minerals mined in the world, such as copper, gold, silver, lead and zinc, are associated with magmas found deep within the roots of extinct volcanoes located above subduction zones. Oil and natural gas are the products of the deep burial and decomposition of accumulated organic material in geologic basins that surround mountain ranges formed by plate-tectonic processes. Geothermal energy can be harnessed from the Earth's natural heat – from active volcanoes or geologically young inactive volcanoes that are still giving off heat deep underneath the surface.

Another interesting theory to emerge recently concerns one of the greatest of all mysteries – the origins of life on Earth. Earlier predominant theory held that life had its origins in warm ponds or similar small bodies of water protected from the harsh environment of the early Earth and far away from the escaping heat of the deep sea floors. Recently, scientists have discovered organisms that thrive in these hellish conditions, and appear to have been around long before the earliest known organisms previously known. Could the hot vents at mid-ocean ridges have been the incubators of life on this planet?

Further reading

Hallam, A., 1989, *Great Geological Controversies*, 2nd edn, Oxford University Press.

WEBSITES

University of California, Berkeley, with animations of plate movement:
www.ucmp.berkeley.edu/geology/
tectonics.html
USGS publication 'This Dynamic Earth', detailing plate tectonics:
http://pubs.usgs.gov/publications/text/
dynamic.html
Information on Alfred Wegener and his theories:
www.ucmp.berkeley.edu/history/wegener.html

SEE ALSO

earthquakes

POLAR ENVIRONMENTS

The global polar environments comprise the regions north of the Arctic Circle (66° 33′ north), which includes the Arctic Ocean, Greenland and the extreme northern parts of Alaska, Canada and the Russian Federation, together with the continent of Antarctica, almost all of which lies within the Antarctic Circle (66° 33′ south). With permanent darkness in the winter, temperatures plummet to −50°C or less. In the summer, despite the 24-hour sunlight, the sun is at a very low angle in the sky and temperatures struggle to reach above freezing point. Precipitation, which mostly falls as snow, is light enough to enable some areas to be officially classed as 'desert'.

Much of the area is permanently covered by snow and ice. The Antarctic ice sheet, several kilometres thick in places, represents more than 90 per cent of the world's fresh water. Conditions are so intensely cold that, apart from a few research stations, the continent is unfit for human habitation. Almost all of Greenland is covered by an ice sheet too, with human settlement only possible along the coast where proximity to the sea moderates the otherwise extremely low temperatures. The Arctic Ocean is almost permanently covered by pack-ice, which is at its greatest extent during the winter and then breaks up and contracts in its extent during the summer months. Icebergs are common during the summer.

Those areas not permanently covered by ice and snow include great swathes of land in North America and the Russian Federation. These are areas of permafrost – permanently frozen soil – where winters are harsh and summers short. The vegetation is tundra – extraordinarily delicate yet well adapted to the cold dry winters and the short summer-growing season. Tundra supports a wide range of animals and the food web is rich and complex. Perhaps the best-known animals are the polar

Case study

Oil and the polar environment: Alaska, USA

The USA consumes some 25 per cent of the world's oil and when huge reserves were discovered at Prudhoe Bay, northern Alaska in 1968, the Government was keen to exploit them. Pack-ice meant that oil could not be transported by tanker from Prudhoe Bay, so an ambitious plan was formulated to build a 1200 kilometre pipeline to the southern port of Valdez. At the time, it was the most expensive privately-funded construction project in American history. It involved crossing almost 1,000 km of frozen ground, three mountain ranges, 800 rivers and streams and several areas where seismic activity often occurred.

continued over

Environmentalists expressed considerable concern. The tundra vegetation is very fragile and might take decades to recover from any damage by construction vehicles. The pipeline was due to cut across important migration routes, for example those of caribou. Since it would be carrying warm oil, it was likely to melt the uppermost layer of permafrost, rendering the pipeline unstable. The region is frequently hit by earthquakes and people were concerned about the possibility of pipe fractures and oil spills.

While some environmental damage was inevitably caused by the large-scale developments, the pipeline itself was constructed with great care, taking account of people's concerns. In places it was raised 6 metres above the ground to allow animals to migrate underneath. The above-ground portion is fixed in place by 78,000 vertical supports driven deep into the permafrost and, since the pipeline is insulated, it does not cause surface thawing. The supports on which it is held have been designed to withstand earthquakes, and the regular pumping stations along its route can control oil flow, should a leak occur. For the remaining 625 kilometres the pipeline was placed underground, but once again care was taken to minimise the environmental impact. Small pipes were embedded in the ground close to the pipeline, each filled with liquid brine, to keep the permafrost solid and stable. Concrete jackets encase the pipe as it crosses rivers and, where it spans areas prone to earthquakes, flexible pipes are used, with regularly spaced valves designed to close automatically in the unlikely event of a tremor or quake that may cause the pipe to crack.

In 1980 the north of Alaska received a special status on account of its ecological importance – it is now known as the Arctic National Wildlife Refuge (ANWR). This designation should in theory enable developments, particularly oil, to be strictly controlled. However, in 2002 President Bush let it be known that he favoured further development in the Prudhoe Bay area for oil, to reduce dependency on foreign sources of oil. Environmentalists are understandably very concerned that any further developments could have far-reaching effects on Alaska's delicate ecosystems.

After extensive surveys, including the tracking of caribou migration routes, work began on the construction of the pipeline in the winter of 1974–5. It was completed within two years at a cost of $8 billion. Around 214,000 litres per month passes at speeds of around 10 km/h through the 120-centimetre diameter pipeline on its six-day journey. Since 1977, over 14 billion barrels of oil have been transported to the terminals at Valdez, and the Prudhoe Bay Field continues to be one of the most productive as the USA tries to meet its rising demand for energy.

bears, which roam for great distances over the pack-ice, hunting primarily for seals.

For hundreds of years indigenous peoples such as the Sami (Lapps) in northern Europe and the Inuits in North America have survived by hunting and gathering, fishing and herding reindeer. They have survived for so long in such a fragile environment solely because they have exploited the resources in a sustainable manner; it has been in their own interests to conserve and protect their environment.

Recently, however, polar lands have been under considerable pressure from governments keen to exploit the regions' rich energy resources, in particular, oil. In Siberia, poor techniques of oil recovery and transportation have resulted in extensive leaks, which have caused massive environmental damage.

WEBSITES

Arctic National Wildlife Refuge (ANWR): www.anwr.org

Canada's polar environment: www.arctic.uoguelph.ca/cpe

Roald Amundsun Center for Arctic Research: www.arctic.uoguelph.ca/cpe/

Trans-Alaskan pipeline: www.alyeska-pipe.com

SEE ALSO

Antarctica

POLLUTION

Pollution occurs when the quality of the environment is disturbed, impaired or contaminated in some way.

Case study

The urban pollution model: Tokyo

Environmentalists have shown that many urban areas have gone through a clear series of pollution stages. Different types of pollution, and different substances, were prevalent at different stages in the economic growth of industrial settlements.

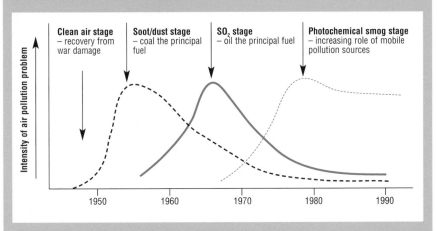

For example, in the aftermath of the Second World War much of the Japanese city of Tokyo had to be rebuilt. There were few factories and little industrial activity – levels of pollution were low. However, as factories were rebuilt and energy was needed for homes and industry, so the levels of pollution rose, principally from the burning of coal. During the 1960s soot and dust pollution decreased, being replaced by oil-related pollutants. Thus the soot/dust stage became superseded by the sulphur dioxide stage. This, too, declined during the 1970s, to be replaced by car-exhaust-related pollution, principally photochemical smog.

Over time overall pollution levels remain fairly consistent although the contributory pollutants themselves change, reflecting new sources of energy and changes in peoples' lifestyles.

Visual pollution includes intrusive buildings, or structures such as mobile telephone masts, which spoil peoples' views. On a larger scale, wind farms are considered by many to be ugly and unnatural structures that impair the aesthetic appeal of the landscape.

NOISE POLLUTION is excessive or unwanted sound and vibration, such as that experienced by people living close to main roads or under aeroplane flight paths. For example, in the UK people are objecting strongly to plans to expand the airports around London, such as at Heathrow and Stansted.

People living close to LANDFILL sites often complain of bad smells – this is olfactory pollution. Other examples of olfactory pollution include main roads and sewage works.

Air pollution is when gases and other, possibly harmful, substances are released into the lower atmosphere, or troposphere. There are many sources of air pollution, including car exhausts (nitrous oxides), landfill sites (methane) and power stations (SULPHUR DIOXIDE). Air pollution can affect peoples' health directly, as was the case during the 1952 London smogs when over 4000 people died. It can also affect people and environments indirectly, through ACID RAIN and global warming.

Water pollution occurs when liquids or solids are deposited in rivers or lakes. This can take the form of solid waste dumped in rivers, such as household items, paper and supermarket trolleys, or more sinister and dangerous pollution involving toxic chemicals from mine waste or leachates from landfill sites. Either way, aquatic environments become damaged as drinking water becomes unsafe and biological habitats are impacted.

Pollution has a greater impact on some natural systems and environments than others. Once released into the environment, chemical pollutants follow particular routes (pollution pathways) and become concentrated in specific places or parts of a system. These places are called pollution sinks and it is here that the impact of pollutants is most apparent. For example, artificial fertilisers consisting of soluble nitrates and phosphates are applied to farmland. These are then leached from the soil or washed off by run-off and eventually enter water courses such as ditches and streams. They then become concentrated in these places by evaporation, especially during hot weather and dry spells. As water percolates down through the soil and bedrock, the pollutants will also enter groundwater stores. In this way, water courses and groundwater stores become pollution sinks. The high concentrations in these sinks can cause further environmental problems, such as encouraging excessive growth of aquatic plants and algae. Eventually these organisms die and huge amounts of oxygen are required from the water as they decay. Gradually, the demand for oxygen will exceed the amount of dissolved oxygen in the water. When this happens, fish and other aquatic organisms will die. The whole downward process is known as EUTROPHICATION.

WEBSITES

World News Network on pollution:
www.pollution.com
US Environmental Protection Agency's
Pollution Prevention Program:
www.epa.gov/p2
UK Royal Commission on Environmental
Pollution:
www.recep.org.uk

SEE ALSO

clean air acts, noise pollution, vehicle exhausts, visual pollution

POPULATION

Daddylonglegs
Photograph by Joseph Kerski

A population is all of the individuals of the same species at a particular time and place. Biological systems in the environment are organised in a hierarchy of levels of integration, from the microscopic and even submicroscopic levels of biomolecules and cellular organelles and cells to the macroscopic level of the biosphere as a whole.

In the environment, every level in the hierarchy is essential to understanding every other level. And yet, of the levels of biological integration, the population stands out as 'first among equals'.

Species are defined by their actual or potential reproductive continuity and by reproductive isolation from other such genetic units. Hence, populations are reproductive units. It is this reproductive function that makes populations so central to the biological view of the world. Populations are of transcendent importance because they are the units of biological evolution. Changes in the genetic composition of populations over time are the basis for biodiversity at all other levels of integration. The genetic continuity of populations across time is the basis for the unity of life, from microbes to humankind. In order to understand the environment, we must understand populations.

Populations have several characteristics that are peculiar to the population as a level of biological integration. These attributes arise from the interaction of individuals and cover the following areas: distribution, DISPERSION, SEX RATIO, SURVIVORSHIP, REPRODUCTIVE STRATEGY, POP-

ULATION DENSITY AND CARRYING CAPACITY, POPULATION REGULATION AND CONTROL, FERTILITY, MORTALITY, MIGRATION, POPULATION INCREASE and POPULATION STRUCTURE.

Human population ecology (demography) is a special case of population ecology because the human population is the only biological population that is (potentially or actually) under the control of cultural norms and behaviours.

Population growth

Biological populations have a tendency to grow. Human populations are a special case of this general phenomenon. Growth in human populations is the cause of a substantial part of human suffering and human population growth has tremendous implications for politics, economics and the environment.

In simplest terms, population growth results from an imbalance between recruitment and loss. An individual is born (or hatches, or is otherwise recruited to the population) at some particular time. A population is made up of individuals and so it is characterised by a birth rate, the numbers of births per unit time. A rate, by definition, is change in a quantity per time. Because populations differ greatly in size, for comparative purposes, birth rates usually are expressed as a ratio, such as births per hundred (%) or births per thousand (symbolized as ‰). The ratio may be an overall ratio of births per thousand in the total population, or it may be a ratio of births per thousand females of reproductive age (the so-called fertility rate). Because birth rates are expressed in several different ways, great care must be taken to avoid erroneous comparisons.

In some cases, a birth rate may be a misleadingly simple statistic. If we want to know how a population will grow, we need to consider all additions to the population, not just births. By definition, births occur at age zero, while immigration occurs at some greater age. Births plus immigration are defined as recruitment, which is the more general term for additions to a population, but there is an important difference between births and immigration. Immigrants are usually closer to reproductive age than newborns; hence,

immigrants have the potential to contribute sooner to population growth.

A population loses members through death and emigration. The sum of deaths and emigrations is termed simply 'loss'. Knowing the age at which an individual is lost to the population is important to knowing how the population will respond. If an individual is lost at reproductive age, its loss will depress population growth. If an individual is lost at pre-reproductive age, population growth in the future may be depressed. If an individual is lost at post-reproductive age, the effect on population growth may be nil.

If the rate of births (or recruitment) and the rate of death (or loss) in a population are known, then a rate of growth can be established. This can be expressed crudely as a formula:

birth rate $-$ death rate = growth rate, or
$b - d = r$

The quantity r is called the biotic potential of a population. The r value may be thought of as a characteristic of the species population. The upper limit of r is the difference between maximal birth rate and minimal death rate. The actual rate of increase is probably always less than the theoretical rate.

The balance between birth rates and death rates determines the rate of population growth. If $b - d = 0$, then we will see a stable population, with gain or recruitment equal to loss. If $b - d < 0$ (that is, any negative number), then the population is in decline. If $b - d > 0$, a positive number (however small), we will observe population growth. In many populations, population growth is exponential, like compound interest.

Factors that limit population growth include food and water availability and purity, light, space, predators, diseases, parasitism and natural hazards. Many of these factors depend on population number and density. For example, for a given population, food and water may be abundant, but if the density or number increases, food and water may no longer be plentiful enough to sustain the population. In another example, in terms of predators, higher densities of a prey population attract more predators, and as the number of prey increases, so does the number of predators. On the other hand, if the number of prey decreases, so does the number of predators.

SEE ALSO

Population increase

POPULATION DENSITY AND CARRYING CAPACITY

Density of individuals in a population is simply the number of individuals per area (or volume of three-dimensional habitat). Density varies within a species from place to place. For example, in interior and southern Alaska, the density of wolves is about one per 65 to 194 km^2; on the west and north coast, it is one per 388 km^2. The density of wolves in Minnesota is one for 1,153 km^2. In short, some parts of a species' range may be occupied more densely than other parts.

Ecologists sometimes distinguish *ecological density* from *raw density*. This distinction is important. Consider the figure, which indicates a study area of 4 km^2, with eight pheasant nests. That yields a *raw density* of two pheasant nests per square kilometre. However, note that there is a 2 km^2 pond in the study area. So the *ecological density* – numbers per area of suitable habitat – is higher: four nests per square kilometre.

For human beings, population density is the number of persons per unit of area. When density figures exclude such areas as inland water, uninhabitable land, or non-arable land, the density is really an *ecological density* for humans.

World overall human population density presently averages 42 persons per km^2, but an enormous range exists. The ten cities with a

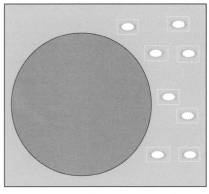

The raw density of pheasant nests on this 4 km^2 study area is 2 per km^2; however, the pond (shaded) occupies 2 km^2, so the ecological density is 4 per km^2

population above two million with the highest population density are Hong Kong (China), Lagos (Nigeria), Dhaka (Bangladesh), Jakarta (Indonesia), Mumbai (India), Ahmadabad (India), Ho Chi Minh City (Vietnam), Shenyang (China), Bangalore (India) and Cairo (Egypt).

Population densities are not just high in cities, but in rural areas as well. Population densities among countries range from over 16,000 per square kilometre in Monaco to 1.7 per square kilometre in Mongolia. The most densely populated large country is Bangladesh, where 134 million people live in a highly agricultural area around the lower Ganges River, with a national population density in excess of 900 people per km^2. High densities make it difficult for the land to support such populations, because of the water, energy and other natural resources required. In addition, any species that is densely packed in a small area is more susceptible to natural hazards, such as floods, hurricanes and earthquakes. In some cases, the environment has reached or exceeded its carrying capacity.

Carrying capacity is the 'full mark' of the environment, the point beyond which there is 'no vacancy'. Carrying capacity is usually defined as the maximum population of a given species that can be supported indefinitely in a defined habitat without permanently impairing the productivity of that habitat. Carrying capacity (conventionally abbreviated as K) is a property of a species' habitat that constrains a **POPULATION**.

All habitats are finite. This is because habitats are subdivisions of a finite resource system, the ecosphere. Carrying capacity is difficult to measure as environments are not neatly calibrated. And, of course, carrying capacity for one species varies with changes in populations of other species.

Stock-growers (those who manage grazing animals) have understood carrying capacity for millennia and they may speak of carrying capacity as 'stocking rate'. They know that if the stocking rate is exceeded, the range will be degraded. They also know that carrying capacity varies from pasture to pasture, season to season and year to year. They know, too, that stocking rates are somewhat higher if competitors – prairie dogs or jackrabbits or grasshoppers, for example – are absent.

As humans, we may feel that we have advanced beyond the need to worry about a carrying capacity for our own species. True, we seem to have had the ability over the past few hundred years to increase our own carrying capacity by eliminating competing species, by importing locally scarce resources, and through technology. However, many caution that shrinking carrying capacity may soon become the single most important issue confronting humanity. These same individuals define carrying capacity not as a maximum population but rather as the maximum 'load' that can safely be imposed on the environment by people. Human load is a function not only of population but also of per capita consumption. Per capita consumption is increasing more rapidly than the former, due, ironically, to expanding trade and technology. Take a look at the size of a typical house built during the 1960s versus a home built since 2000. The same could be said for the size of cars or even the size of soft drinks. In other words, as people consume more, we stretch the carrying capacity as much as the fact that there are more people on the planet. For example, in 1790, the estimated average daily energy consumption by Americans was 11,000 kcal. By 1980, this had increased almost twenty-fold to 210,000 kcal/day.

Furthermore, the number of households in the UK, USA and many other countries has increased faster than the population increase due to demographic changes such as more single people, fewer children and more divorces. More households place pressure on the environment because of increased demand for water, energy, land and waste management. As a result, load pressure relative to carrying capacity is rising much faster than is implied by mere population increases.

WEBSITES

The Carrying Capacity Network is a clearinghouse for information on sustainability:

www.carryingcapacity.org

One means of understanding carrying capacity from an individual and national standpoint is to calculate an 'ecological footprint', as illustrated on:

http://dieoff.org/page110.htm

Human population density information from Wikipedia:

http://en.wikipedia.org/wiki/Population_density

POPULATION INCREASE

Every second, five people are born and two people die, for a net gain of three people each second. That means that 12 people were added to the world's population in the time it took you to read the previous sentence. The world is adding about 78 million more people every year: the population of France, Greece and Sweden combined, or a city the size of San Francisco every three days.

How did the population reach its current level? For most of human history, population growth was very slow. Although there were high birth rates, death rates were also high, resulting in little annual growth. It was not until some time in the seventeenth century that growth began to accelerate. The population of the world reached one billion people around 1820. Though it took all of human history to reach the first billion in 1820, it only took 110 years to double and add another billion. It only took another 45 years for the population to double again to 4 billion in 1975. By 1987 the world's population had added another billion and since that time until the present we have added more than a billion more. The slow growth that characterised much of human history gave way, first gradually, then much more rapidly, to increased rates of population growth.

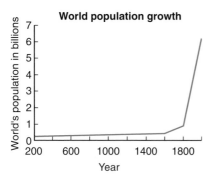

World population growth

World population has grown rapidly only in the last 200 years

Today's population situation is different to any other time in history because the base population is the highest the world has ever seen, and the rate of increase is still near the highest in human history. Advances in medicine, a better standard of living, and safer food and water have led to better health and a lower death rate, so enabling the world to support a larger population.

Most of the environmental concerns facing us today – water quality, resources, landfills, air pollution, urban growth and others – are magnified because of the constant and increasing pressure of population on natural resources. For example, the population of the USA tripled during the twentieth century, but the US consumption of raw materials increased 17-fold. This type of consumption is not sustainable for the long term. Another concern is the equity of human access to basic human services. The richest 20 per cent of the world's people consumes 86 per cent of all goods and services, while the poorest fifth consumes just 1.3 per cent.

What does the future hold in store for the world's population? Most scientists agree that the world's population will not continue to grow at its present rate. Societal changes such as urbanisation, reduced desire for large families and even AIDS are bringing down the birth rate. Another question is whether the Earth would be able to support such a large population in the future. A continual increase in the world's population will put more pressure on natural resources and the carrying capacity of the Earth – the capacity of the

Thomas Robert Malthus

Pessimists about population growth are usually referred to as Malthusians or neo-Malthusians. Thomas Robert Malthus (1766–1834) was a population theorist who was both a clergyman and an economist. He has had a profound impact upon people's thinking about the nature of population growth. In 1798 he wrote his famous 'Essay on the Principle of Population'. According to Malthus, population would always increase at a geometric rate (1, 2, 4, 8, 16, 32 and so on), but food production could only be increased at an arithmetic rate (1, 2, 3, 4, 5, 6 and so on). Therefore it would not take long for the number of people to be greater than the food supply, leading to increased starvation and a rise in the death rate. Humans would always press up against the limits of natural resources, primarily food. Malthus believed that this 'Principle of Population' was inflexible, inexorable and inescapable. His dismal view of the future of humankind was instrumental in helping to saddle economists forever with the description of their discipline as the 'dismal science'.

Today, shortcomings can be seen in Malthus's theory. History has shown that the food supply could be increased as fast or, in some cases, faster than population growth. Because of his religious orthodoxy, Malthus also failed so see how people could limit their family size through contraceptive practices.

The Industrial Revolution and modernisation in the nineteenth and early twentieth centuries reduced some pressure on population, namely, through a decrease in the birth rate. However, the latter part of the twentieth century witnessed a rapid population rise, coupled with considerable malnutrition in many areas, so that many neo-Malthusians believe that population growth will still outrun food supply. They maintain that the world will not be able to continue to support a growing population.

Earth to sustain a certain population at a given level of technology. Every 20 minutes the human population grows by about 3000. At the same time another plant or animal becomes extinct (27,000 each year). According to the United Nations, if fertility were to stay constant at 1995–2000 levels, the world population would soar to 244 billion by 2150 and 134 trillion by 2300. It is inconceivable that the world could support too many more billions than the 6 billion living today.

The only way the Earth's population growth will be brought under control is either through a decline in the birth rate, a rise in the death rate, or some combination of the two. Most would agree that a rise in the death rate is the least desirable way to control growth. Changes in FERTILITY rates make a huge difference in the future world population. A fertility rate of 1.85 would produce a population of 2.3 billion in 2300 – a 4-billion decline from the current population. A scenario of two children per woman would result in a population of 9 billion people in 2300.

WEBSITES

Six billion and counting:

www.pbs.org/sixbillion

Population Connection:

www.populationconnection.org

Population Reference Bureau:

www.prb.org

Overpopulation.org:

www.overpopulation.org/faq.html

United Nations Population Information Network:

www.un.org/popin

Malthus's 'Essay on the Principle of Population':

www.ac.wwu.edu/~stephan/malthus/malthus.0.html

SEE ALSO

fertility, mortality

POPULATION REGULATION AND CONTROL

Any population has the capacity to grow. If there is a positive difference between births minus deaths, the population will grow. Indeed, the population will grow at 'compound interest rates'; that is, it will grow exponentially.

Coupled with this is the fact that every habitat, every resource-base for living organisms, is finite. Some habitats are huge, but they are still finite resources.

All viable populations have the raw reproductive capacity to overwhelm the Earth in short order. Darwin calculated that a single pair of elephants could produce 19 million descendants in a mere 750 years. Because resources are finite, populations must be controlled; and all populations *are* controlled, one way or another.

Density-independent regulation vs. density-dependent control

Some populations seem to be controlled mostly by physical factors such as extremes of weather. Imagine a colony of ants living in a desert wash. A flash flood submerges the colony, drowning all of the ants, young and old, larvae and eggs. The flood could drown ten ants or ten million. The effect of the flood is irrespective of density, hence 'density-independent'. This sort of situation is shown in Graph A in the figure, with density-independent control events indicated at the arrows.

Graph B in the figure illustrates the simplest form of population growth under density-dependent regulation. The population grows to K (or overshoots to some extent) and then is regulated at or near that point. Density-dependent regulation can occur by a number of different mechanisms. For example, a variety of social systems have the effect of regulating populations within the limits of resources. They act in a density-dependent fashion. Territorial behaviour, for example, tends to space organisms over available habitat. Those individuals that do not find and defend a suitable territory will produce poor quality young, fewer young, or none at all. The population is thus regulated.

In a certain sense, there is no such thing as population control that is completely density-independent. This is because the density of the population that remains after the control event will determine how that population will recover. A flash flood on a mountain stream can drown 10 beavers or 1000. That is density-independent control. However, the recovery

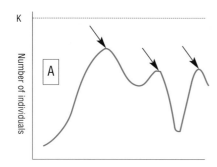

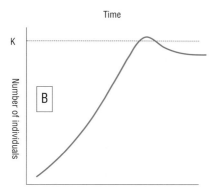

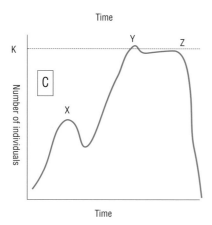

Hypothetical curves of population numbers over time.
A: Density-independent control; B: density-dependent
regulation; C: Combination of density-independent
(points X, Z) and density-dependent control

of the population from the flood will obviously depend on the density of beavers that escaped drowning. Hence, the long-term prospects of the population are in a sense density-dependent.

A given population is likely to be influenced by both density-dependent and density-independent systems. In Graph C in the figure, for example, a hypothetical population is regulated by flooding in spring (Point X) and by a killing frost in autumn (Point Z). In summer (Point Y), food becomes limited and starvation slows population growth. Events X and Z are density-independent and event Y is density-dependent.

Density-independent regulation can result in wide fluctuations in numbers, whereas density-dependent regulation can involve maintenance of populations near equilibrium numbers, with only slight fluctuations.

Intrinsic vs. extrinisic mechanisms

There is another way to classify mechanisms of population control. They may be *intrinsic* or *extrinsic*. As the words imply, intrinsic factors regulate the population from within. Extrinsic factors control the population from outside.

Territorial behaviour is intrinsic to the population and it may regulate population size. Predation and fire are extrinsic to populations and they may exert control. However, sometimes the distinction between intrinsic and extrinsic mechanisms is hard to maintain. Consider lemmings, which tend naturally to avoid one another. When they do encounter each other, they turn around and go in the other direction. The more dense the population, the more often they run into other lemmings, the more they have to work to avoid each other. As they scurry around avoiding each other, they can fall into lakes and wells, march into dense forests (which under better conditions they avoid as unsuitable habitat) and into fields and onto roads. Occasionally lemmings probably even fall off cliffs into the sea. They do all this not as some well-ordered 'suicide march', but as a mostly frantic attempt to avoid each other, to maintain spacing. The end result is that lemmings drown, are eaten by owls, or are run over by lorries, and their numbers are regulated within the carrying capacity of their environment. The movement of crowded lemmings is intrinsic to the population. That is the way lemming genes make lemmings operate. The actual death of an individual lemming, however, is due to some extrinsic cause, biotic (eaten by an owl) or abiotic (drowned in the sea).

Birth rate vs. death rate solutions

There is a final, simple way to classify mechanisms of population control. Have another look at the simple equation $b - d = r$ (birth rate minus death rate equals rate of increase). Mechanisms of population regulation change r. Obviously, r can be changed by working on either b or d, or both. We may think generally of 'birth rate solutions' and 'death rate solutions' as the 'options' that the population has to regulate its numbers. Territorial behaviour often works as a birth rate solution; prospective parents without a territory do not breed successfully. Predation acts as a death rate solution. In most populations most of the time, both kinds of solutions are probably at work.

WEBSITES

Optimal city size and population density article:

www.npg.org/forum_series/

optimal_city_size.htm

Further discussion on density-independent vs. density-dependent effects of population:

www.zoo.ufl.edu/bolker/eep/notes/

densdep.html

Article illustrating the intertwining of politics with immigration and population concerns in the UK:

http://society.guardian.co.uk/environment/

story/0,14124,1294307,00.html

POPULATION STRUCTURE

Population structure refers to the composition of a particular population. Two of the most important characteristics of any population are age and sex. These characteristics are important because they affect many other population variables such as births, deaths, marriage, and international and internal migration. The proportion of children to the elderly affects healthcare, how many schools are needed, retirement benefits and much more.

All these characteristics have an impact on the environment and land use. An ageing population means more retirement communities, golf courses and healthcare facilities, while a younger population means more

schools and day-care centres. Fewer young people means that the population will grow more slowly, resulting in fewer new houses being required, and less demand for water, wood and other natural resources. Working against this trend, however, are rising standards of living and rampant consumerism, which result in increased demand for resources and pressure on the environment.

The proportion of females and males in any given population can reveal a great deal about that population. The universal measure of the sex composition of a population is called the sex ratio. Information about sex is usually both accurate and easy to obtain, and therefore the sex ratio is a relatively simple measure of population structure. The definition of sex ratio is the number of males per 100 females. A sex ratio of 100 would mean that there are equal numbers of males and females in a population; a ratio less than 100 would mean there are more females than males; a ratio above 100 would mean there are more males than females.

The sex ratio for the US population in 2000 can be calculated as follows:

$$\frac{138,053,563 \text{ (number of males)}}{143,368,343 \text{ (number of females)}} \times 100 = 96.29$$

Therefore the USA, with a sex ratio of 96.29, has more females than males.

Three major factors determine the sex ratio:

- the death rate differences between the sexes
- differences in migration rates between the sexes
- the sex ratio of newborn infants.

For the USA, male infants outnumber females, but females generally live longer and have lower death rates than males.

China's sex ratio at birth for 2000 was 116.86. That means that there were significantly more males born in the Chinese population than females. The reason for this difference is usually attributed to the high levels of female infant deaths and the increasing number of sex-selective abortions of female foetuses. Some demographers believe that this high proportion of baby boys will cause significant social problems in the future.

Another important aspect of population structure is the age composition of a population. The proportion of people found in various age groups plays a significant role in a population's social and economic behaviour. A variety of factors determine the age structure of a national population, although the primary factor appears to be the birth rate. In general, a population with a high birth rate will have a significant proportion of young people, whereas a small proportion of young people will be found in areas where the birth rate is low. Migration is another factor that is generally related to age composition. Areas that have experienced significant in-migration usually have an excess of young adults (aged 20–40), whereas areas that have a lot of out-migration usually have more older people.

A world view of age structure reveals some interesting geographic patterns. The youngest populations are generally found in the developing nations of Asia, Africa and Latin America, where many countries have a population with almost half of their inhabitants under 15 years of age. Older populations are generally found in the more developed and wealthier countries in Europe and North America.

A useful method of graphically representing the age and sex structure of a population is to construct a population pyramid. This is a type of bar graph where two bar graphs are placed back-to-back, with the vertical centre line representing zero. The horizontal bars represent the number or percentage in each age group, by sex; and the vertical axis represents the different age groups, usually in five-year intervals. The shapes of population pyramids vary according to the changes in the age and sex composition of a population over time.

A pyramid with a wide base and a narrow top represents a country with a large population of young people and a high birth rate (such as Tanzania), while a pyramid with more of a beehive shape is a country with a relatively stable population with low death rates, low birth rates and a high median age (such as the USA). Examining population pyramids over time shows the effect of wars, famines and changes in fertility behaviour of cultures. For example, Russia's population pyramids during the twentieth century displayed two significant 'dents', for the loss of life experienced by young men and women of army age during the First and Second World Wars. These dents are repeated every 25 years, because when

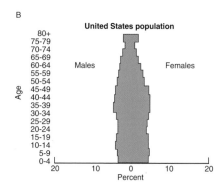

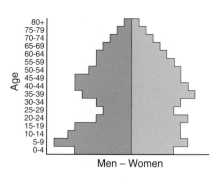

Population pyramid in Germany in 1946, showing the 'dent' from the loss of life experienced by men and women in the Army in their 20s, and the resulting low numbers of children born to this age group

fewer potential parents exist, fewer children result.

WEBSITES

'China's missing girls', describing Chinese sex-ratio problem:
http://app1.chinadaily.com.cn/star/2002/1024/fo5-1.html

Population pyramids and their construction:
http://faculty.washington.edu/~krumme/resources/pyramids.html

Create your own population pyramids for any country:
www.census.gov/ipc/www/idbpyr.html

PRAIRIES

Prairies are low-lying plains that are dominated by grassland. The word *prairie* came into English from French trappers and traders, who were among the first European people to see these lands of broad horizons and few trees. Prairies are found in many parts of the world; they are called pampa in South America, veld in Africa and steppes in Asia. Many prairies are flat because of the action of continental ice sheets, which left behind large deposits of loam (light fertile soils) and sometimes large lakes such as those in Southern Manitoba. Moderate amounts of rainfall, 250–750 millimetres per year, glacial soils and temperatures that vary between –20°C and +20°C create a natural vegetation cover of grasslands across most prairies. Conditions are generally too dry for any tree growth.

Canada's three 'prairie provinces' of Alberta, Manitoba and Saskatchewan cover about one-fifth of the country's territory. In these provinces northern sections are extensive boreal or coniferous FORESTS, but over large areas the natural vegetation is temperate grassland. Over time, large areas of the prairie provinces have been transformed into an extensive wheat-growing area.

Canada's prairies became a focus for settlement and agriculture, especially wheat growing, in the late nineteenth and early twentieth centuries. Settlers arrived by the thousand and the land was laid out in a unique regular grid-iron pattern. Roads were built at 3.2 kilometre intervals, with farms evenly spread along them. A single block of 2.5 square kilometres is called a section. In the wetter eastern prairies farms were a quarter or half section, whereas in the drier west whole sections or larger was more common. The arrival of thousands of settlers, helped by the building of the Canadian Pacific Railway, rapidly transformed the prairies into an area of extensive wheat production, supplying Europe with flour for bread. Saskatchewan became the most important producer of wheat. Today wheat is still a major export product from the prairies, but is usually one of a greater variety of crops.

Wheat output from the Canadian prairies has varied considerably, depending mainly upon physical factors (especially problems of drought and frost) but also in response to demand from Europe and more recently Asia. The 1930s were disastrous years, when drought caused crop failure and led to serious erosion problems – the Dust Bowl. In contrast, there were abundant harvests in the 1960s. Since then the markets have continued to vary and the Canadian Government has guaranteed prices for the farmer, similar to the EU's COMMON AGRICULTURAL POLICY (CAP).

It is not surprising that the Canadian prairies became a focus for wheat growing – the environmental conditions are ideal. In spring the snow thaws quickly and there is ample rainfall. Early plant growth is encouraged by this water and by temperatures that rapidly rise to about 17°C. In summer drier weather and long hours of sunshine help the wheat to ripen. Originally the rich soils required little fertiliser, and the low flat relief and stone-free soils made it easy to use machinery. In addition, efficient railways and good roads and waterways gave a variety of export routes.

However, the northern parts of the prairies are at the very limit of grain growing and frosts are always likely to damage young shoots in spring or the wheat ears just before harvesting – the growing season is almost too short. These are marginal areas for farming, where the threat of drought and soil erosion is high. Rainfall is unreliable and many farmers have to employ strip farming techniques and plant windbreaks to avoid damage from the wind. Strip farming involves planting alternating strips of wheat and grass that reduce areas of exposed soil and the impact of wind.

WEBSITES

US Department of Agriculture, with historical information on farming in the prairies and in other parts of the USA: www.usda.gov/history2/text3.htm

Additional historical reflections: http://collections.ic.gc.ca/exploring/homestead/dprson.htm

PRECIPITATION

Precipitation is the process of water transfer from the atmosphere to the ground. As well as rain, it also includes snow, hail, dew and frost.

Precipitation has a fundamental impact on the environment. Without this water from the sky, our planet would be uninhabitable. Without precipitation, the **HYDROLOGICAL CYCLE** would cease. Vegetation type and consequently the type of animals, microbes and insects that an ecosystem supports are dependent on the type, seasons and amount of precipitation. Precipitation affects erosion and the low river valleys cut into the landscape. **BIODIVERSITY** and **BIOMASS** are affected by precipitation. Precipitation in the form of snow and ice sustains the ice caps in the polar regions. Human impact on the environment is also affected by precipitation. In some parts of the world, people build canals to bring water into an area. In other parts, they build drainage ditches to take water out of an area. Human land use – the type of agriculture and grazing, the type and size of cities, and public works – depend on the environment, and hence on the precipitation.

Clouds are required for precipitation, but not all clouds produce it. In fact, probably less than 10 per cent of clouds create precipitation sufficient to be felt at ground level. Cloud formation requires air to be saturated, which is usually accomplished through some lifting mechanism and the resulting **ADIABATIC COOLING**. Some of the common lifting mechanisms include convection due to surface heating, low pressure, convergence at the surface and/or divergence in the sky and the forced ascent caused by mountain barriers.

As an air parcel is lifted into higher regions with less air pressure, it expands. This expansion cools the air, because the air molecules in it are further apart and collide less frequently. The cooling rate is 10°C per kilometre. Because air near the ground always contains some moisture, if it is lifted high enough it will eventually reach its dewpoint (saturation) temperature. When this happens, clouds form. If the temperature is cold enough, condensation of the water vapour will form ice crystal clouds. Condensation occurs on tiny particles of dust called condensation nuclei, or cloud nucleating agents.

As the cloud becomes thicker, precipitation processes become important. If the thickened cloud remains above freezing, large cloud droplets will form. These larger droplets gain sufficient weight to begin to fall through the cloud. When the pull of gravity supersedes the resistance of the air, the droplet will fall at a steady rate called the terminal velocity. When the larger droplets collide with the smaller ones, the two drops coalesce to form an even larger raindrop. As more impacts occur, the falling drops get larger and fall faster until full-sized raindrops result. The drops continue to grow as they fall through the cloud, but may lose water to evaporation once they are below the cloud base.

If the air in a portion of cloud is below freezing, the Bergeron-Findeison ice crystal process occurs. Some water may remain unfrozen in supercooled droplets down to about −40°C, and provide a source for ice crys-

Snowpack has a significant effect on the local environment, providing water for plants and animals. In many parts of the world, such as the mountains of Colorado, shown on the first of July, above, snowpack lasts until midsummer. This provides water for people, agriculture and industry in the semiarid plains and plateaus below the mountains, impacting the environment through land use far from the mountains themselves

tal growth. Ice crystals form on nuclei through deposition and, as they attract water from the tiny supercooled liquid water droplets, they grow in size. Because the ice has a lower vapour pressure over the surface than liquid water, droplets begin to 'steal' water from the tiny supercooled liquid water droplets that make up the cloud. When the ice crystals become larger and more numerous they eventually become large enough to fall. When they fall to lower, warmer regions of the cloud they may melt or evaporate, or if it remains cold, may fall as sleet or snow. Whichever of these two processes occur, the falling precipitation robs the cloud of water and transfers this water to the ground.

The kind of precipitation that falls depends on temperature conditions within the cloud. If the precipitation originates with the cold cloud process and the air down to the surface is below freezing, snow will fall. If the snow enters a warmer layer aloft and melts to rain and then falls through a cold layer and refreezes, sleet results. If the precipitation is cycled through warm and cold cloud portions repeatedly, particularly if a portion of the cloud is made up of supercooled water droplets and has very active convection, a new layer of ice is 'painted on' every time the particle cycles through the cloud layers, resulting in hail. Such conditions exist in thunderstorm clouds. Under certain circumstances, the precipitation encounters a dry air layer below the cloud base and evaporates, or 'sublimates', before hitting the ground. Such precipitation is detected by radar and can even be seen falling as a rain shaft (or snow shaft) that does not reach the ground. The term given to this is virga.

WEBSITES

Precipitation resources from the University of Illinois:
http://ww2010.atmos.uiuc.edu/(Gh)/guides/mtr/cld/prcp/home.rxml
North America real-time precipitation map:
www.accuweather.com/adcbin/precip_maps?nav=home
Precipitation forecast for Europe:
http://uk.weather.com/maps/lifestyles/intlmountainski/index_large.html
University of East Anglia (datasets):

www.cru.uea.ac.uk/~mikeh/datasets/global
BBC Weather Centre:
www.bbc.co.uk/weather/weatherwise/factfiles/basics/precipitation_whatis.shtml

SEE ALSO

clouds

PRODUCTION/RESPIRATION RATIO

Production and respiration are complementary, antagonistic processes. Production is the process whereby pigments such as chlorophyll in plants and algae capture sunlight and convert it to organic matter and oxygen. Respiration is the process where oxygen is used up to create carbon dioxide, water and cellular energy. The ratio and amount of the two processes is a distinguishing characteristic of the EARTH's various BIOMES.

Most natural biotic communities, from the least productive deserts to the most productive wet tropical ecosystems, fall on the diagonal, which represents equivalence between production and respiration. Environments

---

Production/respiration ratios and fossil fuels

The equilibrium relationship between production and respiration is such an obvious feature of the natural world that perhaps we should unpack its implications for ourselves, as cultural and industrial organisms. Where do human systems fall on this graph? Human cities mostly have a trivial amount of biologic production. Respiration and combustion overwhelmingly prevail. Respiration releases energy from imported food and combustion releases energy from imported fossil fuels.

Fossil fuels come from ancient ecosystems, of course, and peculiar ecosystems in which P/R>1. And as a philosophical and practical matter, humans are therefore almost utterly dependent upon finite stocks of blackened or gaseous leftovers of ancient, aberrant ecosystems in which P>R.

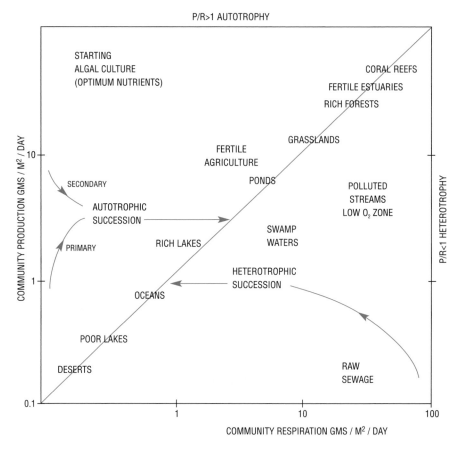

The figure above is a plot of rates of production and rates of respiration in several biotic communities across a wide range of productivity, from deserts to coral reefs. Each of these points represents a ratio between two values, the so-called P/R ratio

along the diagonal represent a 'stand-off' between chloroplasts and mitochondria; a P/R ratio of 1. Deserts have a low production and a low respiration, but these two balance out, leaving deserts on the diagonal. Rainforests and coral reefs have high production, but they also have high respiration, again, tending to balance and placing these environments on the diagonal. Natural environments therefore balance in yet another way.

Below and to the right of the diagonal are communities in which respiration dominates, communities with P/R < 1. These are communities based on the import of resources. Polluted environments have high respiration but low production. Above and to the left of the diagonal are communities in which photosynthesis dominates, communities with P/R > 1. These are systems in which carbohydrates are accumulating, where there is a high production but low respiration.

WEBSITES

Respiration:
http://niko.unl.edu/bs101/notes/lecture7.html

Cell respiration:
www.dccc.edu/homepages/bdadhich/10Bio110cellrespiration.pdf

PROTECTED LAND

NATIONAL PARKS are not the only areas to receive protected status around the world. Wilderness areas, forest lands, wildlife refuges, and state, provincial and local parks are other areas that are set aside by governments, as well as the United Nations as World Heritage Areas, around the world. They provide important wildlife and plant habitats, and buffers against urban development. Governments, private industry and non-profit organisations have all helped set aside land to be protected.

One well-known international organisation is The Nature Conservancy. Since 1951 The Nature Conservancy has protected more than 47 million hectares around the world. Its method of conservation is to raise funds so that special areas can be purchased, and therefore protected from development. Fundraising is done through over a million individual Nature Conservancy members and through mechanisms such as exchanging individual debt for land, conservation trust funds,

Case study

The Arctic National Wildlife Refuge

Renowned for its wildlife, the Arctic National Wildlife Refuge in the north-east corner of Alaska is inhabited by 45 species of land and marine mammals, from the pygmy shrew to the bowhead whale. Most well known are the polar, grizzly and black bear, wolf, wolverine, Dall sheep, moose, musk ox and the animal that has come to symbolise the area's wildness, the free-roaming caribou. Thirty-six species of fish occur in Arctic Refuge waters, and 180 species of birds have been observed on the refuge. The refuge was established in 1960 and enlarged in 1980. Three of the 8 million hectares of the Arctic Refuge are designated wilderness, and three rivers are designated wild rivers. The refuge encompasses the traditional homelands and subsistence areas of Inupiaq Inuits of the Arctic coast and the Athabascan Indians of the interior.

Perhaps the most unique feature of the refuge is that large-scale ecological and evolutionary processes can continue, free of human control or manipulation. A prominent reason for establishing the Arctic Refuge was that this single area encompasses an unbroken continuum of arctic and sub-arctic ecosystems. It is possible to traverse the boreal forest of the Porcupine River plateau, wander north up the rolling taiga uplands, cross the rugged, glacier-capped Brooks Range, and follow any number of rivers across the tundra coastal plain to the lagoons, estuaries and barrier islands of the Beaufort Sea coast, all without encountering an artefact of civilisation.

Since 1980, federal studies have been conducted to assess the amount of oil and gas that may exist in the 0.6 million hectares of coastal plain of the refuge, referred to as the '1002 area'. However, it was not until international tensions of the late 1990s, coupled with falling oil output from nearby Prudhoe Bay, that the controversy heated up: should part of the wildlife refuge be used to drill for oil and gas? The oil companies believe 5 to 16 billion barrels of oil could be recovered in the area, and many Alaskans are eager for the revenue that exploration would generate for their state. Others have resisted drilling in the area because the required network of oil platforms, pipelines, roads and support facilities, not to mention the threat of oil spills, would wreak havoc on wildlife. The coastal plain, for example, is a calving home for some 129,000 caribou. During June 2004 the Drilling Bill was brought before the US House of Representatives, but was shelved, for the time being.

However, the issue is far from dead. With US oil production at nearly a 50-year low, and oil reserves in the country shrinking, President Bush made the Arctic National Wildlife Refuge oil development a key part of his energy package. Congressman John Sununu of New Hampshire tacked an amendment to the Energy Bill, limiting the drilling to just 810 of the 0.6 million hectares along the coast plain. Opponents pointed out that the 810 hectares do not have to be contiguous and only the space of the equipment touching the ground is included in the figure. Each drilling platform can take up as little as 4 hectares.

As the USA and other countries remain dependent on petroleum, areas such as the Arctic National Wildlife Refuge will remain zones of controversy.

ecosystem services payments, resource extraction fees, public finance campaigns and other innovative methods.

A wide variety of protection applies to these areas around the world, and the lands are subject to differing land use and management practices. For example, in the USA limited grazing and clear felling may be done on national forest lands, but not on national park lands.

Another example of protected land is the US National Wildlife Refuge System, made up of 717 units managed by the US Fish and Wildlife Service (FWS). The system includes 500 national wildlife refuges, 166 waterfowl production areas, and 51 wildlife coordination areas, which together cover more area than in the national parks. The refuges were established to preserve natural ecosystems and the diversity of fauna and flora. Some refuges have been established specifically to furnish habitat for a single (often endangered) species. Since 1924 hunting has been allowed, and in some cases encouraged, on certain refuges. FWS views hunting (and fishing) as an effective resource management tool necessary to control populations and maintain proper ecosystem balance. Many animal welfare activists and some conservationists, however, consider it ironic that hunting is allowed on lands set aside to protect and enhance wildlife, and seek to promote legislation that would outlaw hunting on these lands.

WEBSITES

United Nations Environment Programme of Protected Areas of the World, by country: www.wcmc.org.uk/data/database/un_combo.html

National Academies Press report 'Cumulative environmental effects of oil and gas activities on Alaska's north slope': www.nap.edu/books/0309087376/html

Arctic National Wildlife Refuge: http://arctic.fws.gov/index.htm

US Secretary of Interior Gale Norton's testimony before the House Committee on Resources on the Arctic Coastal Plain Domestic Security Act, with the history behind current interest in energy reserves in Alaska:

www.doi.gov/secretary/speeches/030312anwr.htm

US National Wildlife Refuge hunting policy: www.csa.com/hottopics/ern/99dec/pub-6.html

The Nature Conservancy: http://nature.org

RADIATION BALANCE

All the energy that affects the atmosphere comes to the Earth from the Sun. The Sun is very hot – about 5600°C at its surface and 15,000,000°C at its core. Therefore, it emits a great deal of solar radiation in short wavelengths. Most of the Sun's energy is in wavelengths in the visible part of the electromagnetic spectrum – this is the light we see. Travelling through the atmosphere, about half the energy is absorbed, scattered or reflected by clouds. The other half of the energy reaches the Earth's surface, where it is reflected or absorbed. The energy that is absorbed heats the ground surface and is then returned to the atmosphere as long-wave ground radiation. But because the Earth is not a very hot object, the amount of energy returned is much less than is emitted from the Sun, and it is in long wavelengths. Short-wave radiation penetrates the atmosphere more efficiently than long-wave radiation. On the journey back towards space, some of the long-wave ground radiation is absorbed by the atmosphere and reradiated back to the ground surface. This is the so-called 'greenhouse effect', which moderates temperatures on our planet and without which there would be no life on Earth.

There is a balance between the energy entering the Earth and the atmosphere system from space and that which leaves the system. If this were not so, the Earth would either be getting hotter and hotter or colder and colder. Despite the overall balance, there are imbalances between the latitudes. The low latitudes, where the Sun's energy is concentrated throughout the year, have a net surplus of radiation. The poles, however, receive much

The Bowen ratio instrumentation at Lake Eildon, Australia, used to measure the fluxes of heat and moisture to the atmosphere. Image from Dr Jason Beringer, Monash University

less radiation, with the Sun being lower in the sky during the summer and totally absent during the winter. These polar regions have a net deficit of radiation. However, despite these apparent imbalances, the low latitudes are not getting hotter and hotter and the poles are not becoming colder. This is because heat is transferred poleward by balancing mechanisms such as winds and ocean currents. It is these balancing mechanisms that have a huge effect on our weather. In the UK the warm and cold air masses travelling from the south and the north respectively affect our weather from day to day. The warm North Atlantic Drift ocean current, which originates from the Caribbean, has a massive influence on the UK's climate, accounting for the predominantly mild and moist conditions.

Radiation balance occurs on the Earth by day and night, and particularly by hemisphere and season. As winter approaches, the half of the Earth that is tilted away from the Sun receives less and less energy, as the length of day gets shorter and the angle of the sun becomes smaller. More energy is going out at this time than is coming in. As the seasons change to summer, and the tilt of this part of the world is now toward the Sun, more energy is coming in than is going out, and things warm up. Exchanges of heat energy between the hemispheres help moderate this effect. In addition, places near the equator almost always maintain a rough balance between what comes in and what goes out or is redistributed poleward, so temperatures at the equator do not vary much.

While there is an overall radiation balance, short-term changes do occur. There are long and short-term variations in the amount of energy emitted by the Sun, which affect the amount of shortwave radiation received at the top of the atmosphere. The amount of energy that reaches the surface of the Earth may be greater if the atmosphere is more transparent to the short waves, such as when there is less cloud cover, or where stratospheric ozone is depleted and not as much ultraviolet radiation is absorbed. The opposite occurs if there are aerosols in the atmosphere that absorb, scatter or reflect the incoming radiant energy to a greater degree. The 1991 eruption of Mount Pinatubo in the Philippines sent large amounts of sulphur particles and particulates into the stratosphere, where they dispersed around the world and continued to reflect sunlight back to space for over a year. This resulted in lowered temperatures globally for a couple of years.

Changes in the surface reflectivity (called the albedo) can affect the radiation balance. If more of the Earth is covered with snow and ice, as is the case during an ice age, more sunlight is directly reflected back to space without effectively heating the Earth. Similarly, clearing of forests may increase temperatures, because bare ground and many agricultural plants absorb more energy than does natural vegetation. The melting of many of the world's glaciers may be due to radiational balance changes caused by land-use changes nearby, rather than by global warming (as is commonly assumed). The amount of water covering the Earth (versus land areas) also affects this balance, because water absorbs the sunlight to greater depths than land, and water also stores it more effectively. Water also mixes the heat energy through convection and a great deal of energy is utilised to evaporate water. This allows this energy to move great distances with the resulting water vapour, before it is once again released when the water changes back to liquid or ice.

The radiation balance can also be affected during the passage of the long-waves emitted by Earth back to space. If there is a greater amount of 'greenhouse' gases such as carbon dioxide and methane in the atmosphere – the ones that absorb certain wavelengths and then radiate energy back to the Earth – it will get warmer. Increasing carbon dioxide and methane effectively 'store' heat in the Earth's system, causing temperatures to rise. Water vapour is the most effective greenhouse gas on the Earth. If there is little water vapour in the atmosphere, such as in a desert, night-time temperatures plummet rapidly. Clouds also absorb long-wave radiation, so that on a cloudy night temperatures usually fall more slowly than on a crisp clear night.

WEBSITES

NASA radiation balance resource:
http://climate.gsfc.nasa.gov/~cahalan/Radiation/RadiativeBalance.html
Earth and Space Research Organization's radiation balance resource:
www.esr.org/outreach/climate_change/basics/basics.html

University of Wisconsin's information on the Earth's radiation balance:
http://cimss.ssec.wisc.edu/wxwise/homerbe.html
Professor Michael Ritter's resource on the Earth's radiation balance:
www.uwsp.edu/geo/faculty/ritter/geog101/textbook/energy/radiation_balance.html

RAIN SHADOW

A rain shadow is an area of land on the leeward or sheltered side of a mountain range where rainfall totals are relatively low. Where a mountain range lies across the path of the prevailing winds, air is forced to rise up and over it. As it does so, it is forced to cool, and condensation may occur. Clouds then form and rain may result. When the air sinks down on the leeward side of the mountain range, it warms and becomes drier. Clouds dissipate and rainfall becomes less likely. This is the rain shadow, which is sometimes referred to as the orographic effect, meaning 'pertaining to mountains'.

In the UK the Pennine Hills run down the spine of England from north to south. With the prevailing winds coming from the South West, they form an effective barrier to the warm moist air from the Atlantic. The western or windward side of the Pennines receives large amounts of rainfall, accounting for the extensive grasslands and pastoral farming in north-west England. In contrast, the land to the east of the Pennines – in the rain shadow – is much drier and tends to be used for arable farming.

Significant 'lift' occurs as air is pushed into mountains. As air is lifted, it cools adiabatically, usually to the dewpoint temperature. When air reaches its dewpoint temperature, it reaches saturation and clouds form. Clouds will continue to form as long as the air keeps rising and cooling. As condensation proceeds, latent heat is released into the air. Water vapour is lost as precipitation occurs on the windward side, and runs back into the oceans.

The rain shadow effect explains why leeward sides of mountains – the sides away from the predominant windflow – are usually cloud-free and dry. Some of the world's most famous deserts are at least partly 'rain shadow'

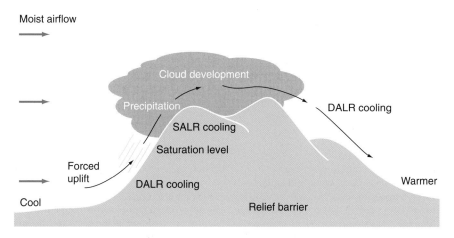

Moist airflow

Cloud development

Precipitation

DALR cooling

SALR cooling

Saturation level

Forced
uplift

DALR cooling

Warmer

Cool

Relief barrier

Diagram illustrating the rain shadow effect

deserts, including the Kalahari in Africa, the Atacama in South America and the Gobi in Asia. In western USA and Canada, rain shadow is the major reason why the environments of the western versus the eastern sides of the Rocky Mountains are vastly different. The western sides are much wetter, supporting lush stands of enormous fir and spruce, but also rarer myrtlewood and redwood trees. The eastern sides are drier, with bristlecone pine, juniper and pinon trees dominating in many areas. This vegetation influences the types of animals the two sides can support.

In north-western USA, the Cascade Range runs north–south in a wide band, paralleling the Pacific coast, beginning about 200 kilometres inland. Some of the world's most lush temperate rainforests are supported by their location on the windward side of the Cascades, with towering Douglas fir, hemlock and Sitka spruce among the main species. Hoquiam, Washington, on the windward side of the Cascade Range, receives 174.5 centimetres of precipitation annually, yet Ellensburg, on the leeward side of the Cascades, receives just 23.1 centimetres annually. On the leeward side, grasslands, yucca, piñon and juniper dominate.

The orographic pattern is repeated the world over, having a significant effect on the environments where mountain ranges and prevailing winds occur – their precipitation and hence their plant life, soils and animal life.

Arid lands on the leeward side of the Sierra Nevada Range, along the California-Nevada border.
Photograph by Joseph Kerski

WEBSITES

Rain shadow effect from World History.com:
www.worldhistory.com/wiki/r/
rain-shadow.htm
Flash animation of rain shadow:
www.mhhe.com/biosci/genbio/tlw3/
eBridge/Chp29/animations/ch29/
rain_shadow_formation.swf
Schematic diagram of rain shadow effect:
www.sp.uconn.edu/~geo101vc/Lecture19/
sld020.htm
Research on the rain shadow effect:
www.weatherpages.com/rainshadow

Washington State's rain shadow:
www.komotv.com/weather/faq/
rain_shadow.asp
Washington's precipitation map:
www.ocs.orst.edu/pub/maps/Precipitation/
Total/States/WA/wa.gif

SEE ALSO

mountains, precipitation

RECYCLING

How often have you sat at Christmas surrounded by your presents and a veritable mound of wrapping paper and packaging? Often the pile of waste cardboard and paper is larger than the gifts themselves.

Research has shown that packaging makes up 25 per cent of the 27 million tonnes of rubbish families in the UK currently produce. No surprise therefore that in the past few years recycling has become an increasingly popular option in many more economically developed countries, especially as it reduces the amount of waste material being dumped.

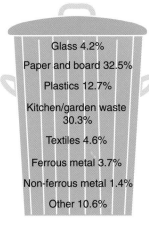

Glass 4.2%

Paper and board 32.5%

Plastics 12.7%

Kitchen/garden waste 30.3%

Textiles 4.6%

Ferrous metal 3.7%

Non-ferrous metal 1.4%

Other 10.6%

The contents of the average family's dustbin.
Source: www.recycle-more.co.uk/

Recycling covers the collection and separation of waste materials, and their subsequent reuse or processing into a usable product. Some recycling processes reuse materials for the same product – an example being old aluminium cans and glass bottles. Other examples convert waste items into new products – such as old tyres becoming rubberised road surfacing.

There are four distinctive areas in which recycling takes place:

- Pre-consumer recycling: industrial recycling that is part of the manufacturing process before the product reaches the customer.
- Product reuse: products such as glass bottles can be returned and then used over and over again.
- Primary recovery: the collection of materials, such as waste paper, which reduces the need to manufacture new materials from virgin raw materials, in this case trees.
- Secondary recovery: waste is used to produce heat, fuel and electricity through 'waste-to-energy' incinerators and the collection of landfill gas.

Recycling rates in north-west Europe

Country	Glass %	Aluminium %	Steel cans %
France	39	*	30
Italy	40	29	**
Netherlands	53	*	45
Norway	6	60	**
Sweden	22	82	**
UK	23	12	19

*Negligible percentage
**Data not available

Source: K. Byrne, 1997, *Environmental Science*, Stanley Thornes.

There are many advantages to be gained by recycling, especially the fact that it helps to conserve finite resources and reduce energy consumption. For example, recycling one tonne of aluminium saves four tonnes of bauxite and 700 kilograms of petroleum coke. In 2002 360 million plastic bottles were recycled in the UK and this helped to save between 50 and 60 per cent of the energy that would be needed to make brand new bottles. In the USA recycling, including composting, diverted 68 million tonnes of material away from landfills and incinerators in 2001. Back in 1990 this figure had been 34 million tonnes.

Case study

SWAP – The Somerset Waste Action Programme

The residents of the county of Somerset in south-west England generate 300,000 tonnes of household waste each year. In 2003 20 per cent was recycled – well above the national average of 12 per cent but SWAP aimed to increase the level of recycling to 28 per cent by April 2004.

Created in 1992, SWAP includes representatives from the County Council, as well as the various District Councils, such as Sedgemoor and Mendip, plus Taunton Deane Borough Council. An important aspect of the scheme is to educate the residents, especially the schoolchildren, so among the SWAP activities are workshops, lectures and activities in schools, educating the next generation of consumers about the importance of recycling.

However, there are economic costs to be considered. Often the recycled products have to compete with brand new products that may be subsidised in price. For example, a plastic recycling scheme in Sheffield, South Yorkshire, UK, between 1989 and 1992 cost £100,000 a year to run, but the resale of the plastics made less than £50,000 per year. It is perhaps for this reason that many schemes require sponsorship or corporate backing.

Some people also feel that recycling fails to address other important issues, such as the limited lifespan of products. Each year in the UK a total of 2 million cars, 6 million large kitchen appliances and 3 million vacuum cleaners are thrown away, and much of this cannot be easily recycled. In many European countries greater emphasis is therefore being given to making products last longer. For example, cars in Sweden are designed to last for an average of 17 years, compared with 12 years in the UK.

WEBSITES

Earth911: www.earth911.org

Environmental Protection Agency:

www.epa.gov/recyclecity

SEE ALSO

reduce, reuse, recycle

REDUCE, REUSE, RECYCLE

Each person in the USA generates about 1.8 kilograms of rubbish each day – 660 kilograms each year! About one-third of this is just packaging materials like wrappers and bags, which end up in a landfill after only one use. What we throw away will be with us for a long time, too. Plastics and aluminium take about 500 years to decompose. If we are going to avoid being overrun with landfills, we need another plan.

There are several strategies in waste management that attempt to lessen the amount of refuse to be buried in landfills or incinerated, combating the 'throwaway mentality' of modern society. They are, in order of priority: reduce, reuse, recycle.

International symbol showing the linkages for reduce, reuse, and recycle

The first strategy is source reduction – to reduce the number of products manufactured. This is the most effective form of waste management since it reduces the amount of waste to be managed and conserves resources used to make new products. Source reduction can be done by 'precycling' – choosing products with minimal impact on the environment,

Case study

Refillable bottles in Prince Edward Island, Canada

The province of Prince Edward Island, Canada, has banned the use of non-refillable beer and soft drink containers. The ban came about incrementally during the 1970s and into the early 1980s as a way to combat littering.

A small deposit is paid on each beer or soft drink container purchased. The consumer can then return an empty soft drink bottle to the retailer or to one of 14 province-wide bottle depots for a refund of the deposit; beer bottles must be returned to one of the bottle depots. Major retailers refund the entire deposit, while smaller retailers often choose to refund a minimum deposit amount set by the Government. Today about 98 per cent of soft drink and beer bottles are returned and reused in Prince Edward Island. Since 1980 1,115,240,000 containers have been reused. The average bottle is refilled 17 times before being 'retired' and recycled.

This reuse of beverage bottles cuts down on the use of non-renewable resources such as aluminium and oil (used to make plastic), and uses less energy, water and landfill space. Reusing the bottles also produces less pollution than a non-refillable system. Prince Edward Island further benefits by reduced litter and an enhanced image for ecotourism.

such as those with little or no superfluous packaging. People can also reduce waste by minimising unwanted junk mail by opting out of mailing lists.

The second strategy is to reuse products wherever possible. An example of this is refilling beverage bottles. Ironically, almost all beverages came in refillable bottles until the rise of plastic bottles and aluminum cans in the early 1980s. Some areas of the world require a small bottle deposit when beverages are purchased. The consumer gets the deposit back when the empty container is returned to a retailer. The used containers are then sent back to the manufacturer where they are washed and refilled, or recycled. In either case, the container does not end up in the waste stream. The case study from Prince Edward Island, Canada, is an excellent example of what can happen when a community commits itself to reuse.

The third strategy is to recycle – to make new products out of discarded products (see **RECYCLING**).

WEBSITES

Grassroots Recycling Network:

www.grrn.org

National Recycling Coalition:

www.nrc-recycle.org

UK Recycling Consortium:

www.recyclingconsortium.org.uk

Prince Edward Island refillable bottles programme:

www.c2p2online.com/documents/donjardine.pdf

Earth facts from Planet Pals

www.planetpals.com/fastfacts.html

Recycling resources from the US Environmental Protection Agency:

www.epa.gov/recyclecity

REMOTE SENSING

Much of our awareness of the environment is by perceiving a variety of signals, either emitted or reflected from objects that transmit this information in waves or pulses. We hear disturbances in the air carried as sound waves, experience sensations such as the heat of a summer's day, react to chemical signals from food through taste and smell, and recognise shapes, colours and relative positions of exterior objects and materials by seeing visible light issuing from them. All sensations that are not received through direct contact with an object are remotely sensed, or sensed from a distance. To understand something as complex as the environment, other sensors besides the human senses are increasingly used. These include cameras, lasers, radio frequency receivers, radar, sonar, seismographs, gravimeters, magnetometers and scintillation counters (to detect microscopic particles).

These devices are frequently mounted on aircraft, spacecraft and ships, and measure force fields, electromagnetic radiation or acoustic energy, often producing an electronic image of the phenomenon being sensed. The science of remote sensing includes the acquisition of these images, analysing the images and applying the information to help understand a problem or issue.

Remote sensing is possible because of the predictable behaviour of electromagnetic radiation. Electromagnetic radiation is energy that travels and spreads out as it goes. It is composed of an electrical field and a magnetic field, and includes every type of energy, from gamma rays (short wavelength, high frequency) to radio waves (long wavelength, low frequency), and everything in between. Visible light, X-rays, microwaves, ultraviolet energy and infrared energy are other types of electromagnetic radiation, each of which can provide useful information about the environment. In short, we can understand the Earth better from space.

To acquire useful information about the Earth's surface through remote sensing, an energy source is needed that illuminates or provides electromagnetic energy to the target of interest. Natural colour images use the Sun as an energy source. Radar images, which can provide pictures of the Earth at night, through rain and through clouds, are produced by sending a radar beam to the surface and sensing the reflected beam that is returned to the device. Remote sensing is possible because some energy is transmitted through the atmosphere, some is reflected from the surface and some is absorbed. Scientists studying the images use these properties of the atmosphere to detect patterns, linkages and trends in the environment. Particles and gases in the atmosphere can affect the incoming light and radiation. Because these particles and gases absorb electromagnetic energy in very specific wavelengths, they influence in which wavelengths we can 'look'. Those useful wavelengths of the spectrum are not severely influenced by atmospheric absorption and are called 'atmospheric windows'.

There are dozens of satellite systems that actively record information about the planet. These include government-sponsored programmes of several countries, such as Landsat, Earth Observing Satellite, NASA's Terra, the 'SPOT' – or Satellite Pour l' Observation de La Terre (the French Earth Observation Satellite) and India's Remote Sensing Satellite. Some of these satellites, such as Landsat, have been continuously recording the Earth since 1972, providing valuable information about how the Earth is changing over time. In addition, commercial systems such as Radarsat, Digital Globe, Space Imaging and Orbimage provide terabytes of data each day, even down to sub-metre resolution. It is not possible to see people's hair colour on these images, but the type of fields, wetlands, buildings, transportation, stream sediments and other surface features can easily be identified.

Remote sensing is an increasingly vital tool to help understand and protect the environment. Remote sensing is applied in such

Case study

The Interagency Vegetation Mapping Project

To monitor the health of forests in north-west USA, the Northwest Forest Plan requires comprehensive and consistent maps of existing and potential vegetation. Remote sensing makes this possible. The US Forest Service and the US Bureau of Land Management (BLM) create vegetation data sets through the Interagency Vegetation Mapping Project (IVMP). IVMP provides maps of the total percentage of vegetation cover, conifer cover, broadleaf cover and conifer size for the range of the NORTHERN SPOTTED OWL. This owl, *Strix occidentalis caurina,* is a threatened species largely because of loss of old-growth forests to logging. These threats are made even greater by wildfires and windstorms. As a result of declining habitat, there are fewer than 100 pairs of northern spotted owls in British Columbia, Canada, 1200 pairs in Oregon, 560 pairs in northern California, and 500 pairs in the state of Washington. Remote sensing and the resulting maps can help protect this species.

diverse ways as wildfire management, detection of beetles in pine forests, early warning for famines based on agricultural assessment, urban growth models, sea upwellings and eddies for fisheries use, and detection of underground ore bodies.

WEBSITES

GetMapping.Com, one of the UK's largest aerial photograph companies:
www1.getmapping.com/home.asp
European Space Agency:
www.esa.int/esaCP/index.html
Canada Centre for Remote Sensing:
www.ccrs.nrcan.gc.ca
NASA's remote sensing tutorial:
http://rst.gsfc.nasa.gov/start.html
Ohio University's remote sensing slide-based tutorial:
http://dynamo.phy.ohiou.edu/tutorial/tutorial_files/frame.htm
The Global Land Cover Facility, the largest source of free Landsat data on the internet:
http://glcf.umiacs.umd.edu/index.shtml
Interagency Vegetation Mapping Project:
www.or.blm.gov/gis/projects/ivmp.asp

RENEWABLE ENERGY

The global demand for energy rose dramatically during the twentieth century. In 1925 the world's energy consumption was the equivalent of 1485 million tonnes of coal. By 1970 this figure had risen to 6821 and in 2000 it was estimated to be close to 15,000 million tonnes, over 10 times what it had been 75 years earlier. The current rate of increase in world energy consumption stands at 5 per cent per annum, and there is no sign of any decline in the next few years. As living standards continue to improve around the world during the twenty-first century, more and more energy will be demanded to ensure that a decent quality of life is maintained and sustained.

But whereas demand is on the increase, the same cannot be said about the supplies of traditional and conventional energy sources such as the non-renewable fossil fuels. Supplies of coal, oil and gas are finite, and

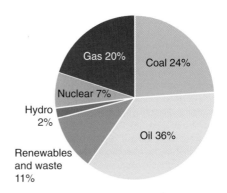

World energy use – 1996 (Source – International Energy Agency)

their increased usage is also accompanied by some rather unwelcome environmental side effects. Global warming and acid rain are just two of the problems caused by burning fossil fuels to produce energy, leading many people to ask whether it would be better to meet our increasing energy requirements by using other, cleaner ways. One very attractive alternative is to use renewable sources.

The term renewable energy covers a variety of sources, including solar, wind, biomass and tidal. Some geothermal sources are regarded as renewable because they are derived from enormous energy sources deep within the Earth's interior. These sources are so large that the rate of depletion by a geothermal energy extraction project is negligible.

Renewable sources are important because they can be exploited without causing major environmental problems. Several highly successful renewable energy companies have been created, such as the RES Group in the case study. These companies have been very active in the creation of successful wind farms. However due to the high costs of development, incentives have had to be introduced to encourage the use of renewable energy. In the UK the introduction of the Non-Fossil Fuel Obligation (NFFO) has meant that electricity companies must generate a certain percentage of their electricity from non-fossil fuels. By 2000 a total of 331 renewable energy projects were up and running as a result of the NFFO in the UK, covering a range of energy sources, including wind, hydro and waste. The Government hopes that 10 per cent of the UK's electricity supply will come from

Case study

The Renewable Energy Systems (RES) Group

The RES Group is one of the world's leading wind energy companies, designing, building and running wind farms that will help to meet many nations' renewable energy targets. Formed in 1981, the RES Group is based in the UK and is a member of the Sir Robert McAlpine Group, one the UK's foremost engineering companies. The RES Group is now active in 15 countries and across four continents, and by the end of 2003 it had built over 775 megawatts of onshore wind capacity around the world. In 2001 it was involved with the creation of what at the time was the world's largest wind farm, or more correctly wind ranch, at King Mountain in Western Texas. The complex contains 214 turbines producing 278 megawatts of energy.

In February 2001 the RES Group also helped to form Offshore Wind Power Limited, which is actively developing offshore wind farms around the coast of mainland Britain.

renewables by 2010, subject to an acceptable cost to consumers.

WEBSITES

Websites that look at the wider issues associated with renewable energy:

www.nrel.gov

www.renewableenergy.com

www.solardome.com

SEE ALSO

biomass, geothermal energy, solar power, tidal power, wave power, wind power and wind farms

REPRODUCTIVE STRATEGY

Differing patterns of **SURVIVORSHIP** are contrasts between life histories and what some ecologists call 'reproductive strategies'. The term 'strategy' in this context can be misleading, as it can connote something consciously done. However, in this case, a reproductive strategy is not some scheme that an individual or a population conspires to follow; rather, it is a part of the fitness-generating, adaptive 'kit' that has accumulated genetically over evolutionary time.

Put simply, there are two sorts of reproductive strategies, the maple–barnacle sort and the elephant–human sort. Barnacles and maples have a *quantitative* strategy: produce lots of poor-quality offspring with minimal parental involvement or investment on the chance that a few will survive to carry your genes into the next generation. Humans and elephants have a *qualitative* strategy. They produce their young mostly one at a time, each young representing a large parental investment, with the parents (usually) taking very good care of those offspring. The end result is the same as with the quantitative strategy. The parents make offspring that carry genes into the next generation. If the parents do a good job with either strategy, their fitness, i.e. their contribution to the genetic heritage of the next generation, is assured.

The reproductive strategy of most species is, of course, somewhere in between the extremes just described. Consider the example of the deer mouse whose story may be simplified slightly for the purposes of illustration. Female deer mice (*Peromyscus maniculatus*) on the Great Plains of North Eastern Colorado, USA, produce an average of five litters of young per year. Each litter contains an average of four young, for a total of 20 little deer mice per year. In the high Rockies, 100 kilometres to the west and 1800 metres higher in elevation, the same species averages only about four litters per year (the growing season is much shorter, after all). Each of those four litters has, on average, five young, again, a total of 20 little deer mice per year.

Deer mice are abundant animals, but they are not as abundant as one might presume from knowing that a pair of deer mice produces 20 young per year. There is not a tenfold increase in the population annually; there is not even a doubling. Deer mice have high mortality, and from one year to the next, populations are fairly stable. Late snows or

Life history and ecological variables of populations summarised and contrasted as *r*- and *K*-'strategies'

Source: After Pianka, E. R. 1970. On *r*- and *K*-selection. *American Naturalist*, 104: 592–7.

Variable	*r*-strategy	*K*-strategy
Mortality	Often catastrophic, non-directed, density independent	More directed, density dependent
Survivorship	Often type III – species that have many offspring, most of whom die early in life	Usually type I (high survival rate of young, live for expected span, die old) and type II (constant death rate throughout life)
Population size	Variable in time, non-equilibrium; usually well below carrying capacity of environment; unsaturated communities or portions thereof; ecological vacuums; recolonisation each year	Fairly constant in time, equilibrium; at or near carrying capacity of the environment; saturated communities; no recolonisation necessary
Competition	Variable, often lax	Usually keen
Selection favours	Rapid development	Slower development
	High maximal rate of increase	Greater competitive ability
	Early reproduction	Delayed reproduction
	Small body size	Large body size
	Single reproduction (semelparous)	Repeated reproduction (iteroparous)
	Many small offspring	Fewer, larger offspring
Lifespan	Short, usually less than a year	Longer, usually more than a year
Emphasis on...	Productivity	Efficiency
Successional stage	Early	Late, 'climax'

cold spring rains may kill some of the fragile pink nestlings. Weasels, foxes, coyotes, owls and snakes eat them. Twenty young turns out to be just about enough to replace the parents, a couple to convey the parents' genes into the future, and 18 to feed the FOOD WEB.

What would happen if a mutation occurred that caused mountain females to produce litters of ten young? Perhaps the individual young would be too small to survive; perhaps ten would be too many for the mother to nurse, or, when weaned, too many for the resources of the habitat. In any event, ten young would not necessarily increase the mother's fitness (her success at becoming a grandmother). And what if a mutation occurred that caused females to have only one litter per year? Unless her young were of distinctly higher quality than those of

her neighbours, for example, young better able to avoid predators, disease, or to survive foul weather, the mother's fitness would be reduced. Thus, we can envision that a population may, by evolution, change its position on the continuum of reproductive strategies toward the quantitative end or toward the qualitative end. However, it can only do so if individuals that make the shift manage to leave more offspring in the reproductive population of the next generation. The reproductive strategy of a population at a particular time is a compromise, summarising the individual reproductive performance of parents. How reproductive strategy evolves depends on the relative fitness of different genotypes.

The primary effect of a reproductive strategy, therefore, is to promote the 'fitness' of parents. But there is a very important secondary effect as well. The reproductive strategy also happens to function to regulate numbers of individuals in a population. Parents who overproduce or underproduce lower their fitness. Population regulation is such an important by-product of a reproductive strategy that some ecologists have argued that reproductive strategies have evolved for the good of the species. This turns out to have been a central problem in evolutionary biology, a problem raised by Darwin himself.

Reproductive strategies are sometimes summarised with respect to two fundamental features of populations and their environments: r, the biotic potential or intrinsic rate of increase versus K, the carrying capacity. A so-called 'r-strategist' achieves fitness by maximising reproductive output; this is basically a quantitative strategy. In contrast, a so-called 'K-strategist' achieves fitness by producing a limited number of young, generally 'sensitive' to the carrying capacity of the environment; this is basically a qualitative strategy.

These two strategies are compared in the table above. This is a very useful way to think about various populations and their life history patterns. The term strategy is a metaphor to describe broad types of life histories, allowing us comparative insight into population ecology of species, but does not imply any deliberate or conscious foresight.

WEBSITES

Reproductive strategies:
http://curriculum.calstatela.edu/courses/builders/lessons/less/biomes/breeding.html
Slides on reproductive and life history strategies:
http://eeb.bio.utk.edu/weltzin/GenEcol03/Lecture/lc_9-25-03.ppt
Checks on population growth with additional discussion on r- and k-strategies:
http://users.rcn.com/jkimball.ma.ultranet/BiologyPages/P/Populations2.html

RESERVOIRS

A reservoir is an enclosed area for the storage of water. Unlike a lake, which is a natural feature, a reservoir has been constructed from the building of a dam or modified from an existing lake. Reservoirs change the run-off, plant type, animal, fish and invertebrate habitat, and even the air temperature of areas that range from a few hectares in size to

Reservoirs in the UK

There are two kinds of reservoir. The first is a natural lake, which has been adapted to store water. The second is a totally artificial body of water that involves the flooding of land that was not previously inundated. Some of the natural lakes in the Lake District have been adapted as reservoirs. Ullswater, for instance, is a main water source for the conurbation of Greater Manchester, over 160 kilometres south. The lake had to be deepened to increase its capacity, with the loss of homes, farms and landscape. Wales has a number of reservoirs that supply the English West Midlands urban areas. Elan Reservoir in Powys and Lake Vyrnwy in central Wales are examples.

With most demand for water being in the South East, some reservoirs have been artificially constructed. Bewl Water Reservoir is the largest body of inland water in southeast England, covering close to 300 hectares. The site was chosen over others because of its suitable contours, clay soils (fairly impermeable) and relatively cheap land. It had been medium-grade farmland with few houses and little wildlife of special interest. Clearance began in 1973 and the 800-metre-long dam built. Roads were rerouted and houses rebuilt elsewhere.

Unlike Ullswater, Bewl Water was also built with the deliberate intention of providing a recreational amenity. The reservoir was therefore filled with trout; footpaths, bridleways, cycle and nature trails were marked out; sailing club buildings and jetties were constructed, as was a playground, car park and café. The sailing club is one of the largest in the country and the trout fishery has a high reputation. The total site area is 480 hectares, of which 127 are set aside as a nature reserve. People visit Bewl from the whole south-east region.

Case study

China's Three Gorges Dam

The Chinese Government is currently promoting the Three Gorges Dam on the Yangtze River as the world's largest hydroelectric scheme. At a cost of US $70 billion it will take up to 20 years to complete. For the Chinese Government it is a symbol of China's development, impending 'superpower' status and high level of organisation. On the other hand, perhaps one could argue that it is only a Communist government such as China's that could organise and impose upon its people such a scheme as this. This plan displaces 1.2 million citizens and moves them from some of the world's best arable land to some of its most challenging. Farmers are being forced to leave their productive river-fed alluvial soils in the 'rice bowl' of China for higher, steeper and significantly poorer land, far from their home region. They simply have no choice in this. Their land will be flooded and rezoned.

Logic dictates that with a still growing population (albeit much more under control due to the one child policy), China needs this crucial food supply. It will not be easy to replace. But the same time, China's industry is growing very rapidly and there is an increasingly urgent need for power. Standards of living and, even more importantly, aspirations, are flying. The Chinese people are demanding the means to make a better living and the electricity to make it happen. They are prepared to work for it – look at how many Western-consumed goods are now 'Made in China'.

The Three Gorges Dam is already underway. It will happen, despite the objections of the environmentalists. People will be moved and land will be inundated. Pollution levels may increase as water ponds up behind the dam and the impact on wildlife is likely to be serious – there are several threatened species including the Siberian crane. On the positive side, however, floods should become less common and navigation should become safer. Some 16 million people should eventually benefit from the electricity produced by the scheme.

hundreds of square kilometres.

Reservoirs are found throughout the world. Britain has several reservoirs but they tend to be relatively small. Many of Australia's reservoirs are located in the Snowy Mountains in the South East, relatively close to main urban areas. California has dams and reservoirs of all sizes. Some countries have huge projects – multipurpose schemes not only for water supply but also for hydroelectric power, irrigation and flood prevention. The Damodar Valley in India and Lake Nasser on the Nile are good examples. Even these will, however, be dwarfed by the Three Gorges Dam currently under construction in China.

The physical characteristics sought for a good reservoir site primarily involve relief, rock type and climate. The Lake District in the UK is ideal for such development. Impermeable rocks prevent loss of water through seepage, while the deep, steep-sided glacial valleys ensure maximum storage capacity for a minimum amount of dam construction. High altitude and a westerly location in the UK means that there is a high rainfall, which maintains supplies.

WEBSITES

Three Gorges Dam information from Public Broadcasting Service:

www.pbs.org/itvs/greatwall/

Three Gorges Dam information from China Online:

www.chinaonline.com/refer/ministry_profiles/threegorgesdam.asp

Bewl Water Outdoor Centre:

www.bewlwater.org

Eca-Watch:

www.eca-watch.org/problems/china/racechina.html

SEE ALSO

dams, lakes

REUSE

While recycling was one of the success stories in the USA in the late twentieth century, the US Environmental Protection Agency believes that even more can be done in producing less

Case study

Reuse in action – the Surplus Exchange in Kansas City

Since 1984 the Surplus Exchange has been playing a leading role in reuse activities in greater Kansas City, linking industry, charity and the environment. The organisation's warehouse is filled with multiple floors of business furniture, equipment and electronics, which are then made available to the non-profit community and the general public. Surplus Exchange trucks run daily routes throughout the metropolitan area to collect equipment from area businesses. The Exchange sets prices to allow organisations and individuals to purchase the items they need, that were collected by the trucks, at the most reasonable prices. Non-profit organisations can also become members of the Surplus Exchange and thereby receive an additional 25 per cent discount on purchases.

waste. It is all about the 3 Rs – reduce, reuse and recycle:

- reduce the amount and toxicity of trash discarded;
- reuse containers and products – repair what is broken or give it to someone who can repair it;
- recycle as much as possible, which includes buying products with recycled content.

Not only does reuse prevent solid waste from entering landfill sites, it can be socially beneficial by taking useful products discarded by those who no longer want them and passing them on to those who do. Simple examples of reuse include refilling bottles, using empty jars and containers for storing leftover food, purchasing refillable pens and pencils, and saving carrier bags after visiting supermarkets. In fact some retail chains give shoppers incentives for reusing bags, either by giving them cash, or making 'long-life' bags cheaply available.

The EPA favours reuse over recycling because a reused item does not need to be reprocessed before it can be used again. It is therefore a cheaper option. There are more than 6000 reuse centres in operation in the USA, ranging from specialised programmes for building materials to local programmes such as Goodwill and the Salvation Army. In some cases, reuse schemes also support local community and social programmes while providing the donating businesses with tax benefits and reduced disposal fees.

Many of the reuse schemes have evolved from local solid waste reduction plans, because reuse requires fewer resources, less energy and less labour, compared to recycling.

It also provides an environmentally preferred alternative to other waste management methods, because it reduces air, water and land pollution, and also limits the need for exploiting new natural resources, such as timber.

WEBSITES

Environmental Protection Agency: www.epa.gov/epaoswer/non-hw/muncpl/reduce.htm

Re-use Development Organisation: www.redo.org

Surplus Exchange in Kansas City: www.surplusexchange.org

SEE ALSO

reduce, reuse, recycle

RICE

The custom of throwing rice at a bride and groom derives from the traditional link made between rice and fertility. In India rice is always the first food offered by a bride to her husband to ensure fertility, and children are always given rice as their first solid food. In Louisiana custom has it that the test of a true Cajun is whether he can calculate the precise quantity of gravy needed to accompany a crop of rice growing in a field. Hence even today rice is important in cultural traditions, as well as being the staple food for hundreds of millions of people around the world. The world's major rice-growing countries include China, India, Bangladesh, the Philippines and Indonesia.

Many historians believe that rice is also one of the world's first cultivated foods, grown as far back as 5000 years BC. Although its precise home cannot be identified, the first recorded mention originates from China in 2800 BC. The Chinese Emperor Shen Nung appreciated the importance of rice to his people and established annual rice ceremonies to be held at sowing time. Today the Chinese celebrate rice by dedicating one day of the New Year festivities to it.

Travellers who carried the seeds with them introduced rice into Europe and the Americas. The UK has never been able to grow rice, due to the inappropriate climate, although parts of France, Spain and Italy gave rise to a thriving European rice industry. Some historians believe rice travelled to America in 1694 in a British ship bound for Madagascar. The ship was blown off course into Charleston, South Carolina, where friendly colonists helped to repair it. In return the Captain, James Thurber, presented Henry Woodward with a quantity of rice seed. Some years later the British made a serious error when, during the American Revolution, they occupied the Charleston area and sent home the entire rice crop, failing to leave any seed for the following year's crop. However, the American rice industry survived, thanks to President Thomas Jefferson who broke an Italian law by smuggling rice seed out of Italy during a diplomatic mission in the late eighteenth century. The rice industry then moved from the Carolinas to the southern states of the Mississippi Basin.

Rice cultivation is unique from other grains in that it must grow in a flat field that is saturated periodically with standing water. The cultivation of rice varies greatly around the world. In many Asian countries cultivation may still rely on primitive methods that take advantage of the large availability of human labour. Simple wooden ploughs are used, drawn by oxen or water buffalo. The ground is naturally fertilised by river floods that deposit fertile alluvium on to the flat flood plains. Logs are often then dragged over the surface to smooth it ready for planting. The seedlings are first planted in nurseries and after about 40 days are transplanted by hand into the fields that have been flooded by rain or river water – these are the paddy fields. During the growing season the fields are kept flooded by hand watering or by primitive irrigation techniques. These involve the transfer of water along channels and the subsequent breaching of the bunds or dikes to allow water to flow into the fields. The fields are then drained before harvesting takes place. Huge amounts of labour are used to harvest the rice that is picked by hand. Since the 1960s the **GREEN REVOLUTION** has led to changes in some of the methods of rice cultivation. By encouraging multicropping and introducing new varieties of seed food, production has increased in many Asian countries.

In contrast, Californian rice – grown mainly in the Sacramento Valley – is highly intensive and mechanistic. It does not rely on

Chris Martin rocks the USA for Make Trade Fair

Chris Martin and his chart-topping British band Coldplay turned up the volume on the Make Trade Fair campaign by inviting Oxfam to promote trade justice at concerts in their spring tour of 2003. Oxfam fielded over 150 volunteers at 14 concerts across the USA and collected over 10,000 postcards calling on George Bush to stop dumping cheap, subsidised exports on poor countries. The team reached hundreds of thousands of concert-goers across the country, with key messages about the Make Trade Fair campaign.

Chris Martin travelled with Oxfam to the Dominican Republic and Haiti in 2002, where he saw first-hand how unfair trade rules affect the lives of the people there. He visited rice-growing areas where farmers once grew enough to provide for Haiti's population. But due to cheap imported rice dumped on the market, they are now facing ruin. Chris explains in his trip journal:

This rice dumping is an example of what Oxfam means when it talks about unfair trade. Haiti has been forced to drop all restrictions on imports, making it one of the freest markets in the world ... so it is flooded with surplus rice grown by heavily subsidised farmers in the USA, and many of its own rice farmers are now moving to the already overcrowded slums in the cities in search of work.

the seasonal and unreliable rains, as in the Asian producing countries. Computerised laser-guided land levelling is used, along with recirculating irrigation systems. These allow farmers to increase yields and reduce the amounts of water required. The water depth is maintained at about 6 centimetres during the growing season. Fields are flooded in April and May and then seeded by aeroplane; later fertilisers may also be applied from the air. By September the crop is mature and ready to be harvested by machines.

However, the massive production of rice in the USA is causing severe problems for many developing countries, which cannot compete with the subsidised rice being sold on the open market.

WEBSITES

Rice farming in the Philippines, on the Global Eye website:
www.globaleye.org.uk/secondary_autumn 2001/eyeon/case.html
Rice farming in Indonesia:
www.deliveri.org/deliveri/bckgrnd/ bolmong.htm
Rice farming in Malaysia:
www.geocities.com/TheTropics/Shores/ 3187/Rice.html

SEE ALSO

Green Revolution

RIVERS

Rivers, the Lotic (flowing) water **BIOME**, are one of the most distinguishing features of the planet **EARTH**. Although at any one time, rivers and streams contain only 1 per cent of the Earth's freshwater (and 0.0003 per cent of Earth's total water), they are, of course, the critical transportation link between land and sea, carrying organic materials and mineral nutrients to estuaries and thence to the sea.

Primary productivity in flowing waters is mostly from rooted aquatic plants and tends to be low, roughly that of a semi-arid grassland. However, secondary production (by consumers such as insects, fish and bottom-dwelling molluscs) may be fairly high because resources are imported to the stream from the adjacent streamside zone. Rates of metab-

olism can be high too because the system is well aerated. Water chemistry and volume can vary greatly with seasons, especially in watersheds dominated by winter precipitation with pronounced spring runoff.

Rivers are hugely significant to humans for various reasons. Rivers are used for political boundaries. Rivers provide us with food, directly through fish, and indirectly through irrigation of farmland. Rivers provide energy, harnessed through dams to create hydroelectric power. Rivers provide water for industry and recreational opportunities, and more importantly, for drinking. The *riparian zone,* an area near rivers, often provides a unique habitat for plants and animals that do not inhabit areas further away from rivers.

Rivers begin in mountains or hills, where rainwater or snowmelt collects and forms tiny streams called gullies. Downslope, when one gully or stream merges with another, the smaller stream is known as a tributary. It takes many tributaries to form a river. Rivers carve deep and steep V-shaped valleys in mountains as fast-moving water erodes rock. As rocks are carried downstream, they also act as erosional forces, and are broken into smaller and smaller pieces of sediment. The sediment is finer than sand by the time rivers empty into the ocean. By carving and moving rocks, running water changes the Earth's surface even more than catastrophic events such as earthquakes or volcanoes. Leaving the high elevations of the mountains and hills and entering plains, rivers slow down and the sediment is deposited. Floods spread sediments over broad areas, creating some of the most fertile agricultural areas on Earth.

Rivers are very active parts of the environment. As they meander across the landscape, rivers are important agents in weathering and erosion. In fact, the amount and type of river sediments present provide a means of measuring the amount of **SOIL EROSION** in what could be a large area. River features include oxbow lakes (abandoned river channels), natural levees and deltas. Ancient rivers, some of which flowed from continental glaciers, have left their mark on the type of soil and landforms found in a given region.

Actual figures vary depending on the source, but the Nile is considered to be the longest river in the world (6,670 kilometres), followed by the Amazon (6,404 kilometres), the Yangtze

Yampa River flowing westward from the Rocky Mountains, Colorado, USA. *Source:* Joseph Kerski

(6,378 kilometres) and the Mississippi-Missouri (6,021 kilometres). The Amazon is the largest river in terms of volume, followed by the Congo of Africa. During the high water season, the Amazon's mouth may be 450 kilometres wide and up to 14 billion cubic metres of water flow into the Atlantic Ocean daily. The Amazon's daily freshwater discharge into the Atlantic is enough to supply New York City's freshwater needs for nine years. The force of the current, from sheer water volume alone and virtually no gradient, causes the current to continue flowing 200 kilometres out to sea before mixing with Atlantic salt water. Early sailors could drink freshwater out of the ocean before sighting the South American continent. It is calculated that three million cubic metres of suspended sediment are swept into the ocean each day. Majaro Island, formed from this deposition of silt, is the world's largest river island, about the size of Switzerland.

All rivers are part of a larger ecosystem, the watershed. The watershed is an area of land that contains a common set of rivers that all drain into a larger river, lake or ocean. For example, the Mississippi River watershed is enormous, draining all of the tributaries from more than half of the continental USA into the Gulf of Mexico. Yet, the Mississippi water-

shed is composed of hundreds of smaller watersheds.

The health of rivers reflect the health of the environment. Throughout human history, rivers have been important means of commerce – for goods, people, and communication. Communities along major rivers became important world trading centres, such as Cairo on the NILE River, New Orleans on the MISSISSIPPI River and London on the Thames. Therefore, for a long time, rivers have borne the brunt of human impact.

For centuries, and continuing today in many parts of the world, rivers have been convenient dumping grounds for human and industrial waste, and subsequently, sources of disease and terrible odours. The Cuyahoga River in Ohio USA actually caught fire in 1936, 1952 and 1969 because of oil and other contaminants in it (which fortunately prompted massive attention and cleanup).

Dams built for irrigation or hydroelectric purposes have altered sedimentation, habitat and water chemistry downstream. Some rivers have been straightened with artificial levees, which may help for flood control (although some claim that they make floods fewer but more severe), but these same dams deprive backwaters and floodplains of fertile sediments.

As the importance of river commerce (compared to roads and aviation) waned during the twentieth century, rivers in some cities have become neglected places of crime, visual, water and air pollution. Some of these cities are now converting riverfront land into **URBAN GREENWAYS**.

Not all water reaches rivers through direct runoff. In fact, most water in rivers is from water that has infiltrated into the soil, becoming groundwater. Since rivers drain the land from runoff and infiltration, they collect pollutants, not only from cities, but also from pesticides and fertilisers applied to agricultural fields. This pollution eventually reaches the **OCEAN**.

WEBSITES

Freshwater ecosystems resource from University of Otago, New Zealand: http://telperion.otago.ac.nz/erg/freshwater/ Missouri Botanical Garden's river biome site: http://mbgnet.mobot.org/fresh/ EPA's Surf Your Watershed site: www.epa.gov/surf/

RIVER RESTORATION

What a difference a century makes. Not long ago, rivers, in common with many aspects of nature, were thought of as features to be tamed and controlled to suit human desires and needs. 'Progress' meant to dam, dredge, channelise and straighten rivers – to make them conform to when and where people wanted them to flow. Nowadays, however, sound river management includes a growing interest in restoring rivers to as close to their original conditions as possible. This policy is known as river restoration. The ecological and societal benefits of river restoration are many.

Rivers transport water, sediment and nutrients from the land to the sea. They play an important role in building deltas and beaches, they remove pollutants from overland flows, and they regulate the salinity and fertility of estuaries and coastal zones. Surface water such as rivers and ponds provide drinking water to millions of people. Rivers serve as corridors for migratory birds and fish, and provide habitat, food and protection from predators to many unique and threatened

Case study

Restoring Florida's Kissimmee River

Before 1940 human habitation was sparse within the Kissimmee basin in south Florida USA. Land use consisted primarily of farming, cattle ranching and fishing the area's renowned large-mouth bass. Rapid growth and development following the Second World War was then extensively damaged when a severe hurricane hit the area in 1947. Following public pressure for measures to reduce the threat of flood damage within the Kissimmee basin, the State of Florida and the Federal Government designed a flood-control plan for central and southern Florida. The meandering river and flanking floodplain were channelised during the 1960s and 1970s into a central drainage canal. Levees and dam-like water control structures forced the Kissimmee into a series of five relatively stagnant pools. As a result, some 12,000–14,000 hectares of wetlands were lost.

The Kissimmee River restoration initiative began as a grassroots movement during the latter stages of channelisation, when citizens and members of the environmental community voiced concerns regarding the environmental impacts of the project. Since 1992 the US Congress has authorised funds to restore over 100 square kilometres of river, wetlands and floodplain ecosystems. The restoration project is a joint partnership with the South Florida Water Management District and US Army Corps of Engineers. The restoration project will involve the filling in of one-third of the canal, primarily in non-residential areas, and two of the five water control structures will be removed. Water levels in several lakes will be allowed to rise, so that continuous inflows and a more natural seasonal pattern will be allowed.

Already, sand bars, flowing oxbow lakes, regrowth of shoreline vegetation, returning waterfowl and better water quality are signs of improvement in the river's hydrology and the region's environmental health.

species of plants and animals. In addition, river ecosystems have economic values associated with recreational, commercial and subsistence use of fish and wildlife resources.

Globally, hundreds of thousands of kilometres of river corridors and millions of hectares of river wetlands have been damaged or destroyed. River restoration seeks to return some of these ecosystems to their approximate pre-disturbance conditions. It includes increasing environmental awareness and education, legal reform, and the physical acts of restoring riverbank vegetation, removing impervious surfaces such as car parks, removing or modifying dams and levees, and improving river water quality. From 1982 to 1992 a total of 311,083 hectares of wetlands were gained as a result of restoration activities in the USA alone. Likewise, dozens of kilometres of rivers were restored in the nation's watersheds over the same time period.

WEBSITES

UK River Restoration Centre:
www.therrc.co.uk
US Environmental Protection Agency on
river restoration:
www.epa.gov/owow/wetlands/restore
American Rivers Organization:
www.amrivers.org
Kissimmee River restoration from the South
Florida Water Management District:
www.sfwmd.gov/org/erd/krr/

SEE ALSO

groundwater and aquifers, levees and embankments

SALINISATION

Salinisation involves an accumulation of salts (for example sodium, calcium and magnesium) at or close to the surface of soil. The amount of land affected by salinisation is increasing. Currently it affects 7 per cent of the world's land area, especially in the more arid or semi-arid zones. In the most extreme circumstances salinisation leads to the formation of a white crust on the ground surface. The vast majority of plants cannot cope with a salty environment – they simply fail and die because the salt is toxic to them. Moreover, the presence of the salt crystals limits the amount of nutrients and air available to the soil.

Salinisation occurs naturally in hot, dry zones where evaporation and transpiration exceed precipitation. In these conditions water from underground moves up through the soil by capillary action. Salts dissolved from the bedrock are present in this soil water, often in considerable quantities. When water reaches the surface and evaporates, the salts are left behind as crystals on the ground surface and in the upper soil pores. This whole process is more likely to happen if the water table is relatively high because the water does not have very far to rise.

Human activity can exacerbate the problem of salinisation. This is especially true in marginal farming areas where irrigation has been employed to try to improve agricultural yields. If too much water is applied to crops, it dissolves the salts and then deposits them when evaporation happens. The western San Joaquin Valley in California suffers in this way. This agriculturally rich zone is now finding its soils becoming more and more salty. Here, the high water table resulting from irrigation prevents salts being leached away and salts become concentrated in the upper part of the soil. New research is being carried out here and elsewhere to try to make plants more resistant to a saltier environment, though perhaps a reduction in the volume of irrigation might be a better alternative.

Up to 40 per cent of Australia's soils are affected by salinisation. Much is a result of natural processes, but the most salinised areas are in the states of South Australia and Victoria, where a large percentage of land is cultivated using irrigation systems. Desert countries such as Syria are almost compelled to employ irrigation methods to be able to produce any crops at all. Here, 50 per cent of the land is affected by salinisation. In Uzbekistan, 80 per cent of soils are salt-damaged.

Solutions to existing salinisation tend to be both complex and expensive, requiring farmers to adapt to higher levels of saltiness rather than attempting to reverse the situation. Lowering the water table by constructing underground drainage systems is one possible – but expensive – approach that has been employed in the Nile Valley. Changing land use is a cheaper alternative. For example, land can be used for livestock rather than arable production because grasses are often more halophytic (salt tolerant) than other more specialised crops. However, this solution leads to other environmental concerns associated with livestock, such as increased soil erosion due to OVERGRAZING. Changing land use from cultivation to grazing also raises human development concerns because more people can be fed from a hectare of crops than one cow on a hectare of land. Alternatively, some crops such as cotton do grow adequately in slightly salty soils. In more temperate climates, barley and sugar beet have proved quite tolerant of such conditions.

The Australian Government named the 1990s the 'Landcare Decade'. This arose from a popular movement to take better care of the land resource. Technical programmes, scientific research and educational programmes were all directed towards solving issues of soil erosion and degradation of all types.

WEBSITE

The Aral Sea:
www.dfd.dlr.de/app/land/aralsee

SEE ALSO

desalination

SEA-LEVEL RISE

Many people are concerned about the effects of sea-level rise that is a consequence of global warming. Coastal regions are increasingly under threat of inundation as ageing sea defences become less able to cope with higher water levels and their maintenance and extension become simply too expensive.

Sea-level change is nothing new. There is plenty of evidence of past rises and falls in sea level at various times in the Earth's history caused by natural shifts in climates. During the last Ice Age (2 million–10,000 years ago), the climate of much of the northern hemisphere fluctuated markedly between periods of warmth and periods of intense cold. During the cold glacial periods, water fell as snow and accumulated on the ground to form huge ice sheets and glaciers. With less water flowing

into the oceans, sea levels dropped. When the climate warmed up during inter-glacial periods, meltwater poured into the oceans causing sea levels to rise. These relatively rapid changes in the amount of water in the oceans are called eustatic changes.

In addition to eustatic change, there is a second, much slower response to glaciation called isostatic change. This involves the changes to the land as it is compressed when covered by great thicknesses of ice, or rises (a kind of 'breathing out') when the weight of ice is removed. After the last ice advance, the immediate effect of the melting ice about 10,000 years ago was to cause a eustatic rise in sea level. This has been balanced by a slower isostatic recovery as the land has risen, so causing sea levels to fall.

Several coastal landforms exist as evidence of sea-level change. Chesil Beach in Dorset, UK, was formed as the post-glacial rising sea swept glacial meltwater deposits (from what is today the sea bed) onshore. This shingle bank, 29 kilometres long and several metres high, is now a tombolo – a bar linking the Isle of Portland to the mainland. In the south-west peninsula of England, the counties of Devon and Cornwall show the effects of the flooding of river estuaries by these rising seas, widening them out into broad inlets (called rias) forming ideal harbours. When sea levels fall, coastal features such as cliffs and beaches become marooned high above current sea levels. Such features are commonplace in western Scotland and in Antrim, Northern Ireland.

Current concerns about sea level change are not, however, associated with natural climate change. They are associated with global warming brought about – many would argue – by human actions such as the burning of fossil fuels. As temperatures rise, so ice in the form of ice sheets and glaciers melt, adding water to the oceans and causing sea levels to rise. A lesser-known impact is that the water already in the oceans expands with the increase in temperature, and this contributes at least as much to rising sea levels as does ice melt – often more.

Much debate exists about the amount of sea-level rise and the likely impact it will have on coastal communities. A certain amount of scaremongering exists, with a wide variety of statistics regularly presented to the public.

The Intergovernmental Panel on Climate Change (the IPCC) published its first predictions in 1990, estimating global warming of 3°C by 2100 – this figure will almost certainly be altered because it was based on an increase of greenhouse gases at the rate current at that time, which is likely to be proved inaccurate.

Some countries, including the UK, are trying to decrease their emissions of toxic and damaging substances by signing up to agreements such as the Kyoto Protocol. Others, including the USA, are proving more reluctant to go in this direction. On the other side of the argument are those who point to the health benefits of a warmer climate and the expansion of areas suitable for agriculture.

A consequence of ice melting is that environments are changing. Canadian polar bears are suffering severely from the heat, for which their systems are certainly not designed. The treeline in Scandinavia and in Canada has been moving north as the climate warms. Some measures against the impact of sea flooding are already in place – the Thames Barrier keeps central London protected from storm surges from the sea. However, the inevitability of a rising sea level will place pressure on such systems. Some nations are directly in the firing line. Pacific and Indian Ocean island groups are especially under threat from rising sea levels, and most of Bangladesh is below 10 metres above sea level.

Most scientists believe that a rise of a few centimetres by 2050 is possible and coastal managers are certainly planning for such an increase in their schemes for coastal defence. As sea levels rise, the potential costs of enlarging existing sea defences and constructing new ones rise exponentially. Governments have already realised that they simply will not be able to afford to keep up. Managed retreat, a relatively recent concept of coastal management, is therefore becoming an increasingly attractive option, allowing areas of low-lying low-value land to be gradually flooded by the sea. These areas naturally become salt marshes, which in time become a barrier to rising sea levels, as silt accumulates and vegetation colonises the area.

Some scientists say that the major concern for the future will be increased storminess – low frequency, high magnitude events capable of destroying coastal defences and flooding coastal communities. We can only wait and see.

Further reading

Warburton, P., 1996, *Atmospheric Processes and Human Influence*, Collins.

WEBSITES

www.grantchronicles.com/astro45.htm

www.massclimateaction.org/
AntarcticNYT030703.htm

www.southsloughestuary.org/EFS/sealev_3.htm

SEE ALSO

climate change, oceans

SEX RATIO

Implicit in the usual age pyramid of a population is the sex ratio, the proportion of males per 100 females. For people, about 106 males are born for every 100 females, leading to an at-birth sex ratio of 1.06. Males have a slightly greater likelihood of dying in childhood than females, and by the time of teenage years, roughly the same numbers of males and females exist. At age 50 and increasingly onward, more females exist, because they have a longer life expectancy in most societies. The USA sex ratio for people aged 85 and older was 0.42 in 1998. The Earth's sex ratio ranges from 1.14 in Guam to 1.0 in

Liechtenstein. China's ratio of 1.09 is in part due to the selective abortion of more girls than boys.

Underlying the sex ratio is some system of sex determination in the development of an individual. The system of sex determination in humans is the most common system in both animals and plants, the so-called XY system. The fruit fly, *Drosophila melanogaster*, has a particularly simple set of animal chromosomes. Both sexes have three pairs of autosomes (non-sex chromosomes – two pairs of long chromosomes and one pair of dot-like chromosomes), plus one pair of sex chromosomes. The female has two large sex chromosomes, so-called 'X chromosomes'; the male has one X chromosome and a somewhat smaller, 'J-shaped' Y chromosome. In the process of meiosis, the paired chromosomes undergo 'reduction-division' and diploid (2N) adults produce haploid (N) nuclei for sex cells. In the first (reduction) phase of meiosis, one of each pair of chromosomes migrates to a daughter nucleus. That means that daughter nuclei produced by females and destined for egg cell nuclei each get an X chromosome, whereas daughter nuclei produced by males and destined for sperm nuclei may get an X or may get a Y.

The familiar XY system is not the only system of sex determination known. In some organisms (grasshoppers, for example), there is no Y chromosome. Females are XX and males are XO. In birds, butterflies and moths, females are heterogametic (XY or XO) and males are homogametic (XX). That is, males develop from zygotes produced by two gametes that are the same with respect to sex chromosomes.

In the social insects such as ants and bees, there are no sex chromosomes. Females (queens and workers) are produced from fertilised eggs; males (drones) develop from unfertilised eggs. Drones have no fathers (although they do have maternal grandfathers!). The diet of the developing larva determines whether a fertilised egg will develop into a sterile worker or a fertile queen. Royal jelly produces a queen.

In some species, environmental conditions actually determine sex. A particular reef-dwelling fish, *Labroides dimidiatus*, lives in a harem, a male tending several females. If the male dies or is removed, the lead female takes over the male's role in courtship and within a few days begins to produce viable sperm. Other such variations on the general theme of sex-determination help to underscore the importance and 'expense' of sexuality.

To repeat, the most common systems of sex-determination are XY or XO types, which produce equal numbers of females and males. However, such a sex ratio may not be particularly advantageous at the time of reproduction. The 1:1 sex ratio is adapted to monogamy (one mate) but not to polygamy (plural mates), and polygamy is the more common situation in nature. Polygamy is of two kinds. Polygyny (plural females) involves more than one female per male and polyandry (plural males) involves more than one male per female.

If we consider the relative sizes of gametes, egg cells typically are hundreds to millions of times larger than sperm cells (and remember that diversity of offspring is the 'reason' behind sex). Males generate diverse gametes cheaper than females do. So one would expect polygynous systems – systems in which males service more than one female – to predominate in nature, and that is the case. Polygynous systems may involve harem formation (as in many large mammals, such as elk and deer), lek behaviour (where males compete in a social courtship display, as in many grouse-like birds), or the 'catch-as-catch-can' promiscuity of some small rodents. Whatever the case, polygynous populations are adapted to rapid growth with maximal genetic diversity at minimal energetic expense, because they need not support as many males as monogamous or polyandrous systems.

If the XY or XO systems of sex determination yield 1:1 sex ratios, yet a majority of species is polygynous, you probably would predict that many populations face the problem of excessive numbers of males. This frequently seems to be the case. The rancher makes steers or geldings out of excess male cattle and horses, respectively. Game managers often set 'antlered only' seasons to weed out extra males from populations of deer or elk. With excess males removed from the population, more resources are available for females and young and the population can increase more swiftly.

What humans do deliberately and con-sciously is 'built in' to the natural functions of many other populations. In harem-forming species such as the elk (or wapiti), many males do not participate in breeding and instead form bachelor herds. In some ground squirrels and many other rodents, adult males direct more aggression toward male offspring than toward female offspring. Males thus are more likely to disperse. While dispersing, those males are beyond the usual protection of the colony or may be out of an ideal habitat – in corridors between patches of suitable habitat, for example. Hence, males are more susceptible to preda-tion than young females who stay home in the natal colony.

Coastline on North Island, New Zealand.
Source: Joseph Kerski

WEBSITES

Human sex ratios for countries:
www.nationmaster.com/graph-T/
peo_sex_rat_at_bir
Detailed discussion entitled: *Simple Math: Sex Ratios:*
http://econ.ucsd.edu/~jsobel/87f04/ratio.pdf
Example of environmental sex determination in turtles:
www.evolutionary-ecology.com/sample/
iiar1713.pdf
Research on sex ratio in mammals:
http://westgroup.icapb.ed.ac.uk/pdf/
Sheldon&West_04.pdf

SHORELINES

Shorelines (often called the 'littoral biome') represent an **ECOTONE** between land and sea, with characteristics of both realms, including all of the challenges of both. One could argue that shorelines are the most demanding on Earth. The challenge of land is desiccation, or 'drying out'; the challenge of the sea (at least for freshwater and land life) is salinity and the physical force of waves.

Shorelines are continuous around Earth's ocean basins, so they have tremendous linear extent but are usually quite narrow. Their width is defined by tidal flux – the zone between the lowest low tide and the highest high tide (and slightly beyond, in the 'spray zone'). The littoral biome consists of two quite different communi-

ties, depending on substrate: sandy beaches ver-sus rocky shores (headlands).

Sandy shores tend to be much less biologi-cally productive than rocky shores. Unless sta-bilised by rooted vegetation in a salt marsh or coastal swamp (such as mangrove; see below), a beach is too unstable in the face of wind and surf to allow growth of producers. Some ani-mals, such as molluscs and various kinds of worms, make a living below the dynamic sur-face. They live mostly as filter-feeders on organic matter imported from the neritic biome of the **CONTINENTAL SHELF**.

Rocky shores provide a firm substrate, allowing producers (algae, such as kelp, for example) and consumers to stake a claim for permanent residence by resisting wave action. Many forms of life, such as limpets, snails, bar-nacles and mussels, prevent desiccation by retracting into a shell at low tide. Algae may be protected by a sticky (mucilaginous) coat-ing. Biologic productivity of kelp beds is the highest of any littoral community, sometimes as high as productive grasslands or temperate forests.

From the chalk cliffs of southern England, to the sandy shores of the Cayman Islands, to the peaks of the Tierra del Fuego in South America, coastlines are some of the most beloved environments in the world, with great beauty and diversity. They take on forms such as high cliffs, sandy beaches, deltas, lava flows, rocks, estuaries, mangroves, salt flats, tide pools and barrier islands, to name a few. Shorelines are also some of the most vulnera-ble environments. They are the sites of many of the world's major cities, with associated

Mangrove swamps

The shrimp on your dinner plate may be contributing to coastal erosion in Vietnam. In South-East Asia and Latin America, mangroves have been cleared for coastal mariculture, especially shrimp farming, with unintended consequences. Much of the inshore fishery is dependent on mangroves and other coastal wetlands as a 'nursery ground'. Mangroves not only protect young fish from predators; they protect coastlines from the storm surge produced by hurricanes or typhoons.

Mangrove swamps form distinctive intertidal littoral (shoreline) or estuarine ecosystems. Mangroves are so unique that some believe that mangroves should be classified as their own biome, rather than part of the shoreline biome. Mangroves occur worldwide on tropical and subtropical shorelines, but in the aggregate cover less than 1 per cent of the area of the globe. More than 50 species of mangroves exist worldwide. Mangroves are among the most productive of estuarine and intertidal environments, with biologic production in the order of 1100 g/m^2/yr.

The term 'mangrove' has two different meanings. First, it is a type of biotic community. Second, it is a form of plant. Mangrove does not refer to a particular species of tree. Rather, mangroves are classified in about a dozen genera in eight different families. Depending on geography and ecological conditions, mangroves of several families participate in the community, variously adapted to life in saltwater. Mangrove swamps of the tropical western Pacific Ocean have greater species richness than do those in tropical America.

Some mangroves bear fruits that are dispersed by saltwater. Others are actually viviparous ('live-bearers'); their seeds germinating in the mature flower (attached to the mother plant), forming roots and shoots. The 'plantlets' eventually drop into the water to root beside the parent. The eventual result is a dense network of trees and shrubs that serve to stabilise the shoreline.

Mangrove swamps tend to be zoned by species. In Florida, for example, red mangrove occurs on the seaward side of the swamp, supported by prop roots that form a network that captures silt and nutrients. Toward the landward side is black mangrove with its characteristic pneumatophores – these are roots that breathe. Pneumatophores are roots that are some form of upward appendage or extension of the underground root system. Because these roots are exposed for at least part of the day and not submerged underwater, the root system can obtain oxygen in an otherwise low-oxygen environment. Further inland is a zone of a third mangrove, buttonwood. Buttonwood, named because its flowers resemble little buttons, is not very tolerant of salt, but can handle low nutrients.

The story of human impact on mangroves shares a similar sad tale to the ways other wetlands have been treated over the centuries. Drainage, clearance for fishing, and pollution have all taken their toll. Furthermore, rising sea levels due to global climate change cause further damage through increasingly severe storm surges.

pollution and land use pressure. Many coastlines have seen much increase in coastal erosion since the mid-1900s. Coastlines have also been soiled by oil and other hazardous material spills.

WEBSITES

Shoreline biome resources from the Missouri Botanical Garden:

http://mbgnet.mobot.org/salt/sandy/indexfr.htm

The Mangrove Action Project, dedicated to understanding and protecting mangroves worldwide:

www.earthisland.org/map/

Mangrove Learning Center:

http://216.156.75.137/mangrovewa.html

SOIL EROSION AND CONSERVATION

People usually think of erosion as something that happens along streams as the water scours the banks, carrying the soil downstream. This is certainly a major form of erosion, but there are others as well. Wind, for example, can lead to erosion as topsoil is blown off open, dry land. On a grander scale, erosion can be caused by ice as glaciers scour the landscape. Much of the mid-western part of North America and Northern Europe was scoured by ice tens of thousands of years ago. Glaciers formed glacial plains, but also hills and ridges known as drumlins, moraines and eskers. In the 'Finger Lakes' region of New York State in the USA, each lake is the result of glacial erosion.

These different forms of erosion all have the common element of relocating rocks and soil as a result of forces such as water, wind or ice. Some erosion is beyond human control. For example, stopping a glacier is not a reasonable option. However, other forms of erosion are under our control, and steps can be taken to maximise soil conservation and minimise environmental damage. Along river banks, planting trees can be used to anchor the soil in place, much as is done by farmers on open land. Also, by restoring natural, meandering river flows (see RIVERS), the speed of the water is reduced because it is not being channelled into a narrower path. This reduction in the speed and volume of water passing a given location helps to reduce the scouring effect on river banks.

Erosion poses several environmental challenges. For agriculture, loss of topsoil reduces the land's ability to support crop growth. If fewer and less healthy crops are grown, it becomes increasingly hard to meet a region's food demands. This is particularly true in developing nations where the need is already great – loss of topsoil makes a difficult situation worse. Erosion is usually exacerbated by deforestation, because when forest cover is removed, soil that was held in place by the tree roots washes into rivers and seas.

Erosion can also be linked to several water quality problems. The added sediment in the water changes its chemistry, which in turn can

Case study

Deforestation and soil erosion in Madagascar

Madagascar, located 400 kilometres east of Africa, is the world's fourth largest island. Its isolation from the African continent for millions of years has preserved and produced thousands of species of plants and animals found nowhere else on the Earth. Of the 10,000 species of plants catalogued on the island – and the list is still growing – 8000 of them are endemic (unique to the area). The density of endemic plants is such that some individual mountaintops have 150–200 endemic plants.

The Indian Ocean near Madagascar is red stained as soil is washed out to sea. NASA

This large number of unique species makes deforestation a great threat to global biodiversity. Deforestation on the island began with colonisation in the late 1800s, but prior to 1950 the inroads into the island's forest ecosystems were generally small. As more and more of the island's best farmland was brought into coffee production for export, the residents of the island have had to clear more forest for their personal agricultural needs. From 1950 to 1985 one half of Madagascar's forests disappeared. By 1985 only 34 per cent of the original forests existing in Madagascar remained. As the island is one of the poorest countries in the world, citizens often must resort to exploitation of their natural resources to find income. However, over time losses of topsoil of this magnitude will exacerbate already pressing environmental challenges.

affect what species live in the water. In agricultural regions, where the run-off from fields includes fertilisers, the changes can be dramatic. More subtle, but no less important, is the problem of soil darkening the water, which increases the amount of sunlight it absorbs. This in turn increases the temperature of the water, altering the habitat and the aquatic species it can support.

Because of these problems, it is very important to take steps to minimise erosion and promote soil conservation. In large industrialised nations, such as the USA, this is carried out by the Natural Resources Conservation Service (formerly the Soil Conservation Service). In developing nations, programmes such as the United Nations Division for Sustainable Development also make soil conservation a priority, since topsoil loss reduces agricultural productivity and sediments in the water make it more difficult to provide clean drinking water. To minimise wind erosion, farmers often plant groundcover on land lying fallow. The roots of these plants help to anchor the soil and minimise erosion from wind and water. One study in Ivory Coast, Africa, found that uncultivated land lost more than 50 per cent more topsoil through erosion than did similar land that had cultivated crops on it. CONTOUR PLOUGHING, which follows the land's natural terrain, helps minimise water-caused erosion on fields. In any nation, high-quality soil is an essential part of a healthy living environment.

Further reading

Tarbuck, E. J., and Lutgens, F. K., 2000, *Earth Science*, 9th edn, Prentice-Hall.

WEBSITES

R. Butler, 'Erosion and its effects':
www.mongabay.com/0903.htm.
US Natural Resources Conservation Service:
www.nrcs.usda.gov
New Zealand's Seafriends Organization:
www.seafriends.org.nz/enviro/soil/erosion.htm
'Landcare field guide' from Australia's Department of Natural Resources and Environment:
www.netc.net.au/enviro/fguide/soil1.html

Deforestation in Madagascar from American University:
www.american.edu/projects/mandala/TED/madagas.htm

SEE ALSO

deforestation

SOILS

Soil can loosely be defined as 'the stuff that plants grow in' – difficult to describe but crucial to life. It is a thin layer of loose material composed of disintegrated rock particles, humus (dead organic matter), water and air. It clothes the planet and provides the basis for our survival – the grasslands on which animals graze and the material in which crops grow. Hence soil plays a vital role in agriculture and is an important determining factor in the nature of agriculture in a location. The soil is also farmers' single most important natural resource – they must look after the soil if they are to be successful.

The first stage in soil formation is the weathering of rock – the so-called parent material. This inorganic mineral matter, called regolith, only becomes a true soil when rotted vegetation increases the organic matter. The soil is rather like the filling in a sandwich, between the underlying rock below and the vegetation on top. Over many hundreds or even thousands of years a fully developed soil is formed, with distinctive layers or horizons evident. In the UK and much of North America the mature soils have largely developed since the end of the last ice advance, some 8000 years ago. Soils develop most rapidly in tropical regions, where the high rainfall and temperatures promote rapid chemical weathering of the parent material and encourage lush vegetation and rapid plant decay.

Globally, different soil types called zonal soils can be identified – each has developed in response to the climate and vegetation in an area. Similar climates have similar soils, for example brown earth soils form in areas with a humid temperate climate. At a more local scale, rock type, drainage and relief all have a role to play in the creation of what are called azonal soils. One example of an azonal soil is a 'raw soil', which is a very thin and immature soil that develops on calcareous rocks such as chalk.

Soil types

Soil type	Characteristics	Location	Climate	Vegetation	Impact of human activities
Podsol	A thick layer of humus at the surface. The soil is acid, pH about 4, because of the acid vegetation and little decomposition due to the cold temperatures. Below this lies a bleached grey horizon. The grey colour is a result of the iron and aluminium being washed out due to high rainfall causing leaching.	Between 55 and 80° north of the equator.	Typically cool and wet but also typical of cold climates in northern Canada and Europe. Temperatures range between −5°C to +15°C and rainfall may exceed 1000 mm per year, more typically less than 750 mm.	Boreal forests – the northern coniferous forests with acid vegetation.	Podsols are of limited agricultural value due to their lack of fertility. Most podsol environments are used for sheep grazing and forestry in coniferous plantations; they are also popular environments for wildlife and tourism. Current issues include acid rain damage, afforestation and tourism management.
Brown earth	Well-drained deep fertile soils that are only slightly acid: pH between 5 and 6.5. The upper layer is rich in humus and well mixed by earthworms. There is little leaching, so high levels of fertility.	Between 30 and 55° north of the equator. Extensive in Western Europe, the east coast of the USA, parts of Central Europe and Asia.	Regions with a humid temperate climate with few extremes of rainfall or temperature – rainfall totals below 800 mm per year. Temperatures average 4–17°C.	Deciduous woodland and grassland vegetation.	Demand for its fertility has seen large areas of original woodland cleared for agriculture and settlement. Good yields are possible, with only limited use of fertilisers or drainage. Current issues include damage by acid rainfall, soil erosion as a result of intensive farming, afforestation with conifers that threaten to change the soils from brown earths into podsols, and urbanisation – all leading to the total loss of the soil to the ecosystem.

Soil types *continued*

Soil type	Characteristics	Location	Climate	Vegetation	Impact of human activities
Chernozem	Sometimes called the black earth: a fine, fertile black or dark brown soil developed on loess, a wind-blown glacial deposit. High concentrations of humus and calcium make it rich in plant foods. Well mixed by earthworms.	Extensive in southern Russia and parts of Hungary and Romania, as well as a similar soil belt from Canada to Texas.	Light rainfall totals, often below 500 mm but high evaporation rates so only mild leaching. Temperatures may range from −20°C to +20°C.	Naturally occuring temperate grasslands such as the steppes, prairies and pampas. When the grasses die back in winter they add large amounts of organic matter to the soil.	Large areas have now been utilised for commercial agriculture, both crop growing and animal rearing. Current issues include soil erosion and the threat from urbanisation.
Latosol	Very deep soil, often over 10 metres, because of intense weathering in the hot, wet climate. There is heavy leaching due to high rainfall totals so any nutrients in the soil are quickly washed out. Silica is the main element washed out, leaving iron and aluminium behind, giving the latosol its name and strong red colour. It is low in organic matter because it is rapidly decayed.	The humid tropics, including the Amazon Basin, the Congo Basin and parts of Indonesia and Malaysia.	Rainfall is plentiful and well distributed, often with over 2000 mm per year. Temperatures are hot with an average of about 27°C.	Rainforests: huge trees called emergents with a canopy, shrub and ground layers below. The lush dense nature of the vegetation suggests fertile soils but in fact about 90 per cent of the nutrients are held in the trees themselves.	Rainforest soils are very much at risk from human activities of deforestation such as plantation agriculture, ranching, timber extraction, mining and slash-and-burn farming. The soils are fragile and once the trees are removed soil erosion occurs, with the secondary effects of gulleying, landslides and increased threat of flooding.

Without careful management, soils can easily be degraded. Overcultivation and over-grazing can lead to soils becoming exhausted – losing their fertility and structure. Soil erosion may ensue, by wind and water. The use of chemical powdered fertiliser can result in soils becoming dry and friable, also making them vulnerable to erosion. During the 1930s in the USA the infamous 'Dust Bowl' resulted from poor soil management and massive wind erosion. Ploughing up and down slopes can increase surface run-off and result in enormous quantities of fertile topsoil simply being washed away to the sea.

The sustainable management of soils is a vital aim for all societies, rich and poor. Soil conservation measures include contour ploughing, the use of natural fertilisers, the promotion of less intensive organic farming and the careful management of stock levels.

WEBSITES

AGRIFOR UK's soil information:
http://agrifor.ac.uk/browse/cabi/5684eded
5c836f0b3868ea280538475b.html
UKOnline soil information:
http://web.ukonline.co.uk/fred.moor/soil/
formed/f0108.htm
World Association of Soil and Water
Conservation:
www.landhusbandry.cwc.net/abwaswc.htm

SEE ALSO

contour ploughing, soil erosion and conservation

SOLAR POWER

On a bright, sunny day the Sun shines approximately 1000 watts of energy per square metre of the planet's surface. This means that in the space of just 30 minutes the Earth receives the equivalent of all the power used by humans in one year. These remarkable facts explain why the thought of harnessing the energy from the Sun has been the dream of scientists for many years.

In 1839 Edmond Becquerel discovered the concept known as the photovoltaic effect, yet it was not until 1954 that the first solar cell was created. Subsequent research meant that by the 1960s solar panels had been invented that could convert the solar energy into domestic usage. (Modern photovoltaic cells convert solar radiation into electricity via a simple chemical reaction in the silica-based cells). Soon, estimates were predicting that domestic fuel bills could be reduced by 50 per cent if solar panels were installed on house roofs. The solar panels also received the support of environmentalists who showed that during its lifetime a single photovoltaic cell will prevent 30 kilograms of carbon dioxide and 88 grams of sulphur dioxide being released into the troposphere. For a while, it looked as if solar power would take off, using the greatest and cheapest source of energy available to humans. But the solar boom never happened, largely because of the sheer cost of installing the panels and the relatively long payback time – the length of time before costs were recovered. Once again, economic factors seemed more important than environmental ones.

Nevertheless, several successful solar power schemes have been developed in Europe and the USA. For example, all 70 houses in Nieuw-Sloten, a suburb of Amsterdam, the Netherlands, have their energy demands met by solar panels. Solar energy was also used to heat the water for the swimming events at the 1996 Atlanta Olympic Games – perhaps the winners of these events should have been awarded green medals, not gold medals.

Solar power in the USA
Many states in the USA receive high amounts of solar radiation. Texas alone receives enough solar energy to supply one and a half times the world's current annual energy consumption – that is if the entire state, all 680,000 square kilometres, was covered with solar cells! The potential has been recognised by the US Government, which in the past 25 years has invested $1.5 billion into solar research. In 1997 President Clinton announced the 'Million Solar Roofs Initiative' which would see a million solar systems installed across the country by 2010.

Some of the world's largest solar panels are situated in the Mojave Desert in Southern California. Built between 1984 and 1991, they generate enough power for half a million people and displace the need for 2 million barrels of oil a year.

Research over the last few decades has shown that there are other ways of harnessing solar energy. These include photothermal methods, whereby solar radiation directly heats water, usually in heating systems. Mirrors are also used to concentrate the Sun's rays on to water-filled, highly absorbent, black-coloured pipes. Another form is passive solar energy, in which buildings with large glass windows capture and store solar energy – a similar concept to a greenhouse. Research is also taking place into harnessing warm water in tropical areas where water has been heated by the incoming solar radiation. Ocean thermal energy conversion, or OTEC, directs warm water under a container holding liquids with low boiling points, such as ammonia. The vapour that is produced is then directed into a turbine to generate power.

Even more lavish schemes have been suggested, including a geostationary solar power satellite. This would involve a huge panel up to 10 kilometres long by almost 3.5 kilometres wide that would orbit the Earth and directly absorb solar radiation for 321 days of the year. It would then beam energy down to the Earth via a giant microwave antenna. However, the cost would be astronomical and there would be a huge risk of either the panel falling out of control back to the Earth, or the microwave beams being used by a terrorist organisation.

At present, harnessing solar power for the bulk of the world's energy needs remains the stuff of science fiction. But this will not deter scientists in the coming years from trying to devise schemes that utilise the largest source of free energy, and the one on which all life on Earth ultimately depends.

WEBSITES

www.solarelectricpower.org

www.ases.org

www.bpsolar.com

www.eia.doe.gov/kids/renewable/solar.html

SEE ALSO

renewable energy

SOLID WASTE MANAGEMENT

People all over the world produce waste from worn out and discarded objects, packaging and other sources. Ultimately, this waste has to go somewhere. In the past people have burned their waste, dumped it in the ocean, or buried it in holes in the ground. We now know that these practices lead to environmental problems, including pollution, contamination of soil and water, infestation of vermin such as rats, and other problems.

To avoid these problems, a number of strategies have been implemented to manage solid waste more effectively. One of the best strategies is to reduce the amount of waste in the first place. By reducing consumption, reusing materials instead of throwing them away, and recycling, it is possible to reduce considerably the total volume of material that must be handled as waste. However, these approaches will not completely eliminate waste, so environmentally responsible solutions need to be put into place. Two of the current preferred options are sanitary landfills and incinerators.

Created to replace old-fashioned 'dumps', modern-day sanitary landfills are regulated to prevent contamination of soil and groundwater. A new landfill is first lined with plastic and clay to prevent seepage. Drains are installed to collect and carry away leachate (water containing dissolved chemicals) for treatment. Refuse is alternated with layers of earth to prevent the wind from blowing the wastes around and to discourage foraging animals. When an area is filled, it is covered with clay and earth, and then it is landscaped, often to be used for farming, since plants help avoid erosion.

Covering the waste in this manner helps to contain it, but it also prevents oxygen from reaching the waste, since oxygen promotes decomposition. Slower decomposition means that it takes longer for the waste to break down and less room is available for additional waste. Also, without access to oxygen, decomposition is anaerobic and creates a by-product called biogas. This highly flammable gas is a combination of methane, hydrogen and carbon dioxide. To prevent possible explosions in nearby homes caused by methane seepage, gas wells can be built to collect the methane. The collected methane can then be released more safely into the atmosphere, or even burned to generate electricity.

The major alternative in the solid waste management field – incineration – involves the burning of wastes. Incineration has real benefits, including a reduction in the volume

Case study

The Garbage Project

The Garbage Project, founded in 1973 by William Rathje at the University of Arizona, has studied patterns in the waste people produce, as measured by the contents of old dumps and more modern landfills. As a result of its research, facts about how waste decomposes have been discovered based on empirical evidence – not just someone's opinion. For example, a surprising conclusion is that organic material may not break down in a landfill if there is insufficient oxygen. The Garbage Project dug up foods such as hot dogs and pastries that were buried for as long as 15 years, but were still recognisable. Many newspapers were readable; in fact, they were used to verify when the waste was buried, and grass clippings were often still green. Biodegradable waste made up more than 56 per cent of the waste in the landfills, by volume, and changed little, even after being buried for more than ten years. Paper and cardboard made up more than 50 per cent of the volume of waste in a landfill, domestic waste averaged about 5 per cent, and food about 1 per cent.

A number of more specialised studies related to solid waste (or 'garbage') have been conducted, such as a study of baby food preferences based on the foods discarded in landfills. While the study of garbage is certainly an unusual (and messy!) field, careful study of our solid waste can give us important perspectives on our culture and environmental practices.

of wastes (by 80–90 per cent), the ability to neutralise some hazardous wastes, and the ability to produce electricity. While these are undoubtedly benefits to the environment, there are drawbacks as well. Incineration can pose a threat to human health by releasing metal particulates into the air, which over time can cause birth defects or cancer. Also, while the total volume of waste is considerably less, incineration creates ash, which has to be put somewhere. Since the ash may contain toxins, it is considered a hazardous waste. In addition, incinerators are expensive to build and operate since they must be fitted with expensive scrubbers to clean the smoke before release.

Both landfills and incinerators share one problem – where they should be located. Few people want waste processing in their neighborhood – the 'not in my backyard' (NIMBY) refrain is often heard when a new solid waste management facility is proposed. Unfortunately, in today's crowded, industrialised nations, there are few options for a facility that does not intrude on someone's space.

While today's approaches to solid waste management are certainly better than the old days when rubbish was literally thrown out the window, problems remain. Good environmental stewardship requires that we do what we can to keep things from becoming waste in the first place.

WEBSITES

United States Environmental Protection Agency's municipal solid waste website: www.epa.gov/epaoswer/non-hw/muncpl

Zero Waste's landfill information: www.zerowasteamerica.org/Landfills.htm

Environmental Literacy's landfill page: www.enviroliteracy.org/article.php/63.html

'How landfills work', from the How Stuff Works Site: www.howstuffworks.com/landfill.htm

Nottinghamshire County Council (UK): www.nottinghamshire.gov.uk/home/environment/recycling/factsheets/landfill.htm

SEE ALSO

incineration, landfill

SPECIES AND ORGANISMS

Species are the fundamental units of biological diversity. A species is a group of interbreeding or potentially interbreeding individuals that are reproductively isolated from other such groups. Species are thus defined by reproductive continuity and distinguished from other species by reproductive discontinuity.

Species are the different kinds of organisms. They are grouped in genera, and designated by two-part Latin names such as *Homo sapiens*, or 'wise human', the name of our own species. We are the only living species of our genus, *Homo*, though at least two other species existed in the past. On the other hand, there are dozens of species of the oak genus, *Quercus* in North America, as different as the California live oak, *Quercus agrifolia*, and the eastern white oak, *Quercus alba*. Species are not generally capable of interbreeding, or the hybrids between them are sterile, or are only rarely formed in nature, whereas the individuals of a given species can, of course, cross successfully with one another and produce fertile offspring.

Scientific classification refers to how biologists group and categorise extinct and living species of organisms. The groupings (taxonomy) from most general to most specific are: kingdom, phylum (animals) or division (plants), class, order, family, genus, and species.

Local **POPULATIONS** represent subsets of species – all of the individuals of a particular species at a particular time and place. Species populations may be organised as 'metapopulations', with subpopulations localised in patches of suitable habitat and linked together by immigration. For many purposes, biologists use the terms 'species' and 'population' interchangeably.

Species are groups of individuals that look alike; that is, the concept of species is 'morphological'. Because of their reproductive continuity, members of a particular species share similar genes (genotypes), so it is hardly surprising that they tend to be similar in appearance (phenotype). However, the morphological species concept is sometimes inadequate. There are species, for example, in which there are several different shapes, colours or patterns within a reproductive unit, or even in the same batch of young. In the tassel-eared Abert's squirrel of the Southern Rocky Mountains and Colorado Plateau of the USA, for example, some individuals are black, a few are brown and some are salt-and-pepper grey with white bellies.

Other species vary widely over their geographic ranges. Geographic variation in the human species in such features as skin pigmentation, facial details, distribution and tex-

Aspen tree in the Colorado Rocky Mountains, USA.
Source: Joseph Kerski

ture of hair, or bodily proportions is well known. Our own variation is probably overemphasised, as it is fairly trivial compared with that of some other species. Similarly, the amount of sexual dimorphism – structural difference between sexes – in humans is minor, relative to elephant seals, weasels and nearly all birds. Even the degree of maturational change is slight in contrast with many other vertebrates, let alone with such organisms as beetles or butterflies.

Another frequent problem with the morphological species concept is similarity between species. Such similarity may be due to close genetic relationship (sibling species resulting from recent speciation), mimicry (as in the case of the monarch and viceroy butterflies) or evolutionary convergence.

Because there are so many interesting exceptions to the morphological species concept, the concept of species as reproductive units has proved very useful, but it is not without problems. One problem is 'hybridisation'. Reproductive isolation between species is not an absolute. Coyotes and dogs form hybrids ('coydogs'); so do bison and domestic cattle ('beefalo'). Examples, both in wild and domestic species of plants and animals, are legion. This does not mean that coyotes and dogs, or cows and bisons, are the same species. Often human intervention, deliberate or not, leads to crossing of populations that act as 'good species' under natural conditions.

Another problem with the reproductive species concept is that asexual organisms do not fit. For organisms with no sexual stage in the life cycle, we must rely on some sort of morphological definition. Over part of its range, the quaking aspen does not develop from seed, but reproduces only asexually. An aspen tree is an aspen tree, however, even though reproduction is by vegetative means (root-suckers). Furthermore, there is no genetic communication between clones (a group of individuals reproduced asexually from a single 'parent'), a clone that may date back ten to twelve millennia to Ice Age climates when there was enough moisture to allow germination.

There is another problem with the biological species definition which is to do with evolutionary time. As we 'watch' a species 'move' through time in the fossil record, we can see change ('anagensis') within a lineage. At two fairly distant intervals in time, members of the same lineage are thus distinctive; different enough to describe as separate morphological species. However, there is indirect genetic continuity between them. If we had a complete record of the intervening generations, there would be no distinction between adjacent generations. Clearly, here is a problem with our species concept. As we move back along our family tree from the branch-tips to the trunk, we see the lines coalesce and our definition blurs, until we come to agree with Darwin that species are not absolute but only 'tolerably well-defined'.

We do not know how many species there are on the **EARTH**. Most estimates range from 5 to 50 million. About 1.5 million species of animals, plants, fungi and microbes have been named to date, one of the best illustrations of the **BIODIVERSITY** of the environment.

Organisms are individual living things, 'organic beings'. Organisms represent 'homeostatic systems', maintaining the order of their internal environment at the expense of the external environment. Individual organisms may be considered the basic unit of ecology, following John Muir's formulaic definition: I + environment = Universe.

Organisms are open systems; they exist only in environments that supply energy and the raw materials for life processes: growth, maintenance, and reproduction. Organisms vary in size from bacteria which are a micron (10^{-6} m) long (such that 25,000 individuals laid end to end would be an inch long), through individual humans 1–2 m long, to blue whales 30 m long, coast redwood trees nearly 100 m tall, and honey fungi with an area of nearly 10 km^2.

Sometimes the term 'organism' is applied to species or some other level of biological integration, but the term probably should be reserved for the uniquely important level of individuals. Individuals represent species and individuals add up to populations. Populations interact to form biotic communities, which interact with the physical environment to build functioning **ECOSYSTEMS**. Apart from individuals produced by cloning and developed in identical environments, no two organisms are precisely the same. Hence, organisms are the fundamental unit of biodiversity, as self-contained, homeostatic packages manifesting genetic information and potentially transmitting that information to future generations.

WEBSITES

Article on defining biodiversity that builds on the species and organisms concepts in this entry:

http://darwin.bio.uci.edu/arboretum/diverse.htm

Interesting discussion about the meaning of species:

http://plato.stanford.edu/entries/species/

Explanation of the scientific classification system:

www.fact-index.com/s/sc/scientific_classification_1.html

STABILITY

Stability, an important condition of the atmosphere, affects many other meteorological conditions such as cloud formation, development of severe storms and even air quality. The atmosphere is said to be 'stable' when there is a tendency for air to sink towards the ground, therefore negating cloud formation and promoting dry and sunny conditions. Unstable conditions exist when air is encouraged to rise, such as when the ground is heated, causing rising convection currents. Under such conditions, clouds are often formed and rain occurs.

Under stable conditions, the air at the surface of the Earth stagnates and, if there are sources of pollution present, serious air quality problems may result. This is particularly true under temperature inversions, when cold air is trapped near the surface because it is

Microbursts

Microbursts result from unstable atmospheric conditions. Microbursts are strong, concentrated downdrafts from convective showers and thunderstorms. They have caused a number of commercial passenger jets to crash on attempted take-offs and landings. An intense microburst can result in damaging winds near 270 km/h.

Weather scientist Dr Tetsuya Theodore Fujita coined the term 'downburst' to refer to a concentrated severe downdraft that causes an outward burst of damaging winds at the surface. Dr Fujita defined downbursts as a surface wind in excess of 62 km/h caused by a small-scale downdraft from the base of a convective cloud.

A wet microburst on 20 May 1974, characterized by a well-defined foot shape on the left side of the rain shaft. (Photograph ©1974, C.A. Doswell III)

Downbursts are different from a normal downdraft, just as a severe tropical storm differs from a hurricane in intensity. Small-scale downbursts have caused aeroplanes to crash (called 'wind-shear accidents'). Fujita defined the microburst as a downburst with a maximum horizontal extent of no more than 4 kilometres.

Downbursts occur in regions of a severe thunderstorm where the air is accelerated downwards by exceptionally strong evaporative cooling (a dry downburst) or by very heavy rain that drags dry air down with it (a wet downburst). When the rapidly descending air strikes the ground, it spreads out in all directions in a circle, like a fast-running tap hitting the bottom of a sink. Often, damage associated with a downburst is mistaken for a tornado, particularly directly under the downburst. However, damage patterns away from the impact area are characteristic of straight-line winds rather than the twisted pattern of tornado damage.

Because microbursts are small, their energy is concentrated into tighter wind shear gradients, experienced by aircraft as rapid changes in wind direction. Investigations of Eastern Flight 66 that crashed at J.F. Kennedy International Airport and Continental Flight 469, both in 1975, indicated that microbursts were responsible. Earlier, a court investigating a wind-shear-related accident at Kano, Nigeria, in 1956 during a thunderstorm concluded that 'the accident was the result of a loss of height and air speed caused by the aircraft encountering, at approximately 75 metres after take-off, an unpredictable thunderstorm cell, which gave rise to a sudden reversal of wind direction'. This ruling, perhaps because of its tentative wording, does not seem to have alerted the rest of the aviation world to the dangers of thunderstorm downdrafts to aircraft landing.

Microbursts also pose hazards to small sailboats, which can be capsized by suddenly shifting strong winds, and to those fighting forest fires, who may be suddenly engulfed in a firestorm fanned up in an unexpected direction.

denser than the warm air above it. The cold air fills up with particulates and pollution from activities on the surface, resulting in smog. Prolonged stability can lead to serious drought. High-pressure systems associated with **ANTICYCLONES** are often quite stable, and have led to prolonged drought in the western USA in recent years. Some of the world's driest deserts, for example the Atacama Desert in Chile, occur next to the sea where cold ocean currents chill and stabilise the air, preventing the upward motions that trigger cloud and precipitation development.

Occasionally air that is stable may be forced to rise, for example, when as horizontally moving air is forced to rise over a mountain range. As it cools and condenses, the release of latent heat can then result in the air becoming relatively warm and unstable. This condition is known as conditional instability and is associated with outbreaks of severe supercell thunderstorms. If the air had been unstable in the first place, it is likely the energy in the water vapour would have been distributed evenly and thinly and without much impact, instead of exploding in a concentrated updraft from which the supercell is born.

WEBSITES

University Corporation for Atmospheric Research, with animations:
www.windows.ucar.edu/tour/link=/earth/Atmosphere/tornado/stability.html
Emery Riddle Aeronautical University:
www.erau.edu/er/newsmedia/articles/cont18.html
Pierce College atmospheric stability resource:
www.piercecollege.com/offices/weather/stability.html

'Microburst handbook' from the University of Oklahoma:
www.cimms.ou.edu/~doswell/microbursts/Handbook.html
Microburst information from the US National Oceanic and Atmospheric Administration:
www.erh.noaa.gov/er/cae/svrwx/downburst.htm

STORM SURGES

A storm surge is a rapid rise in sea level, often resulting from an intense storm centred offshore. Eyewitnesses describe a storm surge as a wall of water that reaches heights far in excess of normal high tides. The impact of storm surges on coastal communities can be absolutely devastating as was the case in eastern England during the North Sea surge of 1953.

These rapid rises in sea level are the product of intense low pressure, such as tropical cyclones. For every 10 millibars decrease in air pressure, the sea rises by 10 centimetres.

Case study

Storm surges

The North Sea surge of 1953

Most of the eastern North Sea coast of England suffered severe damage from the storm surge of 31 January/1 February 1953. A deep depression tracked eastwards to the north of the British Isles and then turned southwards, down the North Sea. Strong winds and the funnelling effect as the storm waters were piled up at the southern end of the North Sea created a wall of water some 6 metres high. To make the situation worse the storm coincided with a spring tide, adding even more height to the water.

Charles Matkin was a 25-year-old fireman involved in the 1953 rescue efforts at Hunstanton in Norfolk.

A gale had been blowing for 36 hours and the tide just didn't seem to go out. The firemen had instructions to pump out houses on South Beach Road, but they had no idea how bad the situation really was. They could not get over the level crossing, which was blocked by a train that had run into a bungalow washed on to the railway line by the surge. The few streetlights in the area were out; it was pitch black. When they were totally surrounded by floodwater they had to wade through.

Near the town there was an American bomber base. The Americans brought in motorboats to try to help get people out, but the boats were too large for the narrow streets. Working through the night, the firemen got only one launched.

It was only when daylight came that they realised the full extent of the damage. The death toll from drowning or exposure was 31, including several children.   *continued*

Case Study continued

A total of 264 people lost their lives in south-east England, but many more (1835) were lost on the other side of the North Sea in the Netherlands. Once the coastal dykes in the Netherlands had been breached by the surge, the sea flowed effortlessly over the land behind, much of which is below sea level.

Since 1953 sophisticated sea defences have been constructed to protect the area from surges even greater than the one of 1953. The huge Delta Scheme on the Dutch side and the Thames Barrier downstream from central London both required high levels of technology and investment, but both have saved that investment several times over by preventing coastal flooding in recent years.

The Bay of Bengal

The myriad of coastal islands and low mainland plains around the Bay of Bengal are prone to storm surges of around 8 metres high. In 1985 the Red Cross estimated that some 40,000 people perished in the region following a cyclone storm surge. In the same area, the small village of Muslimpalli was hit by a storm surge associated with the cyclone of October 1999. The event was reported thus:

> The trauma of this tiny hamlet [Muslimpalli] is mirrored in thousands of other villages across Orissa's once prosperous rice belt. Nearly two months after the disaster, up to 15 million people are living precariously, waiting, as government officials and relief agencies struggle to help them. (*Financial Times*, 24 December 1999)
>
> What is already evident is the unpreparedness of both the state and central governments for the disaster . . . [but] . . . since the cyclones are a regular feature in the region, the local people must be aware of what lies in store if they do not take precautionary measures. (*Hindustan Times*, 2 November 1999)
>
> With . . . waves over 10 metres, buildings and other structures common in rural India can hardly be expected to remain intact. But the enormous loss of lives is a direct result of the constraining socio-economic circumstances in India. (*The Times of India*, 2 November 1999)

Sources: A. Bowen and J. Pallister, 2000, *AS Level Geography*, Heinemann; www.environmentagency.gov.uk/yourenv

Tropical cyclones can involve a pressure drop of 100 millibars, causing a 1 metre rise in sea level. This potential rise is increased in an area of the sea that narrows sharply – for example, as at the Bay of Bengal.

WEBSITES

US Geological Survey 'Coasts in crisis':
http://pubs.usgs.gov/circ/c1075
UK Environment Agency:
www.environment-agency.gov.uk/yourenv
UK Met Office:
www.metoffice.com/education/historic/flood.html
Geography Pages:
www.geographypages.co.uk/flood.htm

SEE ALSO

cyclones (depressions)

SULPHUR DIOXIDE

Sulphur dioxide is a colourless gas, but its effects on the environment are very dramatic. People exposed to high levels of sulphur dioxide can suffer from breathing problems, respiratory illness and worsening respiratory and cardiovascular disease. It is also a major element in ACID RAIN, and contributes to the acidification of lakes and streams, accelerated corrosion of buildings and reduced visibility.

Natural sources of sulphur dioxide include emissions from volcanoes, oceans, biological decay and forest fires. The most important sources of sulphur dioxide resulting from human activity are fossil fuel combustion, smelting, the manufacture of sulphuric acid, the conversion of wood pulp to paper and the incineration of refuse. Coal burning is the single largest contributor, accounting for around 50 per cent of annual global emissions.

Case study

The sources of sulphur dioxide: the example of Ontario, Canada

In 2000 approximately 69 per cent of the sulphur dioxide emitted in Ontario, Canada, came from smelters and utilities, especially electrical generation. Other industrial sources included iron and steel mills, petroleum refineries, and pulp and paper mills. Small sources include residential, commercial and industrial space heating.

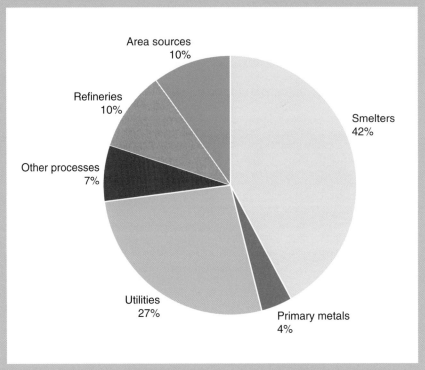

Ontario Sulphur Dioxide Emissions by Sector (Emissions From Human Activity, 2000 Estimates) © Queen's Printer for Ontario. Reproduced with permission

Sulphur dioxide emissions and targets in the UK: 1970–2010

Sulphur dioxide emissions in the UK fell by 70 per cent between 1990 and 2001 to 1125 thousand tonnes per year. This means that the UK is well on track to meet its targets set for 2010 of 625 thousand tonnes under the UNECE Gothenburg Protocol and 585 thousand tonnes under the EU National Emissions Ceiling Directive.

The reasons for the significant drop in sulphur dioxide emissions in the past few decades are reductions in the burning of coal and the decline of heavy manufacturing industry. The use of filtering methods such as FLUE GAS DESULPHURISATION has been very important. Anti-pollution measures have of course also assisted this trend, with the UNECE Gothenburg Protocol and EU National Emissions Ceiling Directive establishing realistic targets for the future.

Sulphur dioxide emissions in the UK (thousand tonnes)

	1970	1980	1990	1995	1999	2000	2001
Large combustion plants	3565	3405	2919	1756	894	899	822
Other sources	2895	1490	800	609	335	289	304
Total emissions	6460	4895	3719	2365	1229	1188	1126

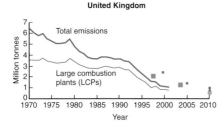

United Kingdom

- 2nd Sulphur Protocol targets for total emissions
- EU targets for LCPs
- EU NECD ceiling for total emissions
- Gothenburg target for total emissions

Sulphur dioxide emissions and targets: 1990–2010

WEBSITES

UK Department of Environment, Food and Rural Affairs:

www.defra.gov.uk/environment/statistics/airqual/aqsulphurd.htm.

Ontario Ministry of the Environment:

www.airqualityontario.com/science/pollutants/sulphur.cfm

Environment Protection Agency:

www.epa.gov/air/aqtrnd95/so2.html

SEE ALSO

acid rain

SURVIVORSHIP

Survivorship is the pattern of mortality at each age in a population. We owe this valuable concept to actuaries, people who calculate life insurance rates. Studying age-specific patterns of mortality in the human population allows actuaries to describe the probability of death of an individual at any given time. If the probability of death at a particular age is known, then the insurance company is willing – for a price – to 'bet' that the 'insured' will not die. If the probability of death is very low, as it is for a 30-year old woman who does not smoke or drink much, the insurance company will 'take the bet' for a fairly modest price. If the individual is a 60-year old man who smokes cigarettes, drinks heavily and partakes of dangerous sports the insurance rate will be correspondingly high. Insurance companies are very shrewd gamblers. In a sense, they do not gamble at all; they do not depend on 'luck'. Rather, they depend on probability; they know the 'odds', which derive from the survivorship curve.

The figure below illustrates four survivorship curves of different shapes. Note first how the figure is labelled. The vertical axis represents the number of individuals that are alive. This could be expressed as the percentage of the original number in the population. Actuaries and other population ecologists usually base their calculations on a cohort of 1000 individuals. A cohort in this sense is a sample of individuals born in the same year whose pattern of survival is being studied. The

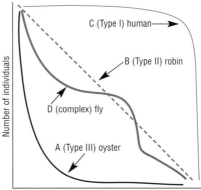

Sketch of four forms of survivorship curves. The classic three 'types', plus a complex curve

Life tables

Survivorship curves are based on life tables, such as is shown in the table below. This particular life table represents one of the classics in the ecological literature and underscores our definition of ecology as 'scientific natural history'. The table was based on a remarkable collection of 608 skulls of Dall's sheep (*Ovis dalli*) collected by naturalist Adolph Murie in what is now Denali National Park, Alaska. The sheep were mostly killed by wolves. Based on details of structure of the skulls and patterns of wear on the teeth, Murie could determine the age at death of a given individual. Note the column headings. Age is tabulated at one-year intervals. This is reasonable, because lambing takes place in a pulse in late spring and early summer, just once a year. Based on age at death, Murie could calculate and tabulate the number dying in an interval (d_x), the number surviving to begin each interval (n_x), the number surviving as a proportion of newborns (l_x), and the expectation of further life (e_x).

Life table for Dall sheep (Ovis dalli) constructed from age at death of 608 sheep killed by wolves in Mount McKinley, (now Denali) National Park, Alaska.

Age interval (years)	Number dying in interval	Number surviving to begin interval	Number surviving as proportion at age 0	Expectation of further life
X	d_x	n_x	l_x	e_x
0–1	121	608	1.000	7.1
1–2	7	487	0.801	7.7
2–3	8	480	0.789	6.8
3–4	7	472	0.776	5.9
4–5	18	465	0.764	5.0
5–6	28	447	0.734	4.2
6–7	29	419	0.688	3.4
7–8	42	390	0.640	2.6
8–9	80	348	0.571	1.9
9–10	114	268	0.439	1.3
10–11	95	154	0.252	0.9
11–12	55	59	0.096	0.6
12–13	2	4	0.006	1.2
13–14	2	2	0.003	0.7
14–15	0	0	0.000	0.0

Source: Based on data in Murie (1944), quoted by Deevey (1947)

There are several ways to develop life tables. The example detailed above is called a static life table, because it is based on a sample of individuals at the time of their death. One can also develop a static life table by investigating the age distribution of a sample of individuals at a particular time. A so-called dynamic life table is based on the observation of the pattern of death of 1000 individuals (a 'demographic cohort') born (hatched, germinated, etc.) at the same time. A survivorship curve is basically a plot of l_x over time, age-specific mortality.

horizontal axis represents time. In this case, time is represented as percentage of maximum longevity. In this way, we can graph together species with very different life spans. Traditionally, survivorship curves are of one of three 'types'. Type I is the human-style curve. Humans are generally not likely to die until they are old. In the USA in 1998 only 1 per cent of all children born alive died by the age of five (compared to 6.7 per cent in Guatemala). However, 300 years ago it was quite a different matter. In the city of York, England, in the seventeenth century, 60 per cent of children had died by the age of five, and only 30 per cent made it to the threshold of reproduction (15 years). Only 20 per cent remained alive by the age of 20. With so few females living to reproduction, only a high **FERTILITY** rate could maintain the population.

Survivorship is *age-specific mortality*. The survivorship curve is characteristic of a particular species. In the figure, take the curve labelled B ('Type II') as the baseline and contrast the other curves with it. Curve B represents roughly constant mortality over the life span. An individual in that population is as likely to die at any one time in its life as at any other time. Probably no species has precisely this curve. Because individuals go through a maturation process, their vulnerability is likely to change somewhat along the way. However, some species do approximate curve B. Robins, for example, seem to be about equally likely to die from exposure when they are nestlings, from starvation when they are juveniles, or from accidents or predation when adult.

Curve A (Type III) represents a fairly common life history pattern. Think of a maple tree – an adult tree sheds several hundred thousand fruits that flutter to the ground. How many of the seeds germinate and how many of the germinated seeds actually produce seedlings? How many of the seedlings survive desiccation to make saplings? How many of the saplings survive attacks by rabbits, bugs and off-road vehicles to make adult trees? The answer in each case is, very few. Once it reaches maturity, however, a maple tree has a good chance of survival.

In many sedentary marine organisms, we see a similar pattern. Oysters or barnacles, for example, have excellent survivorship as adults. They are shelled organisms. Their offspring, however, fare far less well. Oyster and barnacle larvae are

floating planktonic organisms, produced by the tens of thousands and sent off to join the food chain. A minute fraction of the young settle on suitable sites and mature into adults.

The Type III survivorship of curve A is highly concave relative to the origin of the axes. The Type I curve (curve C) is highly convex. Type I is the survivorship curve of humans, elephants and a few other organisms, mostly of large size, which take very good care of their young, avoid predation and climatic extremes, and generally survive until they get old, wear out and die.

Curve D suggests a species with mortality that differs markedly at different life stages. Some populations of flies (and other insects with complete metamorphosis) have this sort of survivorship curve. Eggs are very vulnerable (Type III). Larvae, however, are less so; they live right in their food supply, well protected from drying, and they are fairly mobile (Type I). Pupae fare less well. This is a stressful time physiologically, because the animal is literally digesting itself to recycle the body parts of the larva into the body of an adult. In addition, being immobile, pupae are especially vulnerable to predation (Type III). As adults, however, the flies assume a nearly straight-line survivorship curve (Type II), equally likely to get sprayed, swatted or snapped up by a hungry toad one day as the next. We see in this example that the usual typology of survivorship curves as I, II or III is not quite adequate to many populations in real life. Types I, II and III are probably in reverse order of their prevalence in nature. Type I is so-called not because it is the most prevalent curve but because it is the human curve. The concept of survivorship was brought into general population ecology from the science of human population ecology (demography).

WEBSITES

Survivorship curves discussion and illustrations:
www.life.umd.edu/classroom/biol106h/L29/L29_demo.html

Survivorship and life tables:
www.ento.vt.edu/~sharov/PopEcol/lec6/agedep.html

Human life-table databases for selected countries:
www.lifetable.de/cgi-bin/datamap.plx

SUSTAINABLE DEVELOPMENT

With the world's population now over 6 billion and increasing, the pressure on our planet, its people and resources is intense. Most people agree that we cannot continue to use these resources and abuse our environment as we have done in the past. (WorldAware, UK Department for International Development, 2000)

Sustainable development is an approach now being adopted throughout the world to address the potential problems of resource depletion, pollution and damage to ecosystems. Put simply, it involves development (improving quality of life), which meets the needs of the present without compromising the ability of future generations to meet their own needs.

Sustainable development involves using renewable rather than non-renewable energy sources and promoting reuse and recycling to avoid the need for new resources to be exploited. It involves the use of natural fertilisers in agriculture rather than toxic chemicals, and the use of appropriate technology in industry. It advocates carefully managed programmes of logging that involve selective felling and replanting rather than indiscriminate clear felling. It promotes the use of large mesh nets for commercial fishing rather than the small mesh nets that catch fish of all sizes, including young fish yet to mature. Sustainable development is an approach that many would describe as being 'environmentally friendly'.

In 1992 the United Nations held a conference on Environment and Development in Rio de Janeiro, Brazil. One of the most significant outcomes of what became known as the Rio Earth Summit was a blueprint for sustainable development. Some 180 countries signed an agreement called Agenda 21, which required governments to develop strategies for sustainable development at a national and local level. In the UK every local authority has government targets to reach on aspects such as recycling of domestic waste; as a result, most people have separate waste collections from their homes of paper, tin cans and glass.

If development is to continue without causing long-term damage to the environment, a sustainable approach is not just desirable, but absolutely essential.

WEBSITES

Intermediate Technology Development Group:

www.itdg.org

Case study

Cocoa farming in Cameroon

Cocoa is used in the manufacture of chocolate. It is an extremely important cash crop and is grown throughout much of the tropics. Almost 70 per cent of the world's cocoa is grown in West Africa. The highest yields come from plantations, but the use of chemicals and the practice of monoculture here is harmful to the soil and the environment. A much more sustainable but less intensive (and therefore less profitable) form of cocoa production involves cultivating the cocoa under the canopy of other trees. This is known as 'agroforestry'.

Cameroon produces over 10,000 tonnes (4 per cent of the world total) of cocoa beans each year and many thousand smallholders are dependent upon cocoa for their income. Many have adopted the system of agrofarming because it enables them to grow a variety of crops while providing ideal conditions for the cocoa trees. When land is cleared they deliberately leave some taller fruit and medicinal trees to provide shade for the cocoa. They then plant food crops such as maize and melons with the cocoa trees. Later on crops such as cassava and plantain are planted in between the cocoa. Additional trees such as plum, coconut and oil palm are planted to form, over time, an enclosed canopy, its multistrata system operating in a similar way to that of a natural forest ecosystem. The system is largely self-contained, promotes biodiversity, does not harm the environment and is clearly sustainable.

Centre for Appropriate Technology:
www.cosg.supanet.com/cita.html
UK Government Sustainable Development:
www.sustainable-development.gov.uk/
Sustainable Development International:
www.sustdev.org
New Agriculturalist:
www.new-agri.co.uk/98-5/focuson/
focuson6.html
Soil Association:
www.soilassociation.org/web/
sa/saweb.nsf/printable_library/NT00002F52
Worldaware:
www.worldaware.org.uk

SEE ALSO

biodiversity

SYMBIOSIS

The relationship between the clownfish and sea anemones has always fascinated scientists, divers and underwater photographers. Symbiosis means 'living together'. Clownfish and sea anemone have a symbiotic relationship, with clownfish living in close proximity to sea anemones without falling victim to the stinging cells on the anemone's tentacles. It has been discovered that it is the relationship between the individual fish and its particular host that allows the clownfish to remain unharmed. In aquariums, where a clownfish has been introduced to a new host, it has been observed that the fish exhibits behaviour that is consistent with being stung. Yet, afterwards, the fish returns to the host time after time, swimming in an elaborate 'dance' among the tentacles. Slowly, it allows the tentacles to touch its fins, then gradually the rest of its body, until it appears to be immunised against the sting. The immunity of the clownfish derives from the coating of mucus that covers its body. During its 'dance', it appears not only that the clownfish's mucus is spread onto the anemone's tentacles, but also that the mucus from the tentacles is spread over the fish. The exact mechanisms involved are not understood, but the result is that the fish is afforded complete protection. Although anemones can exist quite happily without clownfish, it appears that the clownfish helps to clean the anemone and, through their fiercely territorial behaviour, the fish helps to protect their host from would-be predators.

A biotic community is the living component of an **ECOSYSTEM**, where all of the species populations occur together at a

Two-banded clownfish (*Amphiprion bicinctus*) in the Red Sea. *Source:* NOAA

particular place in time. Any species is part of the environment of many other species, and it in turn has many species in its own environment. Interaction with other species is unavoidable; this interaction may take many forms. The interaction may be strong and immediate, or quite remote. Speaking very generally, such biotic interaction within a community is called symbiosis. Symbiosis is always 'inter-specific', that is, it occurs only *between* different species. Relationships within a single species, '*intra*-specific', cannot be described as symbiosis.

There are advantages for species to live together. For example, competition exists for food and territory in all environments and to avoid competing with other species, it is expedient for an animal to find a specific **NICHE** within their environment. Another way to avoid direct competition is to form a stable relationship with another species that is not a 'predator–prey' relationship. This allows two species to share harmoniously the same space and/or a food supply.

There are different sorts of possible symbiotic interactions. Consider the simple case of two hypothetical species, species A and species B, and where a beneficial effect is symbolised with +, a detrimental effect with –, and no effect with 0. The possible interactions are shown (and named) in the table below.

**A symbolic classification of symbiosis:
+ = benefit; – = detriment; 0 = no effect**

Species A	Species B	Form of symbiosis
+	+	Mutualism
+	–	Predation/Parasitism
+	0	Commensalism
–	–	Competition
–	0	Amensalism
0	0	Neutralism

When both species involved benefit from the relationship, this is described as mutualism. Clownfish and anemone have this sort of relationship. Commensalism results when one species benefits and the other isn't affected. Parasitism is the term applied when one species benefits and the other is harmed in the process.

There is a fourth, less 'intimate' category of symbiosis known as mimicry, or 'copying', which involves one species imitating another to gain the benefits enjoyed by that species. For example, banded snake eels have been known to mimic a venomous sea snake in order to deter predators.

A species can be involved in two symbiotic relationships at once. For example, it may suffer from parasitism on the one hand, and benefit from mutualistic attention on the other.

Symbiosis terminology may seem rather unscientific as 'beneficial' may seem to imply 'goodness'. However, we can avoid value judgment if we define 'benefit' or 'detriment' in terms of some external, measurable, biological standard, such as increased population size or rate of energy flow.

WEBSITES

Humorous sketches to help keep the different types of symbiosis clear:
www.unibas.ch/bothebel/people/redecker/symbiosis.htm
Symbiosis discussion:
http://users.rcn.com/jkimball.ma.ultranet/BiologyPages/S/Symbiosis.html
Symbiosis discussion from the Starship's 1009-day voyage around the world:
www.ms-starship.com/sciencenew/symbiosis.htm
Notes on symbiosis:
http://homes.jcu.edu.au/~zljes/bz1002/lecture8.htm
Detailed discussion on the relationship between clownfish and anemone:
http://biodiversity.uno.edu/ebooks/ch56.html#ch5

THUNDERSTORMS

Thunderstorms are powerful and sometimes terrifying natural events. They are most commonly associated with extremely unstable warm and moist air, which rises readily to form towering cumulonimbus clouds. While they can occur throughout much of the world, thunderstorms are commonplace in the tropics, where the regular afternoon storm is a way of life. Thunderstorms are often associated with extreme weather conditions, most notably torrential rain and strong, gusty winds. Lightning is a potential hazard, although lightning strikes of people are relatively uncommon. A far greater hazard associated with lightning is wildfire: fires are common in savanna and bush regions of the world, as well as in forested areas, such as parts of California. Another hazard often associated with thunderstorms is hail, which can involve ice crystals up to the size of baseballs. Hailstorms can cause tremendous damage to buildings, cars and crops.

The typical thunderstorm develops from a combination of mechanical lift and free convection when parcels of air are heated from the warm surface of the Earth. The air rises and then is rapidly carried further aloft by lifting mechanisms – free convection alone is unlikely to cause a thunderstorm. Lifting mechanisms are many. One common lifting mechanism involves convergence of air from two directions at the surface, such as when denser air from over the ocean converges on to the land as a sea breeze. This is the common mechanism that initiates the daily afternoon thunderstorms in Florida and in other coastal regions.

Once lifted, **ADIABATIC COOLING** results and cools the rapidly rising parcel to the dewpoint. Condensation begins and latent heat is released, causing further lift. At each stage in ascending, the air loses some of its moisture to condensation, which in turn releases more latent heat, causing more buoyancy, which lifts the parcel higher still. After this process continues, thick tall cumulonimbus clouds rise through the freezing level and continue high into the troposphere. In the cases of the most severe storms, the storm top may ascend a bit into the stratosphere, about 15 kilometres high. Once the cloud levels ascend above the freezing level, some of the water vapour is converted directly into ice crystals through the deposition process. With time, these ice crystals grow larger and larger as they coalesce and, when they reach a certain size, they are no longer held up by air currents and turbulence and so begin to fall.

Strong updraughts cause the ice crystals

Cows killed by lightning. From National Oceanic and Atmospheric Administration.

to rise and fall several times within the cloud, leading to variations in electrical charges within the cloud. Electrons flow from areas of positive charge to negative charge, superheating and illuminating the air with a crack of lightning. The air around a lightning bolt is superheated to about five times the temperature of the Sun. This sudden heating causes the air to expand faster than the speed of sound, which compresses the air and forms a shock wave – which we hear as thunder.

In addition to segregating charges and thus creating lightning, the falling ice and rain can grow in size by a variety of processes. The most important involves falling raindrops or snowflakes colliding to form larger hydrometeors – as falling water in the atmosphere. As larger raindrops fall faster than smaller ones, they collide with and merge with other drops, becoming still larger. This results in the heavy downpours of fat raindrops experienced during a thunderstorm. Hail is also produced almost exclusively by thunderstorms. Ice can loop through the cloud and attract additional layers of ice by passing through thick cloud regions where additional ice layers are 'painted' on. The hail that hits the ground is the result of multiple journeys through the various levels of the thunderstorm.

Even though they may occasionally be destructive, thunderstorms are an important part of the environment in many regions of the world. The Great Plains of the USA is one region where most of the annual precipitation falls from thunderstorms. People in this semi-arid region depend on this precipitation for growing crops; plants and animals in this region also depend on these thunderstorms to survive the long dry spells.

WEBSITES

NOAA National Severe Storm Lab:
www.lightningsafety.noaa.gov/photos.htm
UK Met Office's 'Thunderstorms' leaflet:
www.metoffice.com/education/
curriculum/leaflets/thunderstorms.html
College of DuPage photographs and
diagrams of thunderstorms:
http://weather.cod.edu/sirvatka/ts.html
US National Oceanic and Atmospheric
Administration thunderstorms resource:
www.nws.noaa.gov/om/brochures/trw.htm
Thunderstorm information from the
University of Illinois:
http://ww2010.atmos.uiuc.edu/(Gh)/
guides/mtr/svr/type/home.rxml

SEE ALSO

adiabatic cooling and heating, clouds, hurricanes, precipitation, tornadoes

TIDAL POWER

While King Canute was unsuccessful in controlling the sea, modern engineers have found
ways to harness the power of the tides, which
are generated every 12.5 hours by the gravitational forces from the Sun and the Moon.

Tidal power is another example of a clean
and reliable renewable energy supply that has
grown in popularity over the past 50 years.
The basic requirements are a funnel-shaped
estuary with a large tidal range, and one that
has little shipping or industrial activity. Once a
barrage (dam) has been constructed across an
estuary, the tides will produce a difference in
water levels on either side of the barrage.

There are three basic types of barrage:

- Outflow, or ebb generation: water passes
  through the barrage when the tide is leaving the estuary – from high to low tide.
- Flood generation: water passes through
  the barrage as it enters and fills the estuary
  – from low to high tide.
- Two-way generation: a combination of ebb
  and flood generation. Experiments have
  also been undertaken in some areas to
  pump water into an estuary to raise water
  levels and increase output – in some cases
  by up to 10 per cent.

Who was King Canute?

'Let all men know how empty and worthless
is the power of kings. For there is none wor-
thy of the name but God, whom heaven,
earth and sea obey.' So spoke King Canute
the Great, the eleventh-century Viking King
of England, seated on his throne on the
seashore, with waves lapping round his
feet.

Canute's courtiers believed he was so
great that he could command the tides to go
back. But Canute was a shrewd man, and he
knew his limitations – even if his courtiers
did not. He therefore had his throne carried
to the seashore and sat on it as the tide
advanced, commanding the waves to
advance no further. When they carried on he
had made his point: although the deeds of
kings might appear 'great' in the minds of
men, they were as nothing in the face of
God's power.

Power is then produced by allowing the
head of water to pass through the barrage,
which has a series of sluice gates and turbines
at its base. The sluice gates allow the flow of
water through the barrage to be controlled.
The greater the tidal range, the greater the
energy generated. Tidal barriers can, in theory, be economically viable, if the tidal range is
between 5 and 15 metres.

However, tidal power raises a number of
several environmental concerns. First, areas
behind a barrage may pond up with water and
this could permanently flood low-lying intertidal mudflats where wading birds feed. This
can have a harmful impact on the estuarine
ecosystem (halosere), although replacement
mudflats can be created if 'managed retreat'
strategies are adopted in adjacent areas.

Second, there needs to be regular dredg-
ing in the area behind the barrage. This is
because areas of standing water reduce the
velocities of the river, leading to deposition. If
this suspended material is dumped, the water
behind a barrage will be clearer and there
will be more light penetration in the upper
layers. This can benefit plant life through
increased rates of photosynthesis, but it will
also cause increased micro-organism growth,

Case study

Tidal energy on the Rance Estuary, north-west France

The average difference between high and low on the Rance Estuary, near Mont St. Michel, is 8 metres – at times this difference is as much as 13.5 metres, making it an ideal site for the generation of tidal power.

Following the success of an experiment in 1960, construction began in 1961 of a dam 750 metres long and 13 metres high, behind which was a 22-square-kilometre reservoir of water. At peak periods 18,000 cubic metres of water per second passed through the estuary – a flow that is ten times greater than the River Rhone. Work was completed in 1967 and since then 600 million kilowatt hours of energy have been generated each year – enough to provide energy for a quarter of a million homes. Another bonus for the local residents has been the creation of a dual carriageway on top of the dam wall. The new road has reduced the distance between St Malo and Dinard from 45 to just 15 kilometres, and consequently 26,000 motorists use this road each day.

Apart from the initial development of the plant and its infrastructure, and the inundation of land that was previously dry (caused by the ponding-up of water), there has been a relatively small impact on the environment and the scheme has generally been considered a success. However, proposals for similar barrages elsewhere, for example on the River Severn in England, have raised many environmental concerns, particularly those of altering fragile estuarine ecosystems and interfering with fish migration.

It is an interesting dilemma that an environmentally sound method of electricity generation, favoured by so many, can itself have negative environmental consequences.

for example algal blooms, which can reduce the dissolved oxygen content of water and decrease the number of organisms.

A final consideration is that water will take longer to move through the estuary, causing an increase in flushing time. Water-carrying pollutants from industrial or agricultural sources will therefore remain in the estuary for a longer period of time, increasing the potential amount of damage to the local ecosystem.

The largest tidal range in the world is in the Bay of Fundy in Nova Scotia, where the average difference between high and low tides is 11 metres. The most successful tidal power scheme in Europe is on the River Rance in Brittany, France, where the high tidal range has been harnessed since the late 1960s.

WEBSITES:

Australian Renewable Energy, with graphics about tidal power:

http://acre.murdoch.edu.au/ago/ocean/tidal.html

Energy Educators of Ontario:

www.iclei.org/EFACTS/TIDAL.HTM

Details about the Rance Estuary scheme:

www.edf.fr/html/en/decouvertes/voyage/usine/retour-usine.html

SEE ALSO

renewable energy, wave power

TORNADOES

Tornadoes are among the most feared of weather phenomenon. Hollywood producers of the 1930s exploited this fear when they created the movie of L. Frank Baum's book *The Wizard of Oz*, in which a tornado lifts Dorothy's house completely off its foundation – with Dorothy inside it.

Death and destruction from these tornadoes or 'twisters' is significant in some areas of the world, particularly the USA. All tornadoes in the USA occur in association with thunderstorms, and the largest and most destructive are associated with supercell **THUNDERSTORMS**. While many parts of the world experience thunderstorms, not all thunderstorms spawn tornadoes.

A tornado is formed when several conditions within the parent thunderstorm are met. First, there must be substantial vorticity in the

Idealized view of a 'classic' supercell, looking west

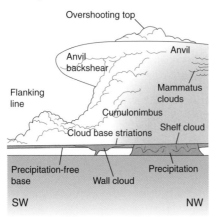

Diagram of a super cloud cell. National Weather Service Storm Prediction Center.

air. Vorticity results when air near the surface is moving more slowly and in a different direction than the air immediately aloft. This creates a tumbling motion in the horizontal, which is lifted and built into a vertical motion by the updraft of the parent thunderstorm. The convection takes on a rotational motion that is transferred to the entire updraft area of the thunderstorm. This is detectible on Doppler radar and is called a mesocyclone.

About half of detected mesocyclones occur in situations where conditions at the Earth's surface constrict and speed up the air circulation. Where areas of cold downdraft in front of and behind a warm, moist updraft push the circulation inwards on itself, air is constricted. This is particularly true if the front downdraft is very cold and accompanied by heavy precipitation, such as hail, and the rear downdraft, although cold, remains unstable. Constriction occurs when the inflow and updraft are particularly strong and the downdraft's circulation partially wraps around the updraft.

Because of the law of conservation of angular momentum, the slowly rotating 30 km/h mesocyclone becomes faster. This is the same physical law that increases the speed of spin on twirling skaters when they pull their arms in. The vertically rotating updraft begins to gather debris if it is near enough to the ground. The lower pressure within the rota-

tion causes a condensing funnel to become visible. To become a tornado, the vortex must touch the ground. Rapid rotation that is visible as a condensation cloud that does not touch the ground is called a funnel cloud. Tornadoes come in all shapes and sizes. The weakest may leave weak damage over a path a few metres wide. The largest – paths of almost a kilometre in diameter – have wiped out entire small towns.

Tornadoes occur in the USA during every month of the year. In January and February most are confined to the coast of the Gulf of Mexico and the southern states, but they quickly spread northwards. More tornadoes occur in April than in any other month, but the pattern of tornadoes is most spread out across the country in August. Thereafter, the pattern shrinks down to the south again, only to repeat the following year. Texas has the most tornadoes, and Oklahoma has the densest pattern of tornado strikes. The tornado causing the most deaths occurred in Michigan, while the tornado causing the most injuries occurred in Texas.

About 1000 recorded tornadoes touch down in the USA every year. Today, only a few are killers, but before warning systems were developed that has not always been the case. About 200 tornadoes have killed 18 or more people. The worst outbreak was from Missouri to Indiana in March 1825, which killed 695 and injured 2027, followed by a May 1840 series in Louisiana and Mississippi, which killed 317 and injured 109. However, tornadoes are not confined to the Midwest. They frequently occur on coasts, spawned by hurricanes. They have occurred in the mountain states, such as during August 1999 when a tornado roared through Salt Lake City, Utah. Tornadoes are not only rural phenomena; they can strike cities – on 16 April 1998, the Tennessee Oilers Football Arena in downtown Nashville, Tennessee was struck by a tornado. Luckily, since it was only under construction, no sporting event was taking place.

While the most powerful and destructive tornadoes occur in the USA, they do occur in other parts of the world. Interestingly, in the UK there are more tornadoes per square kilometre than in the USA – but they are less intense. Tornadoes in the UK are usually associated with active cold fronts, where cold air is rapidly undercutting warmer air.

Torrential rain and strong winds are associated with such conditions.

Tornadoes are ranked according to strength on a scale developed by Dr Theodore Fujita of the University of Chicago. The scale runs from F0, a weak tornado, which produces minimal damage, to F5, capable of producing catastrophic damage over a wide area in the path of the tornado. Such strong tornadoes may have paths over a kilometre wide, and the tornadic circulation include multiple vortices. Since the F scale is determined by damage assessment after-the-fact, many tornadoes in unpopulated areas go unranked. Typical damage defined by the F-scale is as follows:

- F0 (< 117 km/h winds): Light damage. Some damage to chimneys; branches broken off trees; shallow-rooted trees pushed over; sign boards damaged.
- F1 (117–180 km/h winds): Moderate damage. Surface peeled off roofs; mobile homes pushed off foundations or overturned; moving cars blown off roads.
- F2 (181–253 km/h winds): Considerable damage. Roofs torn off frame houses; mobile homes demolished; boxcars overturned; large trees snapped or uprooted; light-object missiles generated; cars lifted off the ground.
- F3 (254–332 km/h winds): Severe damage. Roofs and some walls torn off well-constructed houses; trains overturned; most trees in forest uprooted; heavy cars lifted off the ground and thrown.
- F4 (333–418 km/h winds): Devastating damage. Well-constructed houses levelled; structures with weak foundations blown away some distance; cars thrown; large missiles generated.
- F5 (419–512 km/h winds): Incredible damage. Strong frame houses levelled off foundations and swept away; car-sized missiles blown through the air in excess of 100 metres; trees debarked; incredible phenomena occur.

It is important not to use F-scale winds literally. These precise wind speed numbers are actually guesses and have never been scientifically verified. In addition, different wind speeds may cause similar-looking damage from place to place – even from building to building. Without a thorough engineering analysis of tornado damage in any event, the actual wind speeds needed to cause that damage are unknown.

Tornadoes have an effect on the environment because they are often associated with the colliding of cool dry air masses with warm moist air masses. The intermixing of air masses is an important component of midlatitude climates – keeping them temperate, and bringing precipitation.

WEBSITES

Tornadoes from NOAA:

www.noaa.gov/tornadoes.html

Tornado Project:

www.tornadoproject.com

A tornado chaser's typical day:

www.chaseday.com/tornadoes.htm

Tornadoes resources from the University of Illinois:

http://ww2010.atmos.uiuc.edu/(Gh)/guides/mtr/svr/torn/home.rxml

US National Weather Service Storm Prediction Center:

www.spc.noaa.gov/products/wwa

SEE ALSO

hurricanes, thunderstorms

TOURISM AND ECOTOURISM

The word 'tourism' first appeared in the *Oxford English Dictionary* in 1811, but the concept goes back to ancient Greeks and Romans whose wealthy citizens vacationed at thermal baths and explored exotic places around the Mediterranean. Tourism has since grown to become the world's largest industry, bigger than agriculture or manufacturing, with an annual revenue of almost US $500 billion. Despite disruptions in the world economy, insurrections and health scares such as SARS, tourism is still growing fast, with airline arrivals expected to double by 2010. Leisure is estimated to account for 75 per cent of all international travel. The World Tourism Organization (WTO) estimated there were 694 million international tourist arrivals in 2003, expected to reach 1.6 billion by 2020. In addition, domestic tourism (people going on holiday in their own country) is thought to be

Selva Verde Lodge, Costa Rica

Selva Verde Lodge ('Green Jungle') is an example of an ecotourism destination. Located in the lowland rainforest of north-eastern Costa Rica on a very large, private tract of virgin tropical rainforest and rich second-growth habitats, the lodge is adjacent to Braulio Carrillo National Park and the Organization for Tropical Studies' field station, La Selva. Tourists won't find any water-slides or golf courses here. Rather, they engage in low-impact activities such as hiking, resting in a hammock and swimming in the Rio Sarapiqui, but primarily in viewing local wildlife, including iguanas, howler monkeys, parrots and sloths. The property occu-

One of the buildings in Selva Verde Lodge, elevated on stilts to minimise impact on the environment.
Source: Joseph Kerski

pies over 200 hectares of preserved rainforest, boasts a resident naturalist, local guides, a reference library and the Sarapiqui Conservation Learning Center (SCLC). The preservation is important in the region, where large tracts of land have been given over to banana plantations and other forms of agriculture that many view as unsustainable.

The SCLC began as a public library for local residents and works to foster environmental education in the local community and to support conservation projects to protect the region's rich natural heritage. The SCLC is in partnership with Holbrook Travel, who own Selva Verde Lodge, to provide classes in computer literacy, GEOGRAPHIC INFORMATION SYSTEMS (GIS), environmental education and after-school activities, to facilitate activities so that tourists can interact with the local community; and to lead projects such as water monitoring and recycling. The SCLC is run by a full-time director and a staff of dedicated international volunteers.

four to five times greater than international arrivals. The WTO puts global revenue from tourism in 2003 at US $514.4 billion.

Tourism is also now the world's largest employer. In 2001, the International Labour Organization estimated that globally over 207 million jobs have been directly or indirectly created through tourism. Worldwide, tourism accounts for roughly 35 per cent of exports of services, and over 8 per cent of exports of goods, according to WTO. In the UK alone, 10 per cent of total employment is in the tourism sector. Caribbean countries derive half their GDP from tourism, but they're not alone – 83 per cent of countries in the world have tourism as one of their top five sources of foreign exchange.

The World Travel and Tourism Council (WTTC) predicts that the ten countries that joined the European Union in 2004 will generate up to US $54.6 billion of travel and tourism GDP and create an extra three million jobs. These figures would make tourism

the EU's largest business in terms of income.

Because of the size of tourism, environmental impacts are enormous. Tourism places tremendous pressure on NATIONAL PARKS. Emissions from aeroplanes contribute to climate change. Destinations favoured by tourists, such as beaches and mountains, are often some of the most fragile environments. Wildlife feeding and habitats are destroyed. Tourism also requires other resources to sustain it. A large hotel in Egypt uses as much electricity as 3,600 families, according to UNEP. A tourist in Spain uses 880 litres of water a day, compared with 250 by a local resident, according to the World Wide Fund for Nature. Each year, up to 5,000 hectares (an area the size of Paris) is cleared for golf courses. The planet's 25,000 golf courses also require huge amounts of water – an 18-hole course can consume more than 2.3 million litres of water daily and requires fertiliser and pesticides to produce the smooth, green surfaces that golfers demand.

Because of these large and growing impacts, people, companies and governments increasingly are turning to ecotourism. Associated terms include nature tourism, the visiting and viewing of nature, pro-poor tourism (tourism that benefits poor people in a tourist destination) and community tourism, where small local communities benefit and are involved in the decision-making process. Sustainable tourism is a broader concept than ecotourism, meaning any tourism (including mainstream resort tourism) that does not degrade the environment.

Despite its name, many caution that ecotourism can be as equally damaging to environments as mainstream tourism; that it is often just a marketing scheme to placate tourists' consciences. According to Martha Honey in her book *Ecotourism and Sustainable Development,* true ecotourism involves travel to destinations that are fragile, pristine and usually protected, minimises impact on the local environment, builds environmental awareness for all involved, provides direct financial benefits for conservation, provides financial benefits and empowerment for local people, respects local culture and supports human rights and democratic movements.

Over 50 countries have developed special policies and strategies focused on ecotourism at the national level. The World Ecotourism Summit (WES) was attended by more than 1100 delegates from 133 countries, and 2002 was declared the International Year of Ecotourism by the United Nations Environmental Programme.

Concrete evidence shows that, if managed in a sustainable manner, ecotourism helps conserve biodiversity, alleviates poverty in rural areas and can provide benefits to local and indigenous communities situated near or in officially protected areas. More tour operators, hotel chain owners and others are placing some of their profits into conservation efforts, from energy-saving devices and recycling in their properties to organising volunteer efforts. However, evidence also suggests that many so-called ecotourism sites are actually harmful to the local people and environment.

Tourism a global economic driver. The main challenge for the future is to apply the principles of ecotourism to all forms of tourism development.

WEBSITES

Ecotourism from the UN Environmental Programme:
www.uneptie.org/pc/tourism/ecotourism/home.htm
Detailed report: Ecotourism: principles, practices and policies for sustainability
www.uneptie.org/pc/tourism/library/ecotourism.htm
Paper on monitoring the costs and benefits of ecotourism:
www.uneptie.org/pc/tourism/ecotourism/wes_portfolio/statmnts/pdfs/vefraf.PDF
Selva Verde Lodge, Costa Rica:
www.selvaverde.com

TOXIC AND HAZARDOUS WASTE

Toxic and hazardous waste is waste material that can cause death or injury to living things, even in very small doses. It is usually the product of industry or commerce, but may also come from residential use, agriculture, military forces, medical facilities, radioactive sources and service industries, such as dry cleaning establishments. Toxic and hazardous substances can be released into air, water and land, leading to **POLLUTION**.

Two particularly dangerous types of waste are carcinogenic waste and teratogenic waste. Carcinogenic waste includes asbestos, which can cause mesothelioma – a cancer of the chest and the abdominal lining. Teratogenic waste can cause foetal abnormalities; the most famous example occurred in Minimata Bay, Japan, during the 1950s when an outbreak of diseases, including congenital cerebral palsy, was shown to have been caused by pregnant women eating fish contaminated with methyl mercury – an industrial chemical from a petrochemical and plastic-making company.

Toxic waste treatment and control has proved to be expensive and time-consuming with more resources spent on court battles than on actual clean-up. The disposal of toxic wastes is also a topic of international concern. In 1989 some 50 countries signed a treaty aimed at regulating the international shipment of toxic wastes. In some cases such wastes are shipped to developing countries for cheap disposal, without the informed consent

PCBs

PCBs, or polychlorinated biphenyls were previously widely used in industry as lubricants, coatings and insulation materials for electrical equipment, such as transformers and capacitors. They also found widespread use in consumer items such as hydraulic fluid, fluorescent lights, various appliances and televisions.

More than 680 billion grammes of PCBs were manufactured in the USA, but during the 1970s studies revealed that PCBs had a variety of adverse health effects. They were shown to cause cancer in animals, as well as affect their immune system, reproductive system, nervous system, endocrine system and have other health effects. Studies in humans also gave evidence for potential carcinogenic and non-carcinogenic side effects of PCBs.

Scientists also discovered that PCBs had a long life span in the environment. So, with fears growing about their safety, PCBs were banned from use in 1977. However, PCBs have continued to be released from poorly maintained waste dumps, by the illegal or improper dumping of hydraulic fluids and coolants, through leaks from electrical transformers, and the burning of medical, industrial or city waste.

of their governments. The use of shipping, storage and treatment methods that are often sub-standard endangers human health and the well-being of the environment.

In 1976 the Toxic Substances Control Act in the USA required the Environmental Protection Agency to regulate potentially hazardous industrial chemicals, including halogenated fluorocarbons, dioxin, asbestos, polychlorinated biphenyls (PCBs) and vinyl

Case study

Double standards on both sides of the Atlantic

As protest raged in the UK over 13 US toxic ships due to be decommissioned in the north-east port of Hartlepool, an even deadlier fleet pollutes the seas thousands of kilometres away. 'Why should this country be doing the USA's dirty work?' was the question being asked by the people of north-east England in 2003. The answer was because of tight controls and the fear of litigation in the USA. The UK has an equally infamous record for offloading toxic shipping on to Third World countries.

In November 2003 the *Genova Bridge*, a cargo ship used by the Ministry of Defence in the run-up to the Gulf War, docked in Alang, a port in Gujarat on the north-west coast of India and home to the world's largest ship-breaking yard. As it carried toxic materials, campaigners claimed the *Genova Bridge*'s owners and the UK Government were in breach of international law by not disposing of it in the UK. The people of Alang, including dozens of children who play amid the ship graveyard (ten other ships had been dumped in Alang in 2003), are left exposed to the hazardous waste on board. Cancer-causing asbestos is stripped by impoverished local people using their bare hands, and left to dry in the sun before being sold. Workers, who do not wear protective clothing, are also exposed to mercury, lead, arsenic and chromium poisoning. One in four of the 40,000 workers who are paid 65p (a little over US $1)a day, will die of cancer.

Despite this record, the British Government opposed the plan by Hartlepool firm Able UK to dismantle the 13 US Navy ships. The Environment Secretary said: 'The proposed shipment of these vessels to Hartlepool for dismantling cannot be completed consistent with international rules'. But according to a Greenpeace spokesman: 'This is a classic case of double standards. While the UK authorities don't want US waste in their backyard, they are happy to illegally dump their own elsewhere'. Unfortunately, the net result of this 'export' is that the removal of toxic waste is not undertaken in a way that is safe either for people or the environment.

Case study

The Minamata Disease controversy

The Chisso Corporation (Chisso means nitrogen) began their activities in the town of Minamata, Japan, producing fertiliser, but in the early 1930s they began developing plastics, drugs and perfumes, using a chemical called acetaldehyde, which is produced using mercury as a compound. It has subsequently been estimated that between 1932 and 1968 the company dumped 27 tonnes of mercury compounds into Minamata Bay. Ironically, the dumping only ended when the company's method of mercury production became outdated.

By the mid-1950s people had begun to fall ill with a degeneration of their nervous systems, while others suffered from numbness or slurred speech. Some people had serious brain damage, while others lapsed into unconsciousness or suffered from involuntary movements. There was also an increase in the number of sudden deaths in cats, while birds were strangely dropping dead from the sky. Over 3000 people have been subsequently recognised as having 'Minamata Disease'; as late as the 1990s, the Japanese courts were still resolving suitable compensation for people who had died, suffered from physical deformities, or have had to live with the physical and emotional pain of the disease.

In all, this was a huge cost, so it is ironic that the reason for the opening of the factory, back in 1907, had stemmed from the villagers of Minamata persuading the founder of the Chisso Corporation to build a factory in their town so that they could benefit from the wealth of industrialisation.

chloride. Other US legislation on hazardous wastes includes the Atomic Energy Act (1954), the Resource Conservation and Recovery Act (1976) and the Comprehensive Environmental Response, Compensation and Liability Act, or Superfund Act (1986).

WEBSITES

Environmental Protection Agency:
www.epa.gov/opptintr/pcb/effects.html
Rollins School of Public Health:
www.sph.emory.edu/PEHSU/pcb.htm
American University, Washington:
www.american.edu/TED/MINAMATA.HTM
University of the State of New York (Love Canal Collection):
http://ublib.buffalo.edu/libraries/projects/lovecanal
College of Science Texas A&M University:
http://safety.science.tamu.edu/waste.html
Toxic legacy of Woburn, Massachusetts:
http://home.earthlink.net/~dkennedy56/woburn.html

SEE ALSO

pollution

TSUNAMIS

A tsunami, sometimes incorrectly called a tidal wave, is a wall of water often several metres high that is capable of causing tremendous coastal destruction. The world's largest recorded tsunami was a massive 85 metres high, just over one fifth of the height of the Empire State Building (450 metres). It swept over part of the Japanese island of Ishigaki in 1971. In the 1990s ten major tsunamis occurred, killing a total of over 4000 people. These were dwarfed by a catastrophic tsunami caused by the Sumatra earthquake of December 2004. Tsunamis are not common compared with other disasters, such as earthquakes and landslides, but when they do occur and hit a populated coastal area they can be absolutely devastating.

Tsunamis may strike without warning on an otherwise calm day, triggered by an event a great distance away. They are caused by vibrations out to sea, which are associated with earthquakes, volcanic eruptions or underwater landslides. As tsunamis are propagated across the ocean, their wavelength increases, enabling them to attain significant height as they approach shallow coastal waters. The 2004 tsunami travelled from its source in

Case study

The Papua New Guinea disaster of 1998

Although the sun had set at 6.37 p.m. on 17 July 1998, there was still sufficient daylight for the day's activities to continue. Men were painting a canoe, young people were playing touch football and their elders were moving around in the villages. There was a sound like a heavy helicopter and the sea started to retreat from the shore. People went to the beach to investigate the unusual noise and observed that the sea was 'boiling'. Then there was silence for four or five minutes, followed by the noise of a low-flying jet aircraft. Then the wave approached at great speed.

People ran from the approaching wave but almost all were caught. A few escaped by climbing trees, or by pushing their boats into the lagoon. People in the wave were vigorously tumbled and turned in water that was laden with sand and debris. They were stripped of their clothing, lost skin by sand abrasion, were battered by hard objects and some were cut or impaled by timber and metal objects. Some were buried under piles of logs and debris. Those who were fortunate were carried into the lagoon and clung to floating debris. An infant was deposited miraculously on the floating roof of a house.

Sumatra thousands of kilometres to Thailand, Sri Lanka, Bangladesh and Africa, drowning 200,000 people, dislocating millions more, altering numerous coastal environments and devastating the economies not only of coastal communities but of entire regions.

Ninety per cent of the world's severe tsunamis occur in the Pacific. This is not altogether surprising since their main cause is tectonic activity and the Pacific is ringed with tectonic plate margins. Approximately 400 tsunamis occur in the Pacific each century and around one quarter of these primarily affect Japan or Taiwan. Not all undersea tectonic events result in a tsunami. To cause a tsunami an earthquake needs to be shallow in focus (occur near to the surface) and be at least as strong as 6.5 on the Richter Scale.

Tsunamis have three main ways of causing damage:

- Hydrostatic effects: whole objects or structures can be picked up and moved inland with the onslaught of the wave, or carried out to sea as it retreats.
- Shock effects: debris carried by the wave can at the same time cause damage to other structures.
- Hydrodynamic effects: buildings, bridges and harbours can be torn apart, or their foundations undermined.

Methods are now being developed to help cope with the threats posed by tsunamis. The Pacific Warning System in Honolulu, Hawaii, was established as early as 1948. Seismic stations detect earthquakes and results are interpreted by computer to assess potential tsunami activity. The aim is to alert all areas at risk within one hour of impact. However, whether to issue a warning or not can be problematic. If a warning goes out but the event does not happen, people feel inconvenienced for no reason; local economies may suffer and in the future people may mistrust such warnings. If the opposite occurs – the warning is too late or does not come at all but the disaster happens – then the authorities are simply not trusted again. It can be a no-win situation.

Land use mapping helps to assess the most vulnerable areas by generating relief and topographic information. Public education programmes have also been shown to be effective. The hope is that the use of offshore detectors and land-use zoning will ultimately be effective against tsunamis. Defensive engineering works may also help.

WEBSITES

Tsunami Research Group at the University of Southern California:

www.usc.edu/dept/tsunamis

Public Broadcasting Service:

www.pbs.org/wnet/savageearth/tsunami

University of Washington on tsunamis:

www.geophys.washington.edu/tsunami/intro.html

National Oceanographic and Atmospheric
Administration:
www.pmel.noaa.gov/~tsunami/PNG
2004 Sumatra earthquake and tsunami:
http://earthquake.usgs.gov/eginthenews/
2004/usslav

TUNDRA

The word 'tundra' comes from the Finnish
word *tunturia*, meaning 'treeless plain'. Far
from barren wastelands, though, the tundra
biome is an important one. Like all deserts,
tundra environments are dry (usually receiving less than 25 cm of precipitation annually),
but support many different life forms. Tundra
is the biome not only of high latitudes (Arctic
tundra), but also high elevations (alpine tundra). Climate is cold in winter and mostly cool
in summer. Low biotic diversity, simple vegetation structure, and nutrients in the form of
dead organic material all characterise the
tundra.

At present, tundra covers about 5 per cent
of the Earth's land surface (mostly in the
Arctic) but contributes less than 1 per cent of
terrestrial production. Arctic tundra is the
world's youngest biome. Formed during the
Pleistocene period, between a few 100,000
and 10,000 years ago, tundra spread and
retreated many times in the past, as glaciers
ebbed and flowed from the poles and from
mountain peaks.

Producers include grasses, sedges, cush-
ion-plants and low shrubs. Most plants are
perennials because the short-growing season
does not often allow flowering, sexual reproduction and seed-set. Plant biomass in tundra
is only 0.3 per cent of the global total because
tundra is essentially a two-dimensional environment – plants seldom rise above the modest snow level. Grazers include lemmings,
voles, caribou, Arctic hares and musk ox.
Carnivorous mammals include Arctic foxes,
wolves and polar bears. Migratory birds frequent the tundra, including ravens, snow
buntings, falcons, loons, sandpipers, terns,

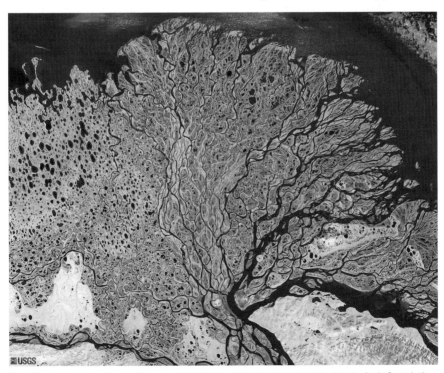

One of the longest rivers of the world, the Lena River forms a large delta where it empties into the Arctic Ocean in the
tundra biome. *Source:* Landsat satellite, USGS

snow birds and various species of gulls. Insects include mosquitoes, flies, moths and grasshoppers. Various species of cod, flatfish, salmon and trout inhabit the inland waters, with sea otters, seals, sea lions and walruses in the seas.

Arctic tundra and alpine tundra are rather similar in structure, but differ in fundamental ways. In particular, in the Arctic, most of the annual light energy budget arrives during the growing season. In alpine tundra at mid-to-low latitudes, light is spread more evenly through the year, at least half of which time is too cold for photosynthesis. As a consequence, Arctic tundra is generally more productive than alpine tundra. In addition, temperate and tropical alpine tundra lacks permafrost – a layer of permanently frozen subsoil (beneath the root zone) typical of Arctic tundra. Permafrost prevents percolation of water into the soil, hence, vast areas of Arctic tundra are seasonal wetlands, as shown in the satellite image above. These wetlands are breeding grounds for many migratory shorebirds and waterfowl, which rear their young on the high secondary productivity (especially insects) of the brief Arctic summer.

For thousands of years, humans have successfully inhabited tundra regions. Anthropologists believe that the first humans came to the tundra in North America from Asia over the Bering land bridge some 11,000 to 40,000 years ago. Later, about 10,000 years ago, another group called the Inuit (popularly called Eskimos) travelled over the land bridge into Alaska. After contact with new cultures in the 1700s and 1800s, the native Inuit population decreased by half.

Environmental impact

Polar regions such as Alaska will be among the first to illustrate the profound impacts of climate change. According to Oregon State University researcher Dominique Bachelet, during the next 100 years, polar regions will experience a massive loss of tundra, as global warming allows these vast regions of cold, dry lands to support forests and other vegetation. This will dramatically alter native ecosystems. Bachelet predicts more precipitation, an overall loss of soil carbon, a probable reduction in forest fires and a likely increase in insect and pathogen attacks on trees. 'The tundra has no place else to go and it will largely disappear from the Alaskan landscape, along with the related plant, animal and even human ecosystems that are based upon it.'

Tundra environments are fragile and especially vulnerable to adverse human impact. While grazing can damage fragile tundra vegetation, a more serious threat has come from dramatic increases in recreation, especially hiking and off-road vehicle use. Trampled and destroyed tundra and can result in severe erosion problems. Heating of buildings and pipelines thaws and destroys tundra.

On a wider scale, global warming will not only affect the Arctic but also have an impact on the entire planet. On the one hand, agriculture might become more viable throughout the regions currently covered in tundra. However, the amount of carbon in plant material in treeless tundra is much lower than in taiga forests. A northward migration of the Arctic treeline due to warming of the climate replaces tundra vegetation with taiga forests, increasing the total amount of carbon dioxide sequestered by plants. Peat bogs in the Arctic presently underlain by permafrost would start to erode away if the frozen peat melts, releasing large amounts of greenhouse gases into the atmosphere. Changes in the amount of freshwater from Arctic rivers flowing into the Arctic Ocean affects ocean circulation patterns, which play a key role in the exchange of heat between tropical and polar regions. Snow cover reflects much of the incoming radiation from the sun back into space, effectively cooling the atmosphere. Tree canopies absorb more solar radiation than tundra plants. Any changes in snow or vegetation cover will affect the global heat budget of the planet. Ozone depletion brings increased ultraviolet radiation, with an unknown effect on the tundra.

Recent industrialisation for the exploitation of non-renewable resources such as minerals (especially nickel) and hydrocarbons in northern Russia and in Alaska has created some large towns, which bring about the added risk of environmental pollution. Air pollution from faraway places has even caused smog over the tundra, contaminating lichen, a large source of food for many animals. However, the largest threat to the tundra biome is probably from the transport of oil. In 1994, up to 100,000 tonnes of oil near Usinsk,

Russia leaked from a pipeline, polluting the tundra over a large area and threatening the ecosystem of the Pechora River basin. New spills have been reported each year, according to Greenpeace.

Threats to the tundra are not new. In the past, caribou, musk ox and walruses were nearly hunted to extinction during the 1800s. Trappers found musk ox to be easy targets for hunting, because when threatened, they refuse to run. In 1930, 34 musk ox from Greenland were reintroduced to Nunivak Island in Alaska in an attempt to save the species from extinction. Today, between 2,500 and 3,000 musk ox live in Alaska, with 125,000 animals worldwide.

WEBSITES

Arctic Studies Center:
www.mnh.si.edu/arctic/index.html
USA long-term ecological research (LTER) site on Niwot Ridge, Colorado:
www.lternet.edu/sites/nwt
The Arctic tundra is represented in the LTER programme by Toolik Lake, in the Brooks Range of Alaska:
www.lternet.edu/sites/arc
Article on the circumpolar environment:
www.dkik.gl/komp/The_circumpolar_enviroment.html
Article on the subarctic forest-tundra: the structure of a biome in a changing climate:
www.zoo.utoronto.ca/fortin/Payette2001.pdf

URBAN SPRAWL

One of the most noticeable changes in the environment in our own lifetimes is urban sprawl. Richard Moe, President of the National Trust for Historic Preservation, defines sprawl as 'low-density development on the edge of cities and towns – poorly planned, land-consumptive, auto-dependent and designed without respect to its surroundings'. Sprawl is familiar to us all: you drive in bumper-to-bumper traffic on a con-

Construction equipment arriving to spread urban development into the desert outside of Las Vegas, Nevada USA. Photograph by Joseph Kerski

gested suburban road past a repeating, endless array of fast-food restaurants, petrol stations and shopping centres, all surrounded by thousands of parked cars. As each season passes, you see construction equipment pave over more and more fields with look-alike housing estates called Oak Meadows or Fox Hollow, whose very names mock what was destroyed.

Sprawl may be one of the most wasteful uses of land ever devised. Every year, 161,000 hectares of rural land in the USA alone are lost to development. During the 1970s and 1980s, for every 1 per cent of metropolitan population growth, land consumption increased by 6 to 12 per cent in the USA. For example, the population of the Chicago area grew by 4 per cent, but the urban area's size grew by 50 per cent. Adding to the problem is that during the same time the number of households increased due to demographic changes that led to more single people and broken-up families.

And sprawl is getting worse. During the 1990s for every 1 per cent of population growth in the USA, land consumption increased by 10 to 20 per cent. If central Maryland's sprawling development continues at this pace, it will consume as much land in the next 25 years as it has in the previous 300 years. Sprawl is not confined to large cities, but also plagues small and mid-size cities. Sprawl is even paving the USA's most productive food-producing region, California's Central Valley, to house a population expected to reach 15 million in 2040 – triple what it is today. In France, long stretches of industrial parks line some roads near the Riviera. Suburban Cairo bumps up against the Pyramids and Athens crowds at the base of the Acropolis.

How did sprawl become such a problem? In the USA, after the Second World War, the Federal Government encouraged housing and highway construction and set off the crush to get to the suburbs. Since the 1950s, federal and state-funded highway construction, including the interstate highways, encouraged the development of once-inaccessible outlying tracts of land. However, some local laws are also to blame. Many municipal zoning laws insist on large residential plots and a strict separation of residential and commercial uses in new communities. Most zoning ordinances separate different types of land uses, establish minimum distances between houses, minimum setbacks from roads, minimum parking-space requirements, minimum road widths and so on, so that the only type of development that can occur is sprawl. The lack of land-use planning and the reliance on zoning ordinances has promoted sprawl. There are no corner stores. There are few buses. You have to drive everywhere. The average American driver spends the equivalent of 55 eight-hour workdays each year behind the wheel. Residents of sprawling communities drive three to four times as much as those living in compact, well-planned areas.

Sprawl's costs include extending new infrastructure to outlying areas; traffic congestion; the loss of fertile farmland, open space and wildlife habitat; increasing air and water pollution; the loss of regionally distinctive communities; and the demise of huge areas of 1950s and 1960s-era inner ring suburbs, just as the construction of those same suburbs sucked the vitality out of city centres during the mid-twentieth century.

One answer to urban sprawl is 'smart growth'. Smart growth is urban planning that includes restrictions on sprawl, promoting pedestrian-friendly communities and mass transit, and reversing government programmes and tax policies that help create sprawl. The US Environmental Protection Agency practised smart growth by denying permits for the proposed Legacy Highway near Salt Lake City that would have destroyed wetlands adjacent to the Great Salt Lake,

increased air pollution and promoted development along it. Smart growth also includes requiring developers to pay impact fees to cover the costs of new roads, schools, water and sewer lines – the 'hidden' costs of sprawl that residents must bear. Smart growth advocates the revitalisation of already developed areas, and the prevention of new development in floodplains, coastal areas and other disaster-prone areas.

Through smart growth, regional plans are encouraged, so one community's pro-sprawl stance does not run rampant over an adjacent city's limits on growth. Some areas are setting up urban growth boundaries to rework government subsidies and to specify the areas in which urban development must be contained, promoting higher-density redevelopment in established neighbourhoods, on vacant plots and on abandoned sites (known as brownfield development). For example, along with the 1979 growth boundary, in the USA Portland voters also established Portland Metro, the country's first elected regional government. One of Metro's successes is its regional transit system, used by 45 per cent of Portland's downtown commuters.

People in areas such as Japan and Europe have historically viewed their land as a scarce, valuable commodity, not as an endless resource to be developed and then abandoned for 'greener pastures'. After the Second World War greenbelts were established around some cities in the UK, and during the 1950s and 1960s the UK, France, the Netherlands and Sweden enacted strict development laws that directed growth into city centres and formed self-contained new towns in outlying areas. Now many European nations are addressing sprawl by tightening their development laws, such as the UK's recent restrictions on superstores, which limit their construction along highways, and aim to protect the countryside and retail activity on village high streets. China has taken the strongest stand against sprawl of any country by designating 80 per cent of its arable land as 'fundamental farmland'. To build on such land requires approval from all levels of government, as well as the State Council.

Anti-sprawl forces are steadily gaining powerful converts in the USA. Many smaller households – couples without children, single parents, empty nesters and others – are choosing to live in downtowns, older neighbourhoods, attached homes and more sensible-sized houses. Some 'new towns' even have shops, housing, schools and recreation all at walking distance. In the 1998 elections in the USA, voters approved 72 per cent of the 240 measures to limit growth or preserve open space. In New Jersey 68 per cent of voters approved the Governor's $1 billion bond issue to set aside half the state's remaining 0.8 million hectares of open space.

In March 1999, in a speech entitled 'Developing common ground', Hugh McColl, CEO of Bank of America, announced his support for smart growth, declaring that the USA could no longer afford sprawl. Even some leading US real estate developers are now jumping on the anti-sprawl bandwagon. In 2004 John Williams, a successful suburban Atlanta developer and Chairman of the Metropolitan Atlanta Chamber of Commerce, announced that he would only build smart growth projects from now on.

But we need to act fast – as the bulldozers continue to work faster than policies or good intentions.

WEBSITES:

Smart growth information:

www.nrdc.org/cities/smartgrowth

Urban sprawl information from the Sierra Club:

www.sierraclub.org/sprawl/factsheet.asp

Farmland Preservation Trust:

www.farmland.org

Sprawl Watch:

www.sprawlwatch.org

VEHICLE EXHAUSTS

Increased personal mobility is one of the greatest changes brought about by modern technology. However, in common with many other changes since the Industrial Revolution, these developments have not taken place without a cost – in this case noxious and potentially dangerous car fumes.

The main emissions from a car engine are:

- Nitrogen gas (N_2): air is 78 per cent nitrogen gas, and most of this passes right through the car engine without being changed.
- Carbon dioxide (CO_2): a product of combustion, resulting from the carbon in the fuel bonding with the oxygen in the air. It is one of the 'greenhouse gases' thought to be responsible for global warming.
- Water vapour (H_2O): another product of combustion, resulting when hydrogen in the fuel bonds with atmospheric oxygen.
- Carbon monoxide (CO): a poisonous gas that is colourless and odourless. It is extremely dangerous and can kill people.
- Hydrocarbons or volatile organic compounds (VOCs): produced mostly from unburned fuel that evaporates. Solar radiation breaks these down to form oxidants, which react with oxides of nitrogen to cause low-level ozone (O_3), a major component of smog.
- **NITROGEN OXIDES**: a group of highly reactive gases that contain nitrogen and oxygen in varying amounts. These contribute to both smog and **ACID RAIN**.

The latter three emissions are the most harmful. They can have a direct impact on peoples' health and they can have an indirect effect on human activity by contributing towards acid rain, which can damage lakes and trees, and by contributing towards global warming. It is no surprise that these are the emissions that recent legislation has sought to regulate and are also the ones that **CATALYTIC CONVERTERS** aim to reduce. However, in many less economically developed countries there is yet another pollutant that is produced. As vehicles in several countries do not use unleaded fuel, lead is still being emitted into the lower atmosphere. In children, lead causes retardation, learning disabilities, hearing loss, reduced attention span, behavioural abnormalities and kidney damage. In adults, it causes problems of hypertension, high blood pressure and increased risk of heart disease.

Case study

The drive for lead-free fuel in Africa

One of the aims of the World Summit held in Johannesburg in 2002 was to increase the use of lead-free fuel in Africa. Within a year of the summit ending, a survey conducted by the United Nations Environment Programme (UNEP) found that Egypt, Libya, Mauritius and Sudan had become fully lead-free and that another 22 countries were in the process of drawing up plans to phase out lead in petrol.

After the summit it was also decided that as a small but symbolic push towards the lead-free goal, the on-site petrol station at the UN headquarters in Nairobi, which had previously sold both leaded and unleaded petrol, would only sell unleaded fuel. But there were many other parts of the continent, especially the Central African nations, where leaded fuel was still being used. As a UNEP official said: 'Much of Africa, mainly for technological reasons, a lack of awareness of the health risks and misconceptions about the impact of unleaded fuels on engines, has lagged behind'.

WEBSITES

Environmental Protection Agency:
www.epa.gov/greenvehicles
BBC Vehicle Emissions:
http://news.bbc.co.uk/1/hi/health/
medical_notes/336738.stm
History of lead in petrol from Radford
University:
www.radford.edu/~wkovarik/lead
History of lead in petrol from the US
Environmental Protection Agency:
www.epa.gov/history/topics/perspec/lead.htm

SEE ALSO

catalytic converters, nitrogn oxides, pollution

VISUAL POLLUTION

Visual pollution refers to those elements of the environment that people find unattractive, including buildings, business signs, street signs, vacant sites, utility poles and wires, graffiti, weeds and litter.

Much visual pollution can be traced to the dependency of modern society on communications (telephone lines and powerlines, for example), industrialisation (powerplants and refineries), waste disposal (incinerators) and particularly motor car transportation. The car helped increase people's mobility and, as new neighbourhoods were built, the ensuing suburban sprawl and cycle of urban decay was a familiar story in most European and North American cities. As car dependency increased, city land became more and more devoted to roads, bridges, car parks, petrol stations, new and used car sales and other car-related use. By the 1960s people realised that the price of the car was indeed a heavy one to pay, and the price was the vitality and visual appeal of the city.

Visual pollution is not confined to urban areas. In country areas people are concerned about the increasing signs of visual blight. Large quarries of gravel and minerals, landfills, mobile-phone towers, electrical sub-stations and everything from refrigerators to old cars on people's land are becoming as common as trees, meadows and other natural vistas.

Some of these issues are addressed through local laws, but it is difficult to enforce what people feel they have a right to keep in their own back yards. Illegal dumping is punishable by fines, although it often takes place at night and the perpetrators are rarely caught. Communities in many rural areas have been devastated by the growth of agribusiness, the loss of family farms and rural depopulation. The resulting small towns that serviced these farms are thus largely boarded up.

In the USA, as people discovered that community after community was becoming an ugly, ubiquitous mess of signs and shopfronts, sign ordinances went into effect across the country in the 1970s. Many businesses resisted what they viewed as their right to put up any type or size of sign of their choice.

Visual blight adjacent to a proposed freeway in Kansas City, Missouri. Photograph by Joseph Kerski

Streetlights, street signs and utility line improvements can be tackled within a community's architectural control efforts, but these measures are often expensive for a city to undertake. Visual pollution is often more difficult to address than other forms of pollution because it is largely dependent on human perception. For example, lawsuits have been brought against wind farms that generate electricity through wind turbines, because area residents said 'Not in my backyard'. In the UK, this NIMBY culture has often thwarted developments in wealthier areas, where local people are more politically aware and able to influence planning policies more effectively.

Early efforts to reduce visual pollution have targeted the older urban cores. Vast urban renewal projects dating from the 1930s to the 1970s replaced old terraced houses with high-rise housing projects. Soon after their construction, however, many of these projects became symbols for everything undesirable about cities – crime, rubbish, poverty and drugs. Many motorways were even built through decaying areas in the hopes of revitalising them, as the Bronx River Parkway in New York had done in the early 1900s. However, motorways had an adverse economic, social and visual impact on the urban

neighbourhoods they bisected, and areas near and under them became one of the last places that anyone would want to be near, much less live in.

Coupled with increasing awareness of urban visual pollution, people realised that older suburbs were becoming nearly as unappealing as the urban centres from which people had fled to the suburbs in the first place. Half-vacant strip development, decaying shopping malls and rows of houses that all looked the same became the impetus for changes in urban planning and development. New infill development is now replacing some of these blighted areas with such amenities as pedestrian-friendly businesses and townhouses with porches and a garage in the back, rather than on the street.

Graffiti – words or signs scribbled on public or private property without the owner's consent – represents a different challenge for communities because it can recur. A report by the city of Albuquerque indicated that the environmental cost to the city of graffiti is $1,000,000 every 18 months. In cities throughout the world bridges and walls are targeted by the graffiti writers. While some graffiti can be witty, humorous and even be classified as artistic, a good deal is offensive and

Case study

A new downtown for Lakewood, Colorado

Lakewood, Colorado, just west of Denver, was typical of many suburbs in attracting economic growth by constructing an enormous shopping mall. In 1966 Villa Italia mall opened to great acclaim, becoming the largest and most successful shopping mall between Chicago and Los Angeles. Developer Gerri Von Frellick was inspired by the Galleria of Milan, and he decorated Villa Italia with statues of Roman figures, terrazzo floors and columns taken from Denver's old US Customs House. For 35 years 'Villa' was viewed as the commercial and social centre of Lakewood. But, like other shopping malls, one by one, the stores closed, and by the end of the century Villa was dead and razed.

The City of Lakewood decided to do something unusual with the Villa site. A suburb from its inception, Lakewood never had a downtown. Now it will. BelMar, the name for Lakewood's downtown, is under construction as a new cultural, residential and commercial district. This district already includes the Lakewood city hall, civic centre, public library, cultural centre, heritage centre and Lakewood City Commons retail centre. Laid out in city blocks with local merchants, benches and a leisurely atmosphere, BelMar contrasts sharply with surrounding big-box, national fast-food chains, and the newest mega-mall 10 kilometres west.

unwelcome. Some communities use partnerships between owners and managers of public properties whereby the city removes the graffiti and is reimbursed by the affected property owner. As most acts of graffiti are committed by 10–15 year olds, many people believe that the best solution is to educate those who are under 10 years of age. They must see that graffiti is vandalism and unacceptable, and learn community leadership and community responsibility. Neighbourhood associations and 'Adopt-A-Block' campaigns in the USA help combat graffiti by creating partnerships to clean specific areas and keep them free of vandal street art.

Further reading

Jakle, J., 1992, *Derelict Landscapes – The Wasting of America's Built Environment*, Rowman and Littlefield Publishers, Inc.

WEBSITES

Visual pollution and its problems:
http://joefrank.tripod.com/derelict.html
Scenic America, an association to preserve scenic beauty:
www.scenic.org/index.htm
Reducing urban blight:
http://www.urbanblight.org/

The history of BelMar and Villa Italia:
http://experts.uli.org/Content/ResFellows/Beyard/Beyard_C06.htm

SEE ALSO

pollution

VOLCANOES

Mount St Helens in Washington State, USA, was long known to be a threatening volcano, having last erupted in 1857. In 1975 Dwight Crandall and Donal Mullineaux of the US Geological Survey had identified Mount St Helens as the most likely of the Cascade Range volcanoes to erupt, and they urged a programme of regular monitoring and civic preparations. So, when the mountain awoke with a mild earthquake on 20 March 1980, the scientific community did too. Sensors were put in place all around the peak to broadcast readings to data-logging computers many kilometres away from the foul gases and shuddering ground. Data were gathered, and accurate maps of the volcano, compiled from laser-ranging measurements, were turned out in mere days.

However, Mount St Helens erupted in a way that no one had foreseen. It was the worst volcanic disaster in the recorded history of the USA. Fifty-seven people died, and the ash from the explosion covered much of the western USA.

The word 'volcano' comes from the little island of Vulcano in the Mediterranean Sea, off the coast of Sicily. Centuries ago, people living in this area believed that Vulcano was the chimney of the forge of Vulcan – the blacksmith of the Roman gods. They thought that the hot lava fragments and clouds of dust erupting from Vulcano came from Vulcan's forge as he beat out thunderbolts for Jupiter, King of the Gods, and weapons for Mars, the God of War. In Polynesia, people attributed eruptive activity to the beautiful but wrathful Pele, Goddess of Volcanoes, whenever she was angry or spiteful. Today, volcanic eruptions

Case study

Environmental recovery after the eruption of Mount St Helens

The eruption of Mount St Helens removed 350 square kilometres of forest, dramatically altering wetlands, forests, meadows and rivers. The recovery in the area has increased our understanding of how natural disturbances regulate the productivity and biodiversity of ecosystems.

This was not the first time the environment had experienced an eruption: Mount St Helens has erupted more than 20 times in the past 4500 years, well within the 500,000 years estimated to produce an old-growth Douglas fir forest.

In the blowdown zone, high trees were snapped off or blown down, and shorter trees were buried under a metre of ash. A year after the eruption, wind-dispersed herbs such as fireweeds, which can survive on nutrient-poor surfaces that retain little water, were the first to colonise the barren surfaces. Isolated patches of plants had survived the blast because they were underneath snow, behind large trees, or behind steep ridges. Many subsequently perished because of the dramatic change in conditions. Low plants that are adapted to living in shade, such as wintergreen and fawn lily, were unable to tolerate the post-eruption conditions of increased light, temperature and dry winds, and soon perished. Surviving plants accelerated the overall recovery process because seeds did not have to arrive from distant sources. Pacific silver fir and mountain hemlock were producing cones by 1993, 13 years after the eruption. Because conifers have heavy seeds and require a symbiotic soil fungi (*mycorrhizae*) to survive, it was thought that they would not re-establish for quite some time. However, conifers are well-established in the entire eruption zone.

In contrast to the slow recovery of some upland vegetation, most riparian areas (along streams) recovered rapidly. Bank erosion re-exposed some buried shrubs and trees such as salmonberry and willow. Fragments of some species, such as willows, were swept downstream from their original locations and then sprouted. Surviving plants produced wind and water-dispersed seeds that colonised wet shorelines.

No bird species survived in either the area where the trees were blown over (the 'blowdown zone') or the area where lava and boulders were ejected (the 'pyroclastic zone'). However, within a year, seven bird species (such as Vaux's swift and the dark-eyed junco) had colonised – these were ground foragers that nest on the ground or in cavities of trees, and species that fly from perches to forage. Both types occur in open landscapes with sparse vegetation. When riparian vegetation grew again during Year 7, new bird species appeared. At Year 15, the blowdown zone in the disturbed forest had 70 per cent of the bird species of the undisturbed forest, but the species composition remains markedly different between the two. In the pyroclastic zone, characterised by undulating pumice hills and gullies, only 46 per cent of the original species richness exists.

One surprise from the eruption was that it created more aquatic habitat than existed before. In streams, by Year 10, 80 per cent of the invertebrates that had existed before the eruption had re-established. In contrast, recovery of trout was only at 5 per cent. Amphibians, such as frogs and toads, recovered more quickly than expected, because ice, snow and cold water at the time of the eruption helped buffer some of the volcano's effects. Four species of frog and toads and one species of salamander had colonised all available lakes within five years after the eruption, even though no dispersal corridors existed between lakes. These animals had apparently travelled great distances over non-forested pumice ground – an amazing feat for such small creatures.

are studied and interpreted by volcanologists.

Throughout the Earth's history, volcanoes have played a key role in forming and modifying the planet upon which we live. More than 80 per cent of the Earth's surface – above and below sea level – is of volcanic origin. Gaseous emissions from volcanic vents over hundreds of millions of years ago formed the Earth's earliest **OCEANS** and **ATMOSPHERE**, supplying the ingredients vital to sustain life. Over geologic eons, countless volcanic eruptions have produced mountains, plateaus and plains, which subsequent erosion has sculpted into majestic landscapes. Weathered volcanic rock has also formed some of the most fertile soils on the planet. These mountains, landscapes and soils have a great effect on the surrounding environment – the plants, animals and climate.

Most of the world's active volcanoes are located along or near the boundaries between shifting tectonic plates and are called 'plate-boundary' volcanoes. (**PLATE TECTONICS**) Therefore, they often occur near zones of frequent earthquake activity, and indeed some volcanic eruptions are accompanied by **EARTHQUAKES**. The eruption of Mount St Helens in 1980 was immediately preceded by a magnitude 5 earthquake that helped dislodge the magma bulge forming on the volcano's side, causing an enormous **LANDSLIDE**. However, some active volcanoes are not associated with plate boundaries. Instead, they are located on 'hot spots' in the Earth's crust, such as in Hawaii, a chain of islands in the middle of the Pacific plate.

Unlike other mountains, volcanoes are not formed by folding and crumpling, or by uplift and erosion, but are built by the accumulation of their own eruptions – lava, rock 'bombs', crusted-over ash flows and tephra (airborne ash and dust). A volcano is most commonly a conical hill or mountain built around a vent that connects with reservoirs of molten rock below the surface of the Earth. The term volcano also refers to the opening or vent through which the molten rock and associated gases are expelled.

When an eruption begins, the molten rock may pour from the vent as fairly calm, non-explosive lava flows, such as what often occurs in Hawaii, or it may shoot violently into the air as dense clouds of lava fragments, such as what happened at Mount St Helens. In a violent explosive eruption, some of the finer ejected materials may be carried by the wind only to fall to the ground hundreds of kilometres away. The finest ash particles may be carried many times around the world by stratospheric winds before settling out. These clouds of ash can affect the climate in some areas by blocking the sunlight, causing **CLIMATE CHANGE** with cooler temperatures than normal.

WEBSITES

Mount St Helens photographs:
http://pubs.usgs.gov/publications/msh/preface.html

Volcano World:
http://volcano.und.nodak.edu/vw.html

South Dakota State University's 'How volcanoes work' resource:
www.geology.sdsu.edu/how_volcanoes_work

USGS 'Volcanoes' booklet:
http://pubs.usgs.gov/gip/volc/

USGS on Mount St Helens:
http://pubs.usgs.gov/publications/msh/title.html

Volcano information from GeoNet Internet Geography, a resource for pre-collegiate British geography students and their instructors:
www.bennett.karoo.net/topics/volcanoes.html

SEE ALSO

earthquakes, landslides, plate tectonics

WASTE DISPOSAL

A recent survey in the USA showed that Americans generate rubbish at a rate of 1.8 kilograms per person per day. This translates to 600,000 tonnes per day or 210 million tonnes per year. For the US Federal Government, this is a very serious problem, especially since these rates are almost twice as much per person as many other more economically developed countries.

In the UK each person produces about a third of a tonne of rubbish each year. Nearly

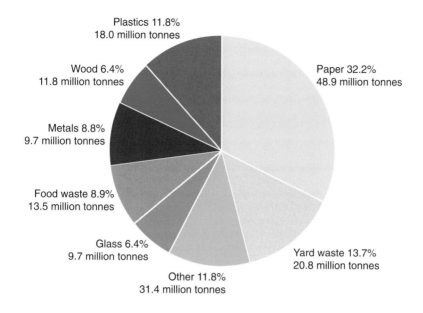

Plastics 11.8%
18.0 million tonnes

Wood 6.4%
11.8 million tonnes

Metals 8.8%
9.7 million tonnes

Food waste 8.9%
13.5 million tonnes

Glass 6.4%
9.7 million tonnes

Other 11.8%
31.4 million tonnes

Paper 32.2%
48.9 million tonnes

Yard waste 13.7%
20.8 million tonnes

Total generation: 151.9 million tonnes

The composition of waste disposed of in the USA by means other than recycling (2000)

Source: Characterisation of MSW in US: 1996 Update, US EPA, Washington, DC

150 million of tonnes of waste from domestic and industrial sources has to be thrown away each year. The typical composition of household refuse in the UK is

- paper and card: 33%
- organic matter: 21%
- debris, dust and cinders: 17%
- plastics: 11%
- glass: 9%
- metals: 7%
- textiles: 2%.

While the UK does not produce as much waste per person as in the USA, this is still a vast amount of rubbish. Part of the problem is that we live in a throw-away society and are happy to throw things away, such as ballpoint pens as soon as the ink runs out, rather than seek refills. In May 2004 thousands of convenience stores introduced the 'disposable DVD', which do not have to be returned, but have an internal timer that expires. They are simply thrown away after viewing. Millions of CDs may soon show up in landfills. But many argue that just because something can be disposable does not mean it should be disposable.

Many people are unaware of how much we consume, and so long as our domestic refuse is collected on a frequent and regular basis, we tend to forget about how greedy our modern societies have become. Only when the rubbish collection is delayed do we start to appreciate how wasteful we are.

The disposal of waste in most more economically developed countries is the responsibility of national and local governments. This is largely to reduce the potential hazards to human health that untreated waste and rubbish could cause. Ideally, waste should also be disposed of in ways that have the least impact on the environment.

There are three main means of waste disposal:

- Landfill: burying waste in holes in the ground (often in old quarries and worked-out sand and gravel pits).
- Incineration: burning the waste, although it can produce other types of waste in the form of smoke.
- Recycling: sorting the rubbish and making it into usable materials.

None of these options is without its problems. Recycling is the most environmentally friendly of the three, but it is a very expensive process and not all waste can be recycled.

Of the 210 million tonnes of rubbish, or solid waste, generated in the USA annually, about 56 million tons (or 27 per cent) is either recycled (mainly glass, paper products, plastic and metals) or composted ('green' or 'yard' and human waste). The remaining refuse is disposed of by either landfill or incineration. Unfortunately large amounts of paper, garden waste and plastics are still not recycled.

In some less economically developed countries the recycling of waste is an important element of the informal economy, whereby residents of the slums and shanty towns scavenge the rubbish tips in search of items that could be sold or converted into a saleable item. In the sprawling Indian city of Mumbai, home to over 16 million people, these rag-picking activities are the only option for hundreds of thousands of poor people to earn money. Men, women and young children gather waste in the streets and then deliver it to middlemen who run warehouses where the waste is sorted according to quality. Paper can then be created, ranging from medicine packaging and sweet boxes to coloured and office paper, with the rag-pickers earning on average around 40 to 50 rupees a day for their efforts – the equivalent of 50 to 60 UK pence.

WEBSITES

Waste Recycling Group:
www.wrg.co.uk
rethinkrubbish.com:
www.rethinkrubbish.com/home
The Times of India:
http://timesofindia.indiatimes.com/
articleshow/412824.cms
Edugreen:
http://edugreen.teri.res.in/explore/
solwaste/segre.htm
The 1992 film Baraka, which includes environmental footage from 24 countries, including gripping footage of rag-pickers:
www.imdb.com/title/tt0103767

SEE ALSO

incineration, land fill, recycling

WATER SUPPLY

'Water, water everywhere, but not a drop to drink'! Only 2.7 per cent of the water on our blue watery planet is actually fresh water. Furthermore, 77 per cent of that small amount of fresh water is 'locked up' in ice sheets, ice caps and glaciers. Of the remaining 23 per cent only 3 per cent occurs in lakes, rivers and in the atmosphere. The rest is stored underground as groundwater.

Despite the relatively small proportion of fresh water, this equates to billions of litres for the world's population. Despite this, hundreds of millions of people have difficulty in obtaining the 5 litres a day needed for survival. The reason for this is that both water and the world's people are unevenly distributed. In the average British city, daily domestic water consumption is 175 litres. This compares with just 45 litres in rural Bangladesh.

Water supply is the provision of water for people to use in their own homes and economic activities. We use water so often and for so many different activities that we forget the massive processes that are involved in finding it, cleaning it and moving it to where we want it to be. Many of us are unaware of the overuse of water in some regions today and consequent risks to future standards of living.

Think of the dozens of purposes to which we each put water every day. The basic ones are obvious – drinking, bathing, washing clothes, cooking and cleaning. These actually use relatively little water. Washing the car and watering the garden take rather more; toilets, sewage disposal and our leisure activities may use huge amounts. On occasions it may not be possible to serve all these from local supplies. The summer of 2003 was abnormally dry in many parts of Europe. Water supplies were put under tremendous pressure, and for some areas, such as in northern Italy, water was actually cut off for several hours a day to conserve supplies.

In the south-western USA, with its growing population and high standard of living, there is a massive demand for water. People want their gardens, parks and golf courses – and they want them looking very green. The volumes of water taken from a precarious supply for uses like these, which could be classified as frivolous, exceed that supplied for the more basic domestic needs. Cities like Tucson and

Case study

Supplying water for California

California has a low annual rainfall (San Francisco 500 millimetres, Los Angeles 300 millimetres), and the major cities lie in some of the driest parts of the state. In the Los Angeles Basin alone, 20 million people live in approximately 6 per cent of California's habitable land with only .06 per cent of the state's streamflow. California is a water-guzzling state, not only due to its large and growing population, but also because of the large agricultural output and the almost matchless standard of living. Despite the existence of many projects to deliver water – the pumping of aquifers, construction of huge reservoirs and transference of water from neighbouring states via canals – more supplies are still needed. Throughout the history of southern California, water provision has been a highly controversial political issue, tied to the development of the area's urban and agricultural areas.

Less well known is the water conservation success story in the Los Angeles Basin. The drought that had started in 1987 suddenly intensified in 1989–90, forcing water rationing for the first time. Water agencies were genuinely concerned about meeting record levels of demand, and so began to fund aggressively and implement water conservation programmes, along with improved groundwater management and water recycling.

The response was dramatic. In 1990 the Metropolitan Water District's water sales peaked at 2.6 million acre-feet; by 1993, these sales had plummeted to 1.5 million acre-feet. Demand has remained low, climbing in the late 1990s to just 1.8 million acre-feet – still far below the 1990 level. The unthinkable has happened: today the district's service area is using about the same amount of water as it used 15 years ago, despite an almost 30 per cent growth in its population.

Phoenix, Arizona, are constantly reassessing their water supply needs and management strategies, just to try to keep pace with demand.

There are several sources of water supply that people can exploit. In the UK people rely largely on **RESERVOIRS**, direct extraction from rivers and the pumping of **GROUNDWATER** from **AQUIFERS**. The main issue in the UK is that the greatest demand for water is in the South and East, whereas the bulk of the UK's rainfall occurs in the North and West. With high rates of evaporation and generally low-lying land, the South and East are poorly suited to the construction of reservoirs. To increase supply, groundwater is being used more intensively. In the future water will probably have to be imported from other wetter regions of the UK, and this will be very expensive. Parts of North America have a similar problem. For example, there are plans to bring Canadian water to California and its neighbouring states. Should this ever happen, it will be a major engineering feat with huge costs.

Increase in the demand for water is apparent in all countries, no matter what their economic level. This increase is fuelled by population increase, climate change, the need to develop and industrialise, as well as the demands made by increasing standards of living. Access to a reliable supply of sufficient clean water is one of the key characteristics that today divides the rich world from the poor. It has been said that the one single act that can improve the health and life expectancy of more people than any other is not improved vaccines, but the provision of clean water to the world's population.

The World Bank and WaterAid are both instrumental in making sure that water supply and sanitation are on the development agenda. In 1992 a World Development Report stated: 'Access to safe water remains an urgent human need in many countries. Part of the problem is contamination; tremendous human suffering is caused by diseases that are largely conquered when adequate water supply and sewerage systems are installed. The problem is compounded in some areas by growing water scarcity, which makes it difficult to meet increasing demand except at escalating cost.' Over 1.1 billion people are estimated to lack clean water supplies. Water supply problems can even disrupt children's education. Walking long distances to collect clean water or waiting in queues for limited supplies takes time, which children should be spending in school. In Tanzania an increase of 12 per cent in school attendance was recorded when water supplies were made available within a 15-minute walk.

WEBSITES

World Bank:

www.worldbank.org/watsan

WaterAid:

www.wateraid.org.uk

BBC World Water Crisis:

http://news.bbc.co.uk/hi/english/static/in_depth/world/2000/world_water_crisis/default.stm

Worldwatch on water supply and population:

www.worldwatch.org/press/news/1999/09/23

New Jersey Water Supply Authority:

www.njwsa.org

Kingsland Water Supply Corporation:

www.kingslandwater.com

World Water Supply Bulletin:

www.independent-media.tv/gtheme.cfm?ftheme_id=21

SEE ALSO

irrigation

WATER USE

Water is an extremely precious resource. While there is enough water to supply our needs many times over, water – like population – is distributed unevenly across the planet. Some areas have huge reserves and are, in many respects wasteful, whereas other areas face critical shortages, impacting on the lives and livelihoods of ordinary people as well as affecting entire economies.

The World Bank divides water use into three categories, as the table illustrates. The huge differences between the various countries listed reflect different levels of economic development and, to some extent, different climates.

Industrialised countries, such as Germany and Belgium, have high industrial water consumption. Some rich countries employ intensive irrigation techniques to maximise crop production, such as Spain and the USA. Australia, the most arid of the countries in the table, uses most water for domestic and agricultural purposes. Poorer countries, with little industrial demand,

How water supply is used in different countries

	Agriculture %	Industry %	Domestic %
Australia	33	2	65
Belgium	4	85	11
Canada	12	70	18
France	15	69	16
Germany	20	70	11
Spain	62	26	12
UK	3	77	20
USA	42	45	13
China	87	7	6
Bolivia	85	5	10
Ghana	52	13	35
Indonesia	76	11	13
Kenya	76	4	20
Malaysia	47	30	23
Mexico	86	8	6
Peru	72	9	19

From WORLD DEVELOPMENT REPORT 1999/2000: by World Bank, © 2000 by the International Bank for Reconstruction and Development/The World Bank. Used by permission of Oxford University Press, Inc.

utilise the vast majority of the water they have in agriculture. Many people remain peasant farmers, tied to the land for lack of opportunity elsewhere and, as rainfall is not always reliable, irrigation can be crucial to food supply. Malaysia, which is the most economically developed of the countries in the lower half of the table, shows the most even balance, illustrating the impact that development appears to have on water use.

For many people in less economically developed countries water supply and sanitation are the most important of all environmental issues. The United Nations designated the 1980s as 'International Drinking Water Supply and Sanitation Decade'. By 1990 it was hoped that all people would have access to both. This target has proved to be much too ambitious. More than 2 million deaths from diarrhoea still occur each year, but could be avoided if everyone had access to reasonable water and sanitation services.

Case study

Overuse of water in Kenya

Korr in Northern Kenya is an example of the overuse of borehole water. This region has not only low but also unreliable rainfall. If the rains fail it is obviously disastrous. A small variation of even 10 per cent below the average can be critical; this part of Kenya lies in the semi-arid Sahel zone of Africa so variability is often up to 30 per cent. Because of the nature of the environment most people were nomadic herders, moving with their animals following the rains and thereby also the pasture.

A Roman Catholic mission was set up at Korr to help meet some of the needs of these migratory peoples. Borehole water was exploited and proved reliable in quantity and quality. Education and health services were also provided, so increasingly people began to give up nomadism and settle permanently at Korr. The demands then placed on the borehole by people and grazing livestock were greatly increased.

At the same time other pressures have been placed on this fragile environment. Firewood is the traditional source of energy. Exploitation by a sedentary population has been too heavy; much of the light woodland and even scrub has been destroyed. The herds have eaten out grass stocks, and this, together with trampling by hooves, has led to soil erosion.

Levels of water in the aquifer have dropped several metres. Pumps have been fitted to give better access, but these only encourage continued overuse of this water resource. This experience shows that water use in much of Africa is an issue that needs addressing seriously and urgently.

One of the main issues regarding water use is waste and overuse. In the UK about a third of water is 'lost' between reservoir and consumer, due to leaking pipes. Poor maintenance over many decades now means that during dry periods people have to endure water bans on certain domestic uses. And this is in a country whose inhabitants feel that it is always raining!

Overuse of water is both wasteful and, in the long term, unsustainable. Overuse occurs throughout the world, in both rich and poor nations. However, if it continues, water supplies may become contaminated or even run dry.

WEBSITES

Water Aid:

www.wateraid.org.uk

Water – Use it Wisely:

www.wateruseitwisely.com

Center for Improved Engineering and

Science Education:

www.k12science.org/curriculum/drainproj

US Geological Survey:

http://water.usgs.gov/watuse

CNN:

www.cnn.com/US/9811/10/water.use.down

WATERFALLS

It was not until I came on Table Rock and looked – Great Heaven, on what a fall of bright green water! – that it came upon me in its full might and majesty. Then, when I felt how near to my Creator I was standing, the first effect, and the enduring one – instant and lasting – of the tremendous spectacle, was Peace . . . Niagara was at once stamped upon my heart, an Image of Beauty: to remain there, changeless and indelible, until its pulses cease to beat, for ever. (Charles Dickens, 1842, *American Notes*)

Waterfalls are one of the world's spectacular natural features. Niagara Falls is perhaps the most famous; it is one of the most visited sites in the world with 18 million people annually travelling to 'wonder' at it. Venezuela's Angel Falls is equally spectacular but is much more remote and, as a result, has far fewer visitors. The huge demand to visit waterfalls such as these in often fairly remote locations can lead to environmental problems. Roads and car parks have to be built, together with facilities for tourists such as visitor centres, toilets, hotels and footpaths to viewing points. All these developments impact upon the natural environment, and wherever large numbers of people gather there will also be problems such as pollution, litter and noise.

The world's highest waterfalls

Falls	Country	Height in metres
Angel	Venezuela	979
Tugela	South Africa	947
Utigard	Norway	800
Mongefossen	Norway	774
Yosemite	USA	739
Mardalsfossen	Norway	656
Tyssestrengane	Norway	646
Cuqenan	Venezuela	610
Sutherland	New Zealand	580
Kjellfossen	Norway	561

Many rivers have waterfalls somewhere along their course. Most commonly waterfalls form where a river flows from a hard band of rock on to a softer band of rock. The more rapid rate of erosion of the softer rock creates a 'step' in the river's profile – this is the start of a waterfall. The erosion at the base of the waterfall can lead to undercutting and the formation of a distinct overhang. Eventually the overhang will collapse and so, over time, as this process is repeated, the waterfall moves upstream.

Five of the world's ten highest waterfalls are in Norway. The reason for this is that glaciation has created features called hanging valleys. A hanging valley is a small tributary valley that was not eroded down to the same level as the main trunk valley. This was because it contained less erosive ice. After the last ice advance, when melting had occurred, tributary valleys were left 'hanging' on the sides of the main valleys, forming sharp breaks of slope over which the present-day waterfalls plunge. Although they are spectacular to view, they are found in very remote areas and, therefore, have few visitors.

Generally considered the tallest in the world, the Angel Falls of the Guyana Highlands of Venezuela are 15 times the height of Niagara and have the longest free fall of water in the world (745 metres). They are named after an American, Jimmy Angel of Missouri, who first saw them in 1933 when he found his plane stranded in marshy ground at the top of the tepui or plateau off which the waterfall drops.

WEBSITES

InfoNiagara:
www.infoniagara.com
Tourism Niagara.Com:
www.tourismniagara.com
Niagara Falls State Park, with excellent pictures:
www.niagarafallsstatepark.com
Salto-Angel.Com, dedicated to Venezuela's Angel Falls:
www.salto-angel.com
Angel Falls, with excellent pictures:
www.angel-falls.com
Ladatco Tours:
www.ladatco.com/ANGL-ABT.htm

WAVE POWER

The highly sculptured and contorted coastlines around the world provide a visible indication of the immense power that is unlocked when waves break. Anyone unlucky to be caught out at sea during a storm will also testify to the sheer force that waves can generate. In the past 30 years various experiments have been undertaken as people have tried to harness this source of renewable energy.

Ocean Power Technologies, a company based in New Jersey, USA, has been experimenting with wave power in Hawaii. In 2002 the company received a $4.3 million grant from the US Navy to create 'powerbuoys' off the island of Oahu. The powerbuoys are anchored up to 4 metres below the surface of the ocean, and inside each is a piston-like structure that moves as the buoys bob up and down, driving a generator that produces electricity that is sent ashore by an underwater cable.

Other experiments have used devices called oscillating water columns (OWCs).

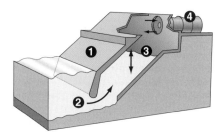

Islay Wave Power Station. 1. Wave capture chamber set into rock face. 2. Tidal power forces water into chamber. 3. Air alternately compressed and decompressed by 'oscillating water column'. 4. Rushes of air drive the Wells Turbine, creating power. From BBC News at bbcnews.co.uk

When waves hit an OWC, the air inside is compressed and forced through air turbines. The first successful OWC device was used in Japan to power a light on the top of a buoy for navigation, while another in Tofteshallen, Norway, was designed to produce 500 kilowatts of electricity. However, the Norwegian OWC was blown out to sea during a violent storm in 1998, and the experiment failed.

British scientists have been more successful and in November 2000 the world's first commercial wave power station went into action on the Isle of Islay in Scotland.

Supporters of wave energy believe that there is enough recoverable wave power around the shores of the UK to exceed domestic electricity demands. They also argue that just 0.1 per cent of the energy within the world's oceans could supply more than five times the current global demand for energy. Today, experiments are taking place to develop the technology that will be essential if this energy is to be harnessed efficiently and economically. If the experiments are a success, the next few years could see the creation of large-scale wave plants in near-shore or offshore environments. Some opponents argue that such vast plants create visual pollution, in addition to the noise pollution that occurs with the generators. But the supporters of wave energy have countered these protests by showing that the onshore or near-shore plants could be designed as part of harbour walls and coastal defences, making coastal areas safer and possibly less prone to erosion.

WEBSITES

Britain's wave power schemes:

www.wavegen.co.uk

The BBC website, with details about the Isle of Islay scheme:

http://news.bbc.co.uk/1/hi/sci/tech/1032148.stm

SEE ALSO

renewable energy, tidal power

Case study

Limpet 500: the world's first wave power plant, Islay, Scotland

Limpet 500 (Land Installed Marine Powered Energy Transformer) was developed after a 10-year project by researchers at Queen's University in Belfast, Northern Ireland, which resulted in the construction of a demonstration plant capable of generating 75 kilowatts of power. With funding from the EU, the scheme was transformed into a commercially viable operation by Wavegen.

Limpet 500 consists of two basic elements – a wave energy collector and a generator that converts wave energy into electricity. The energy collector is simply a sloping, reinforced shell built into the rock face at the shoreline. It has an inlet that is large enough to allow seawater to enter freely and subsequently leave a central chamber. As more and more waves enter the shell, the level of water rises, thereby compressing the air at the top of the chamber. The air is then forced through a 'blowhole' and into the turbine. This has been developed so that it continues turning even when the water inside the shell chamber recedes as the waves outside draw back. This constant stream of air in both directions, created by the oscillating water column, produces enough movement in the turbine to drive the generator that converts the energy into electricity.

WETLANDS

Until recently, the word 'swamp' conjured up negative images. Swamps were seen as undesirable places, unsuitable for agricultural and urban development. These wetlands were therefore filled in, drained, dredged and used for peat for centuries. Yet swamps, marshes and other wetlands are prime habitats for fish, shrimp, crabs and amphibians, providing critical nesting sites for waterfowl and habitat for wildlife. As the transitional zones between land and water environments, wetlands are among the most biologically fruitful ecosystems, rivalling tropical rainforests in fertility. Alive with nutrients and plant debris, they also filter out pollutants and buffer inland areas from storm, wave and flood damage. Wetlands also support food production. Rice, a staple crop for half the world, grows primarily in waterlogged paddies, and coastal wetlands are spawning grounds for commercial fish harvests.

Wetland areas, where the soil is frequently or permanently waterlogged, serve a vital role in maintaining ecological balance. Wetlands such as marshes, swamps, bogs, lakes, estuaries, deltas and floodplains help maintain ecological balance. However, because of their destruction over the centuries, they now cover just 2 million square kilometres of the world's land surface – a small fraction of their original extent. In the USA the 159 million hectares of wetlands that existed in the 1780s were reduced by three-quarters to 42 million hectares 200 years later. New Zealand has lost 90 per cent of its marshy terrain, and more than 70 per cent of European wetlands have disappeared. Even the immense peat bogs of England and Ireland, once thought to be an inexhaustible source of fuel, are 90 per cent depleted.

Wetland vegetation is classified as hygrophytic, or water-tolerant. Most plants cannot survive waterlogged conditions because the lack of air in the soil prevents their roots from sucking up moisture and nutrients. Hygrophytes have special adaptations to allow this to happen.

Wetlands of various types exist all over the world, from equatorial to tundra areas, and the contrasts between them are huge. Each has its own particular ecosystem and food chains. The warmer and wetter the climate, the more productive is the system and the faster the cycling of both energy and nutrients. Sub-tropical wetlands, such as the Florida Everglades, have high rates of growth and recycling of nutrients. Not only are cold wetlands slower in operation due to low temperatures, such as those in the Canadian tundra, but they are also seasonal, virtually closing down altogether during the winter months.

It is possible for wetlands to be managed in an economically productive way while also conserving wildlife. Human use of temperate wetlands has proved that management can increase the number of niches within the ecosystem. For example, wet grasslands have long been part of the traditional farming system in the UK. In the summer, grazing is provided for livestock and hay is grown for winter fodder. Meadows may flood in winter – the Somerset Levels in the South West of the UK is a perfect example. In this way they act as a water store within the **HYDROLOGICAL CYCLE** and so may prevent flooding in more economically vulnerable places elsewhere in the drainage basin. Such traditional management practices have protected valuable habitats, especially for water birds. The grazing animals, usually cattle, are integrated into the ecosystem. Ditches act as 'wet fences' to control stock and at the same time provide a perfect environment for innumerable insects and water mammals.

Intensification of agriculture can, however, destroys wetlands. When land is converted to arable production ditches are drained and filled in and all the specially adapted wildlife is lost. In Europe government policies have consistently promoted farm production. The price support mechanisms of the Common Agricultural Policy have only served to encourage this. A direct contradiction exists between this official promotion of intensive agriculture and the EU's legal obligation to protect wetlands.

Wetland conservation and rehabilitation is increasingly an aspect of leisure and tourism. The Norfolk Broads (the largest remaining wetland in Britain) contains waterways created by peat-digging in medieval times. The richly diverse fauna and flora in this area has been recognised by Ramsar status (wetland of international importance, recognised under the Ramsar Convention). Since 1975 the

Case study

The Florida Everglades

The Everglades National Park is a wetland of international importance and the only sub-tropical reserve in the whole USA. It occupies the whole of the southern tip of the state of Florida, covering 11,655 square kilometres. Containing some tropical ecosystems and some warm temperate ones, it includes a huge variety of ecosystems:

- sawgrass prairies
- mangrove swamps
- cypress swamps
- pinelands
- hardwood forest
- estuarine environments
- marine/coastal environments.

This wetland provides a home to a wide variety and number of birds, in particular the larger species such as roseate spoonbill, wood stork, great blue heron, crane, ibis and several of the egret family. It is the only place in the world where alligators and crocodiles coexist.

Threats to the Everglades include expanding urbanisation from Miami to the east and Fort Myers to the west, agricultural expansion from the north, and the resulting pollution and drainage that accompanies these activities. Nevertheless, success stories abound. For example, the 1990 Florida legislature created Preservation 2000, a $3 billion programme to purchase environmentally sensitive lands. Since its inception, Preservation 2000 has been responsible for the protection of over 162,000 hectares of land throughout Florida. Another example of progress is that wildlife crossings and under-passes have been constructed for the panthers on Alligator Alley and on State Highway 29 in Collier County's Big Cypress Swamp by the Florida State Department of Transportation.

RSPB (Royal Society for the Protection of Birds) management at Strumpshaw Fen (in Norfolk) has aimed to clear dykes of sediment, counteract pollution and reestablish key species. Typical fen plants such as sow thistle and milk parsley now inhabit extensive water meadows. Several fish and reptile species have become well established. The area is open to the public for recreational activities and annual usage has increased steadily.

If wetlands are to survive they must be managed sustainably and may also have to be economically viable at the same time. In Villafafila Lagoons Nature Reserve in the province of Zamora, Castilla y Leon, Spain, environmentally sustainable development is the aim. A mixed farming system supports some of the local population. Local food-processing facilities have been increased to add value to products. Protected by Ramsar status and with an increasing bird population, there has been a growing stream of visitors to the area. Guided visits, pony trekking and bicycle hire are just some of the developing leisure businesses boosting the local economy.

WEBSITES

Wetland loss in the USA:
www.npwrc.usgs.gov/resource/othrdata/wetloss/wetloss.htm
USA National Wetlands Inventory:
http://wetlands.fws.gov
Wetlands International:
www.wetlands.org
USGS National Wetlands Research Center:
www.nwrc.usgs.gov
USGS, 'The South Florida environment: a region under stress':
http://sofia.usgs.gov/publications/circular/1134
Conservation Everglades Restoration Plan:
www.evergladesplan.org

Everglades National Park:

www.everglades.national-park.com

www.nps.gov/ever

Friends of the Everglades:

www.everglades.org

Nova Southeastern University's

oceanographic center:

http://www.nova.edu/ocean/eglades

Everglades.com:

www.everglades.com

WILDFIRES

Every year news headlines capture the drama of wildfires raging out of control. This seemingly wanton destruction plays out an annual rite of summer, consuming millions of hectares of grassland and timber. These large-scale fires often lead to human tragedy, as people die and homes are engulfed in flames. What is less often understood is that these fires are both natural and necessary for the development of a healthy ecosystem. In the Front Range of Colorado, for example, studies of vegetation revealed that wildfires occurred every seven years during the 1800s, before fire suppression was practised.

In 2003 Europe experienced one of the hottest and driest summers on record. Fires broke out widely in southern Europe during July and August. In Portugal 11 people perished, and several people were killed in France by fires that were believed to have been caused deliberately.

Modern efforts to suppress wildfires are necessary if we are to protect lives and prop-

Wildlife in a stream at the edge of a wildfire, taken in Bitterroot National Forest in Montana on August 6, 2000. Credit: John McColgan, a fire behaviour analyst from the Bureau of Land Management in Fairbanks, Alaska

erty, but without periodic fires, new plant growth never has a chance to take hold in an area. Without the heat from periodic fires, many nutrients remain locked in organic matter. In addition, some plants depend on the heat from a fire to support reproduction, for example, the Californian sequoia tree. This occurs in cases where the heat of the fire causes the seeds to be released (such as is true for certain pine species) and in cases where fire clears the land to support the growth of plants that cannot thrive in the shade of the original tree canopy. When the canopy is removed by fire, these plants thrive. Post-fire plant growth is also supported, as the ashes from the fire contribute valuable minerals to the soil. Deadwood also blocks out sunlight from the ground, thereby reducing species diversity. As destructive as they may appear, wildfires are actually quite valuable environmental forces. The challenge is to find a means by which people and fires can co-exist.

Given people's natural reluctance to see their homes and communities burn, we suppress fires by limiting activities such as fires at campgrounds, and work to put fires out as quickly as possible. This approach serves an obvious short-term benefit, but in the long run there is an unnatural build-up of deadwood that would normally be consumed in a periodic fire. This increases both the likelihood of a fire starting from a lightning strike or a spark from a campfire, and the severity of the fire once it starts. With all the fuel present in the form of the deadwood, the fire will probably burn hotter and longer than would be the case in the absence of a history of fire suppression.

To address this dilemma – needing to protect people and their homes while at the same time avoiding an excess of fuel accumulation – many areas practise controlled or prescribed burns. In these burns the area and extent of the burn are determined in advance and the fire is set deliberately, with adequate equipment available to bring it under control as needed. While these burns do have the potential to get out of hand, this happens only rarely. On the plus side, the burns go a long way towards reducing the fuel load that would be present in an unburned area. The fire also leads to the many ecological benefits' as nutrients are released into the soil and new plant growth flourishes.

Case study

Yellowstone National Park: 1988 and today

In 1988, major fires raged in Yellowstone National Park, USA, during a dry summer, scorching almost 324,000 hectares, or more than one-third of the park. Fortunately, there were only two human casualties, neither of which was directly attributed to the fire. Many animals perished, however, and dozens of buildings were consumed in the flames. Due to the intensity of the fires, the National Park Service abandoned its previous policy of allowing wildfires in the camp to burn out of control, but even the best efforts of firefighters and military personnel could not fully extinguish the flames. It took the first snowfalls of the autumn to complete the job.

In the aftermath of the fire, new growth was seen sprouting all over the park. Nutrient-rich soil now supports wildflower growth that would not have been possible under a dense tree canopy. Seeds from the lodgepole pine tree, freed from the cones by the heat of fire, have now begun growing in densities up to half a million plants per hectare. Thus, while the park is different as a result of the fire, it remains a vibrant ecosystem. Change is natural.

Further reading

Raven, P., and Berg, L., 2004, *Environment*, 4th edn, John Wiley and Sons.

WEBSITES

Online near-realtime wildfire mapping site for the USA:

www.geomac.gov

Online near-real time wildfire mapping site for Australia :

www.sentinel.csiro.au/mapping/viewer.htm

USA National Interagency Fire Center:

www.nifc.gov

The 1988 Yellowstone wildfires:

www.lib.niu.edu/ipo/ip881117.html

www.uwyo.edu/vegecology/
Turner%20et%20al.pdf

WIND POWER AND WIND FARMS

Harnessing the power of the wind is not a new idea – the Chinese first began using wind power for irrigation over 2000 years ago. Windmills were a familiar sight across the countryside of many countries (such as the Netherlands) up until a hundred years or so ago, with the energy being used in a very localised way, perhaps for grinding corn. These wind generators and mills still exist in many less economically developed countries, providing a cheap and highly effective form of energy for rural communities, often in remote areas and on isolated farms.

In the past two decades, commercial wind farms have been constructed in more economically developed countries, with tall turbines and rotating blades. From these, electricity can be generated, providing the wind blows at speeds in excess of 16 km/h.

The UK's first wind farm was opened at Camelford in Cornwall in December 1991, on an area of moorland 250 metres above sea level, where average wind speeds are 27 km/h. Enough energy is generated to produce electricity for 3000 homes. There are plans to supplement these terrestrial wind farms with offshore turbines, but there is, at present, insufficient wind across the UK landmass to allow the country to be self-sufficient in wind power. Despite the commitment of the British Government to develop wind power, Britain is never likely to match the amount of wind energy produced in other European nations. For example, in Denmark there are over 6500 modern turbines, and ambitious plans to increase wind power in the next few years.

Wind power is an environmentally friendly form of energy, because it does not generate air or water pollution, and does not involve finite resources. Compare this with the 1350 tonnes of ash produced each day at coal-fired power stations, not to mention the 3 million tonnes of carbon dioxide generated each year, plus the 6500 tonnes of sulphur dioxide.

Supporters of wind power also argue that

Case study

Wind power in Denmark

Denmark is considered to be the European leader in wind power. Since the 1980s thousands of turbines have been installed, on land and at sea, and by 2001 around 20,000 people worked in wind energy – more than in fishing, one of Denmark's traditional industries. By 2003 21 per cent of the nation's electricity consumption came from wind energy, and the Danish Government hope this figure will be close to 50 per cent by 2030. Whereas other nations have focused their efforts on land, the Danes have increasingly looked to the sea; despite the higher construction costs at offshore farms, the energy returns are much greater.

At Horns Rev, some 18 kilometres off the coast of Jutland, the Danish engineers have created their largest wind farm to date. Completed in 2002, the 80 turbines generate 600 million kilowatt hours a year – enough energy to power all the refrigerators used by the 5 million Danes.

greater use of wind farms would reduce the harmful effects or environmental damage of other sources such as nuclear energy or opencast (open-pit) coal mines. Another advantage for countries harnessing wind energy is that winds tend to be stronger during the winter, the period when electricity demand peaks. British farmers opting to create wind farms also gain an extra source of income, at a time when many are cash-strapped – thus the creation of wind farms could boost investment in depressed rural areas.

It would be wrong however, to assume that wind power does not generate pollution – the tall turbines create both noise and visual pollution. Unlike the old-style windmills, they do not blend into the landscape, and their rotating blades can be noisy, disturbing local residents, grazing animals and other wildlife, as well as interfering with radio and television signals.

It is also a misconception that the turbines can harness the power of storms, since turbines usually have to be switched off during powerful storms because there is a danger that the blades may break or become damaged. There are also hazards in cold weather, because ice can form on the blades and this can fly off when they start rotating, causing a hazard for local residents or passing motorists.

There are also economic drawbacks, as wind power is more expensive than other forms of energy – coal or gas-generated power costs £16 per megawatt hour, nuclear power costs £19, but wind power costs £28.80 (2004 values). In addition, the best location for wind farms is often furthest from areas of high energy demands, so expensive transmission grids need to be constructed.

WEBSITES

British Wind Energy Association:
www.bwea.com
European Wind Energy Association:
www.ewea.org
American Wind Energy Association
www.awea.org
Wind power in Denmark:
www.windpower.org/en/core.htm

SEE ALSO

renewable energy

WOODLAND ECOSYSTEMS

A woodland ECOSYSTEM is not just a collection of trees but many other plants, birds and animals that together live in equilibrium with their environment. Woodland ecosystems are vital to the health of the planet. Woodlands help maintain the carbon dioxide balance and add to global biodiversity. They reduce overland water flow that may cause flooding and provide a wide range of resources for people, including food, shelter, products, building materials, fuel and tourism.

Two main ecosystems are the northern coniferous woodlands and temperate deciduous woodlands.

Just to the south of the Arctic summers become warmer and the growing season longer, allowing trees to grow. The climate is still harsh, with prolonged severe winters and cool summers: in many areas temperatures are below freezing for six months of the year and permafrost is common. This zone of coniferous woodlands is called the boreal forest or taiga and formerly stretched in an almost continuous belt across much of northern North America and Eurasia. Taiga is also found on high mountains in lower latitudes such as the southern Rockies and Scottish Highlands.

Evergreen coniferous (cone-bearing) trees dominate this ecosystem, with their needle-shaped wax-covered leaves – witness the traditional Christmas tree. There are only four main species of conifers – spruce, pine, fir and larch – far fewer than in the deciduous woodlands further south. The trees are highly adapted to the environment. Being evergreen they are able to photosynthesis as soon as temperatures allow. They can resist drought from strong winds and frozen soils and remain undamaged by heavy snowfall due to their conical shape and flexible branches. Thick bark reduces water loss and helps protect from frost. The trees can also withstand the relatively acid and infertile podsol soils. Animal life is limited by the severe winters but includes deer, moose in North America, and a variety of rodents and carnivores such as foxes, wolves, lynx, mink and sable.

Traditionally the boreal forests of North America were inhabited by small groups of Indians who lived sustainably using the woodlands for fishing, hunting and gathering. With the arrival of the first Europeans, hunting and fur trading grew in importance. Mining rapidly overshadowed this after the Klondike Gold Rush in 1896, which raised the population of Yukon to 30,000 by 1901 (compare this with the current population of just 22,000). This phase of exploitation has continued, and developed at an even greater pace after the Second World War as new mineral discoveries were made. Timber exploitation also expanded to meet the growth in demand for paper. Timber was particularly concentrated around the Great Lakes where the local rivers could be used to transport logs to the mills; timber extraction has also concentrated in the southern part of the boreal forests since much of the north is inaccessible and the timber grows too slowly. Because of the region's strategic importance, it now has many radar stations. In addition, oil and gas has been extracted, assisted by the building of pipelines.

In contrast to the boreal forests the temperate deciduous woodlands of north-eastern USA and the UK have been extensively cleared. In the USA large areas remain, contributing to the spectacularly colourful autumn – or fall – just before the trees shed their leaves.

Deciduous trees with dense undergrowth of shrubs and flowering plants dominate the woodlands. The trees are 45–50 metres tall and the main species include oak, lime, elm, beech, chestnut and maple. In much temperate woodland one or two species dominate, although diversity ranges from 8 species per hectare in Europe to 40 species per hectare in North America. The leaf fall adds to the fertile brown forest soils that are only mildly acid, and the leaves are well mixed by earthworms and insects. Within the ecosystem the animal life is as varied as the plants. In the USA the fruits, nuts and seeds from the forest provide food for squirrels, opossums and birds, and larger animals include lynx and bear. The animals adapt to the cold winter conditions by a variety of means: some migrate to warmer climates; others hibernate like the bears; squirrels store food; while stoats and weasels change colour to camouflage themselves.

The effect of human activity on these woodlands has been enormous due to the popularity of these zones for settlement and agriculture. The climate is more equable than in the taiga, with rainfall totals of between 500 and 1500 millimetres annually. Winters are below freezing for two to three months per year in north-eastern USA and China, but milder in the UK and Western Europe, where the Gulf Stream operates. Summers are cool, with temperatures between 15 and 20°C.

Today large areas of these woodland areas are protected and conserved to avoid further destruction. Plans abound to promote tree planting and increasingly these involve the planting of the native deciduous species rather than the faster-growing coniferous trees.

WEBSITES

Woodland ecosystems:

www.connix.com/~harry/forest.htm

Boreal Forest Network:

www.borealnet.org/main.html

Boreal forests, from a Canadian website:

www.borealforest.org/index.php

Temperate deciduous forests

www.cnr.vt.edu/dendro/Forsite/tdfbiome.htm

SEE ALSO

forests and forestry